Handbook for Cultura

MW01487632

Larry D. Purnell • Eric A. Fenkl

Handbook for Culturally Competent Care

 Springer

Larry D. Purnell
College of Health Sciences
University of Delaware/Newark
Sudlersville, MD
USA

Eric A. Fenkl
College of Nursing and Health Sciences
Florida International University
Miami, FL
USA

ISBN 978-3-030-21945-1   ISBN 978-3-030-21946-8   (eBook)
https://doi.org/10.1007/978-3-030-21946-8

This Springer imprint is published by the registered company Springer Nature Switzerland AG
The registered company address is: Gewerbestrasse 11, 6330 Cham, Switzerland

Preface

The usefulness of the Purnell Model for Cultural Competence, its organizing framework, and comprehensive assessment guide has been established globally. They are being used worldwide for individual and family health-care assessment and planning and intervention in health promotion and wellness; illness, disease, and injury prevention; and health maintenance and restoration.

The aggregate data, a research principle, in culturally specific chapters is meant as a guide. The content may not apply to every individual in that culture depending on individual preferences, acculturation, and assimilation into the dominant culture. The variant cultural characteristics on the model help prevent stereotyping, assuring patients are individuals within their culture.

Each chapter is organized by the model's 12 domains of culturally competent care and provides key approaches and interventions highlighted in bold type, providing a quick reference for working with each individual. Thus, recommended approaches and interventions may need to be adapted based on patient's and family's personal perspectives and circumstances. The health-care provider must make a concerted effort to set aside any ingrained personal biases and prejudices.

With increased migration around the world, health-care providers are caring for patients whose cultures are foreign to them, making care more complex. Given the worldwide diversity, and the diversity within cultural groups, it is impossible to include the beliefs and practices of all cultures and subcultures.

Specific criteria were used for identifying the groups represented in the guide and were selected based on a variety of the following criteria:

- The group has a large population in North America, such as people of African American, Appalachian, Cuban, Jewish, and Mexican heritages.
- The group is relatively new in its migration status, such as people of Arab, Bosnian, Hmong, Somali, and Vietnamese heritages.
- The group is widely dispersed throughout the world, such as people of Chinese, Cuban, Greek, Filipino, Hindu, Iranian, and Puerto Rican heritages.

- The group has little written about it in the health-care literature, such as people of Brazilian, Haitian, Japanese, Korean, Russian, and Turkish heritages.
- The group was of particular interest to the readers, such as people from Amish, Hmong, and Somali heritages.

Sudlersville, MD, USA   Larry D. Purnell
Miami, FL, USA   Eric A. Fenkl

Contents

Introduction

Caring for culturally diverse patients and families has become commonplace throughout most of North America and worldwide. Geographical areas that previously were largely homogenous, with the exception of possibly one or two culturally diverse groups, no longer exist. This multicultural mosaic being created is adding richness to our communities and societies at large.

Many cultural groups fall into the category of vulnerable populations with increased health and health-care disparities. Even when disparities are not a concern, patient satisfaction may be of concern. If we are to decrease health and health-care disparities and increase patient satisfaction, more attention must be paid to cultural values, beliefs, and practices, which include spirituality and complementary and alternative therapies. If health-care providers and health-care delivery systems do not adapt and include diverse practices, patient and family care are compromised, resulting in not only poorer health outcomes but also an increase in cost.

Recent efforts to promote global recognition of multiculturalism have encouraged health-care providers to become more aware of their own cultures and to the cultural differences among their patients in order to provide effective, acceptable, and safe health care. Each provider and organization must make a strong commitment to providing culturally competent care, regardless of the setting in which the care is provided and the backgrounds of the participants in care. As a result of this commitment, health-care organizations and educational programs have begun to recognize the need to prepare health-care professionals with the knowledge, skills, and resources essential to the provision of culturally competent care.

Culturally competent providers value diversity and respect individual differences regardless of one's race, religious beliefs, or ethnocultural background. The goal of this Guide is to provide a framework to promote transculturally competent care that respects each person's right to be understood and treated as a unique individual.

Educational programs and access to resources necessary for cultural competence are not always readily available in a format that is concise and easy to use. This Guide has been developed to be used in a manner that can be easily accessed and understood. This approach to gaining essential knowledge using aggregate data is designed to help health-care professionals provide the highest quality care for patients and families of all backgrounds. Respect is essential to the development of dynamic interpersonal relationships that promote a positive influence on each person's interpretation of, and responses to, health care in a multicultural environment.

The real challenge for health-care providers is gaining timely access to concise information that facilitates an accurate understanding of cultural beliefs from the perspective of both the patient and the provider. This concise Guide uses the Purnell Model as a framework to simplify health assessments and interventions quickly and accurately, while enabling culturally relevant care. Although the primary focus of the handbook is on the needs of health-care providers in the United States, the organizing framework and major concepts presented in each chapter are relevant to multicultural health care throughout the world. The small size of this Guide and the outline format of its presentation, with recommended highlighted culturally relevant interventions, make it an essential resource for all health-care providers who are committed to providing culturally competent care, regardless of their practice setting.

Chapter 1
Transcultural Diversity and Health Care

1.1 The Need for Culturally Competent Health Care

The need for cultural competence is increasingly recognized by health-care providers and health-related organizations in the United States and globally. The social ideology of a melting pot has been replaced by recognizing that people deserve respect within their cultural framework and as individuals. The literature on health and health-care disparities across ethnic, social, and economic groups continues to demonstrate compelling evidence for health-care providers and health-care organizations to be attentive to cultural diversity and inclusion as well as cultural competency. Major goals of Healthy People 2020 that impact on health disparities, with culture being one force to help eliminate them, are to (a) attain high-quality, longer lives free of preventable disease, disability, injury, and premature death; (b) achieve health equity, eliminate disparities, and improve the health of all groups; (c) create social and physical environments that promote good health for all; and (d) promote quality of life, healthy development, and healthy behaviors across all life stages.

Culture is defined as the totality of socially transmitted behavioral patterns, beliefs, values, customs, lifeways, arts, and all other products of human work and thought characteristics of a population of people that guide their worldview and decision making. These patterns may be explicit or implicit, are primarily learned and transmitted within the family, are shared by most members of the culture, and are emergent phenomena that change in response to global phenomena and technological advancements. Culture is learned first in the family, then in school, and then in the community and other social organizations. Culture is largely unconscious and has powerful influences on health and illness beliefs and treatment. Health-care providers must recognize, respect, and integrate patients' cultural beliefs and practices into health prescriptions to help eliminate or mitigate health disparities and provide patient satisfaction.

This handbook describes aggregate data on the dominant cultural characteristics of selected ethnocultural groups and provides a guide for assessing cultural

© Springer Nature Switzerland AG 2019

L. D. Purnell, E. A. Fenkl, *Handbook for Culturally Competent Care*,
https://doi.org/10.1007/978-3-030-21946-8_1

values, beliefs, and practices. Based on individuality and the **variant cultural characteristics of culture** listed below, aggregate data will **not fit every individual in a cultural group.** Health-care providers who understand their own cultures and their patients' cultural values, beliefs, and practices are in a better position to interact with their patients and provide culturally acceptable care that increases opportunities for health promotion and wellness; illness, disease, and injury prevention; and health maintenance and restoration. To this end, health-care providers need both cultural general and specific cultural knowledge. The more one knows about a cultural group, the more targeted will be the assessment and interventions. Without specific knowledge of cultural groups for whom they provide care, health-care providers might not know what questions to ask to provide culturally competent care. However, any generalization made from aggregate data about the behaviors of a cultural group or individual is almost certain to be an oversimplification; within all cultures are subcultures, ethnic groups, and individuals who do not adhere to all the values of their dominant culture and vary according to variant cultural characteristics as identified later in this chapter. Subcultures, ethnic groups, and ethnocultural populations are groups of people who have experiences different from those of the dominant culture with which they identify. For this Guide, subcultures are defined as a group of people from several cultures who join together for an array of reasons. For example, a support group for substance abusers, bikers, or veterans may have members from European American, Korean, and Panamanian cultures; they create their own subculture within their dominant culture.

Culture has a powerful unconscious impact on health professionals. Each health-care provider adds a unique dimension to the complexity of providing culturally competent care. The way health-care providers perceive themselves as competent providers is often reflected in the way they communicate with patients. Thus, it is essential for health providers to think about themselves, their behaviors, and their communication styles in relation to their perceptions of different cultures. Before addressing the multicultural backgrounds and unique individual perspectives of each patient, health-care professionals must first address their own personal and professional knowledge, values, beliefs, ethics, and life experiences in a manner that optimizes assessment of and interactions with patients who come from a culture different from that of the health-care provider. Self-awareness in cultural competence is a deliberate and conscious cognitive and emotional process of getting to know oneself: one's own personality, values, beliefs, professional knowledge, standards, ethics, and the impact of these factors on the various roles one plays when interacting with individuals who are different from oneself. The ability to understand oneself sets the stage for integrating new knowledge related to cultural differences into the professional's knowledge base and perceptions of health interventions.

Culture and race are not synonymous. However, the controversial term *race* must be addressed when learning about culture. Race is genetic and includes physical characteristics that are similar among members of the same group, such as skin color, blood type, and hair and eye color. Although there is less than a 1% genetic difference among the races, this difference is significant when conducting health assessments, determining genetic conditions, and prescribing medications and treatments, as outlined in culturally specific chapters that follow. People from a given racial group may, but do not necessarily, share a common culture. Perhaps the most significant aspect of race is social in origin; race can either decrease or increase opportunities, depending on the environmental context.

1.2   Variant Characteristics of Culture

Major attributes that shape peoples' worldview and the extent to which people identify with their cultural group of origin are called variant characteristics of culture. Some characteristics may change while others remain stable over life. Variant characteristics include:

- nationality,
- race,
- skin color,
- gender,
- age,
- religious affiliation,
- educational status,
- socioeconomic status,
- occupation,
- military experience,
- political beliefs,
- urban versus rural residence,
- enclave identity,
- marital status,
- parental status,
- physical characteristics,
- sexual orientation,
- gender issues,
- health literacy,
- length of time away from the country of origin, and
- reason for migration (sojourner, immigrant, or undocumented status).

Immigration status influences a person's worldview. For example, people who voluntarily immigrate generally acculturate and assimilate more easily. Sojourners who immigrate with the intention of remaining in their new homeland for only a short time or refugees who think they may return to their home country may not have the need or desire to acculturate or assimilate. Additionally, undocumented individuals (illegal immigrants) may have a different worldview from those who have arrived legally. Some variant characteristics of culture may be fluid and should not be seen as categorically imperative.

The literature reports many definitions for the terms *cultural awareness, cultural sensitivity*, and *cultural competence*. Sometimes these definitions are used interchangeably. However, cultural awareness has more to do with an appreciation of the external or material signs of diversity, such as the arts, music, dress, or physical characteristics. Cultural sensitivity has more to do with personal attitudes and not saying things that might be offensive to someone from a cultural or ethnic background different from that of the health-care provider. Increasing one's consciousness of cultural diversity improves the possibilities for health-care providers to provide culturally competent care. Cultural competence, as used in this book, means:

1. Developing an awareness of one's own existence, sensations, thoughts, and environment without letting them have an undue influence on those from other backgrounds.
2. Demonstrating knowledge and understanding of the patient's culture, health-related needs, and culturally specific meanings of health and illness.
3. Continuing to learn about cultures of patients to whom one provides care.
4. Recognizing that the variant cultural characteristics determine the degree to which patients adhere to the beliefs, values, and practices of their dominant culture.
5. Accepting and respecting cultural differences in a manner that facilitates the patient's and family's abilities to make decisions to meet their needs and beliefs.
6. Not assuming that the health-care provider's beliefs and values are the same as the patient's.
7. Resisting judgmental attitudes such as "different is not as good."
8. Being open to cultural encounters.
9. Being comfortable with cultural encounters.
10. Adapting care to be congruent with the patient's culture.
11. Engaging in cultural competence is a conscious process and not necessarily a linear one.
12. Accepting responsibility for one's own education in cultural competence by attending conferences, reading professional literature, and observing cultural practices.

As of July 2015, the U.S. population was more than 321 million (U.S. Department of Commerce: U.S. Census Bureau 2015). Immigration continues at approximately 11% per year. India was the leading country of origin for new immigrants, with 147,500 arriving in 2014, followed by China with 131,800, Mexico with 130,000, Canada with 41,200, and the Philippines with 40,500. Table 1.1 shows the diversity of the U.S. population.

The census collects data on two categories: race and ethnicity; therefore, these numbers total more than 100%. Race and Hispanic origin are two separate and distinct categories. Race categories as used in Census 2010 include the following:

1. *White* refers to people having origins in any of the original peoples of Europe, the Near East, the Middle East, and North Africa. This category includes Irish, German, Italian, Lebanese, Turkish, Arab, and Polish.

Table 1.1 Diversity of the U.S. population as of July, 2015

White persons	77.1%
Persons of Hispanic or Latino origin	17.6%
Black and African American	13.3%
Asian	5.6%
American Indian/Alaskan Native	1.2%
Native Hawaiian and other Pacific Islanders	0.2%
Persons with two or more races	2.6%
White persons—not Hispanic	63.7%

Data from U.S. Department of Commerce: U.S. Census Bureau (2018). Retrieved from http://quickfacts.census.gov/qfd/states/00000.html

2. *Black* and *African American* refer to people having origins in any of the black racial groups of Africa and include Nigerians and Haitians and any person who self-designated this category regardless of origin.
3. *American Indian* and *Alaskan Native* refer to people having origins in any of the original peoples of North, South, and Central America and who maintain tribal affiliation or community attachment.
4. *Asian* refers to people having origins in any of the original peoples of the Far East, Southeast Asia, and the Indian subcontinent. This category includes the terms *Asian Indian, Chinese, Filipino, Korean, Japanese, Vietnamese, Burmese, Hmong, Pakistani,* and *Thai*.
5. *Native Hawaiian* and other *Pacific Islander* refer to people having origins in any of the original peoples of Hawaii, Guam, Samoa, Tahiti, the Mariana Islands, and Chuuk.
6. *Some other race* was included for people who are unable to identify with the other categories. Additionally, the respondent could identify, as a write-in, with two races (U.S. Department of Commerce: U.S. Census Bureau 2015).

1.3 Reflective Exercises

1. What changes in ethnic and cultural diversity have you seen in your community over the last 5 years? Over the last 10 years? Have you had the opportunity to interact with newer groups?
2. What health disparities have you observed in your community? To what do you attribute these disparities? What can you do as a professional to help decrease these disparities?
3. Who in your family had the most influence in teaching you cultural values and practices? Mother, father, grandparent?
4. What activities have you done to increase your cultural competence?
5. Given that everyone is ethnocentric to some degree, what do you do to become less ethnocentric?
6. How do you distinguish a stereotype from a generalization?
7. How have your variant characteristics of culture changed over time?

8. Distinguish between a stereotype and a generalization.
9. What ethnic and racial groups do you encounter on a regular basis? Do you see any racism or discrimination among these groups?
10. What does your organization do to increase diversity and cultural competence?

Bibliography

Centers for Disease Control and Prevention (2016) https://www.healthypeople.gov/2020/default
Healthy People 2020 (2016) https://www.healthypeople.gov/2020/default
Immigration Policy Institute (2016) http://www.migrationpolicy.org/article/frequently-requested-statistics-immigrants-and-immigration-united-states
U.S. Department of Commerce: U.S. Census Bureau (2015) http://quickfacts.census.gov/qfd/states/00000.html

Chapter 2
The Purnell Model for Cultural Competence

The Purnell Model for Cultural Competence and its organizing framework can be used as a guide for assessing the culture of patients. The major explicit assumptions on which the Purnell Model is based include the following:

1. All health-care professions need similar information about cultural diversity.
2. All health-care professions share the metaparadigm concepts of global society, family, person, and health.
3. One culture is not better than another culture; they are just different.
4. Core similarities are shared by all cultures.
5. Differences exist within, between, and among cultures.
6. Cultures change slowly over time.
7. The variant cultural characteristics (see Chap. 1) determine the extent to which one varies from the dominant culture.
8. If patients are coparticipants in their care and have a choice in health-related goals, plans, and interventions, their health outcomes will be improved.
9. Culture has a powerful influence on one's interpretation of and responses to health care.
10. Individuals and families belong to several subcultures.
11. Each individual has the right to be respected for his or her uniqueness and cultural heritage.
12. Caregivers need both cultural-general and cultural-specific information in order to provide culturally congruent care.
13. Caregivers who can assess, plan, intervene, and evaluate in a culturally competent manner will improve the care of patients for whom they care.
14. Learning culture is an ongoing process that develops in a variety of ways, but primarily through cultural encounters (www.transculturalcare.net).
15. Prejudices and biases can be minimized with cultural understanding.
16. To be effective, health care must reflect the unique understanding of the values, beliefs, attitudes, lifeways, and worldview of diverse populations and individual acculturation and assimilation patterns.

© Springer Nature Switzerland AG 2019   7
L. D. Purnell, E. A. Fenkl, *Handbook for Culturally Competent Care*,
https://doi.org/10.1007/978-3-030-21946-8_2

17. Differences in race and culture often require adaptations to standard interventions.
18. Cultural awareness improves the caregiver's self-awareness.
19. Professions, organizations, and associations have their own culture, which can be analyzed using a grand theory of culture.
20. Every patient encounter is a cultural encounter.
21. Providers who engage in critical self-reflection have a better understanding of their own biases and of patient's cultures.

2.1 The Purnell Model

The Purnell Model for Cultural Competence and its organizing framework can be used in all practice settings and by all health-care providers. Moreover, it has been classified a holographic and complexity grand theory. In fact, even mediators with the American Bar Association have found the model valuable. The model is a circle, with an outlying rim representing global society, a second rim representing community, a third rim representing family, and an inner rim representing the person (Fig. 2.1). The interior of the circle is divided into 12 pie-shaped wedges depicting cultural domains (constructs) and their associated concepts. The dark center of the circle represents unknown phenomena. Along the bottom of the model is a jagged line representing the nonlinear concept of cultural consciousness. The 12 cultural domains and their concepts provide the organizing framework. Each domain includes concepts that need to be addressed when assessing patients in various settings. Moreover, health-care providers can use these same concepts to better understand their own cultural beliefs, attitudes, values, practices, and behaviors. An important concept to understand is that no single domain stands alone; they are all interconnected. The 12 domains are overview/heritage, communications, family roles and organization, workforce issues, biocultural ecology, high-risk health behaviors, nutrition, pregnancy and the childbearing family, death rituals, spirituality, health-care practices, and health-care practitioners. For a more complete description of the domains, the reader is referred to the textbook by Purnell, *Transcultural Health Care: A Culturally Competent Approach* (2013, Fourth Edition, F.A. Davis Company).

2.1.1 Assessment Guide

Following each of the domains and concepts is a box that includes suggested questions to ask and observations to make when assessing patients from a cultural perspective. It is recognized that clinicians are not able to complete a thorough cultural assessment for every patient. The list of questions is extensive; thus, the clinician must determine which questions to ask according to the patient's presenting symptoms, teaching needs, and the potential impact of culture.

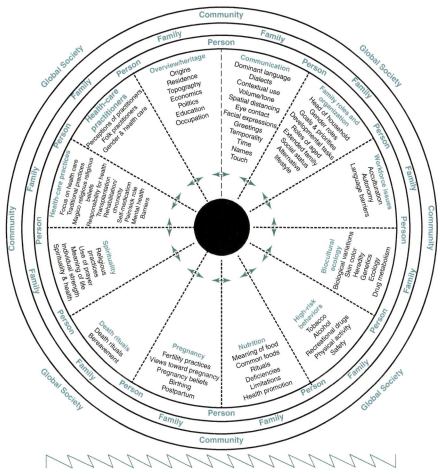

Fig. 2.1 The Purnell model for cultural competence

2.1.2   Domains and Concepts

2.1.2.1   Overview and Heritage

Includes concepts related to the country of origin and current residence; the effects of the topography of the country of origin and the current residence on health, economics, politics, reasons for migration, educational status, and occupations. See table below.

2.1.2.2 Communications

Includes concepts related to the dominant language, dialects, and the contextual use of the language; paralanguage variations such as voice volume, tone, intonations, inflections, and willingness to share thoughts and feelings; nonverbal communications such as eye contact, gesturing and facial expressions, use of touch, body language, spatial distancing practices, and acceptable greetings; temporality in terms of past, present, and future orientation of worldview; clock versus social time; and the amount of formality in use of names. Differences between the language spoken by the health-care provider and the patient, educational level, and health literacy are additional communication add to the communication difficulties. Effective communication is the first and probably the most important aspect of obtaining an accurate assessment. See table below.

2.1.2.3 Family Roles and Organization

Includes concepts related to the head of the household, gender roles (a product of biology and culture), family goals and priorities, developmental tasks of children and adolescents, roles of the aged and extended family, individual and family social status in the community, and acceptance of alternative lifestyles such as single parenting, same sex partnerships and marriage, childless marriages, and divorce. See table below.

2.1.2.4 Workforce Issues

Includes concepts related to autonomy, acculturation, assimilation, gender roles, ethnic communication styles, and health-care practices of the country of origin. See table below.

2.1.2.5 Biocultural Ecology

Includes physical, biological, and physiological variations among ethnic and racial groups such as skin color (the most evident) and physical differences in body habitus; genetic, hereditary, endemic, and topographical diseases; psychological makeup of individuals; and the physiological differences that affect the way drugs are metabolized by the body. In general, most diseases and illnesses can be divided into three categories: lifestyle, environment, and genetics.

Lifestyle causes include cultural practices and behaviors that can generally be controlled: for example, smoking, diet, and stress.

Environment causes refer to the external environment (e.g., air and water pollution) and situations over which the individual has little or no control (e.g., presence of malarial mosquitos, exposure to chemicals and pesticides, access to care, and associated diseases and illnesses). See table below.

Genetic conditions are caused by genes.

2.1.2.6 High-Risk Health Behaviors

Includes substance use and misuse of tobacco, alcohol, and recreational drugs; lack of physical activity; increased calorie consumption; nonuse of safety measures such as seat belts, helmets, and safe driving practices; and not taking safety measures to prevent contracting HIV and sexually transmitted infections. See table below.

2.1.2.7 Nutrition

Includes the meaning of food, common foods and rituals, nutritional deficiencies and food limitations, and the use of food for health promotion and wellness, and illness and disease prevention. Multiple diseases and illnesses are a consequence of this major cultural component. See table below.

2.1.2.8 Pregnancy and Childbearing Practices

Includes culturally sanctioned and unsanctioned fertility practices; views on pregnancy; and prescriptive, restrictive, and taboo practices related to pregnancy, birthing, and the postpartum period. See table below.

2.1.2.9 Death Rituals

Includes how the individual and the society view death and euthanasia, rituals to prepare for death, burial practices, and bereavement behaviors. Death rituals are slow to change. See table below.

2.1.2.10 Spirituality

Includes formal religious beliefs related to faith and affiliation and the use of prayer; behavior practices that give meaning to life; and individual sources of strength. See table below.

2.1.2.11 Health-Care Practices

Includes the focus of health care (acute versus preventive); traditional, magico-religious, and biomedical beliefs and practices; individual responsibility for health; self-medicating practices; views on mental illness, chronicity, and rehabilitation; acceptance of blood and blood products; and organ donation and transplantation. See table below.

2.1.2.12   Health-Care Practitioners

Includes the status, use, and perceptions of traditional; magico-religious, and bio-medical health-care providers; and the gender of the health-care provider. See table below.

2.2   Purnell Cultural Assessment Tool

An extensive cultural assessment is rarely completed in the clinical setting because of time and other circumstances. A seasoned clinical practitioner will know when further assessment is required. Thus, this tool should be used as a guide. *Shaded items in italics are part of any standard assessment. Other items may also be part of a standard assessment, depending on the organization, setting, and clinical area.*

Cultural assessment tool	Comments
Overview, inhabited localities, and topography	
Where do you currently live?	
What is your ancestry?	
Where were you born?	
How many years have you lived in the United States (or other country, as appropriate)?	
Were your parents born in the United States (or other country, as appropriate)?	
What brought you (your parents/ancestors) to the United States (or other country, as appropriate)?	
Describe the land or countryside where you live. Is it mountainous, swampy, etc.?	
Have you lived in other places in the United States/world?	
What was the land or countryside like when you lived there?	
What is your income level?	
Does your income allow you to afford the essentials of life?	
Do you have health insurance?	
Are you able to afford health insurance on your salary?	
What is your educational level (formal/informal/self-taught)?	
What is your current occupation? If retired, ask about previous occupations	
Have you worked in other occupations? What were they?	
Are there (were there) any health hazards associated with your job(s)?	
Have you been in the military? If so, in what foreign countries were you stationed?	
Communications	
What is your full name?	
What is your legal name?	
By what name do you wish to be called?	
What is your primary language?	
Do you speak a specific dialect?	
What other languages do you speak?	

Cultural assessment tool	Comments
Do you find it difficult to share your thoughts, feelings, and ideas with family? Friends? Health-care providers?	
Do you mind being touched by friends? Strangers? Health-care workers?	
How do wish to be greeted? Handshake? Nod of the head, verbal greeting only etc.?	
Are you usually on time for appointments?	
Are you usually on time for social engagements?	
Observe the patient's speech pattern. Is the speech pattern high- or low-context? Note: patients from highly contexted cultures place greater value on silence	
Observe the patient when physical contact is made. Does he/she withdraw from the touch or become tense?	
How close does the patient stand when talking with family members? With health-care providers?	
Does the patient maintain eye contact when talking with the nurse/physician/etc.?	
Family roles and organization	
What is your *marital*/partner status?	
How many children do you have?	
Who makes most of the decisions in your family?	
What types of decisions do(es) the female(s) in your family make?	
What types of decisions do(es) the male(s) in your family make?	
What are the duties of the women in the family?	
What are the duties of the men in the family?	
What should children do to make a good impression for themselves and for the family?	
What should children *not* do to make a good impression for themselves and for the family?	
What are children forbidden to do?	
What should adolescents do to make a good impression for themselves and for the family?	
What should adolescents *not* do to make a good impression for themselves and for the family?	
What are adolescents forbidden to do?	
What are the priorities for your family?	
What are the roles of the older people in your family? Are they sought for their advice?	
Are there extended family members in your household? Who else lives in your household?	
What are the roles of extended family members in this household? What gives you and your family status?	
Is it acceptable to you for people to have children out of wedlock?	
Is it acceptable to you for people to live together and not be married?	
Is it acceptable to you for people to admit being gay or lesbian?	
What is your sexual preference/orientation? (ask only if appropriate and later in the assessment after a modicum of trust has been established)	

(continued)

Cultural assessment tool	Comments
Workforce Issues	
Do you usually report to work on time?	
Do you usually report to meetings on time?	
What concerns do you have about working with someone of the opposite gender?	
Do you consider yourself a "loyal" employee?	
How long do you expect to remain in your position?	
What do you do when you do not know how to do something related to your job?	
Do you consider yourself to be assertive in your job?	
What difficulty does English (or another language) give you in the workforce?	
What difficulties do you have working with people older (younger) than you?	
What difficulty do you have in taking directions from someone younger/older than you?	
What difficulty do you have working with people whose religions are different from yours?	
What difficulty do you have working with people whose sexual orientation is different from yours?	
What difficulty do you have working with someone whose race or ethnicity is different from yours?	
Do you consider yourself to be an independent decision maker?	
Biocultural Ecology	
Are you allergic to any medications?	
What problems did you have when you took over-the-counter medications?	
What problems did you have when you took prescription medications?	
What are the major illnesses and diseases in your family?	
Are you aware of any genetic diseases in your family?	
What are the major health problems in the country from which you come (if appropriate)?	
With what races and ethnic groups(s) do you identify?	
Observe and document skin coloration and physical characteristics	
Observe for and document physical handicaps and disabilities	
High-risk health behaviors	
How many cigarettes a day do you smoke? *Do you smoke a pipe (or cigars)?*	
Do you smoke a pipe (or cigars)?	
Do you chew tobacco?	
For how many years have you smoked/chewed tobacco?	
How much alcohol do you drink each day? Ask about wine, beer, spirits?	
How many energy drinks do you consume each day	
What recreational drugs do you use?	
How often do you use recreational drugs?	
What type of exercise do you do each day?	
Do you use seat belts?	
What precautions do you take to prevent getting a sexually transmitted infections or HIV/AIDS?	

Cultural assessment tool	Comments
Nutrition	
Are you on a special diet?	
Are you satisfied with your weight?	
Which foods do you eat to maintain your health?	
Do you avoid certain foods to maintain your health?	
Why do you avoid these foods?	
Which foods do you eat when you are ill?	
Which foods do you avoid when you are ill?	
Why do you avoid these foods (if appropriate)?	
For what illnesses do you eat certain foods?	
Which foods do you eat to balance your diet?	
Which foods do you eat every day?	
Which foods do you eat every week?	
Which foods do you eat that are part of your cultural heritage?	
Which foods are high-status foods in your family/culture?	
Which foods are eaten only by men? Women? Children? Teenagers? Elderly?	
How many meals do you eat each day?	
What time do you eat each meal?	
Do you snack between meals?	
What foods do you eat when you snack?	
What holidays do you celebrate?	
Which foods do you eat on particular holidays?	
Who is present at each meal? Is the entire family present?	
Do you primarily eat the same foods as the rest of your family?	
Where do you usually buy your food?	
Who usually buys the food in your household?	
Who does the cooking in your household?	
How frequently do you eat at a restaurant?	
When you eat at a restaurant, in what type of restaurant do you eat?	
Do you eat foods left from previous meals?	
Where do you keep your food?	
Do you have a refrigerator?	
How do you cook your food?	
How do you prepare meat?	
How do you prepare vegetables?	
What type of spices do you use?	
What do you drink with your meals?	
Do you drink special teas?	
Do you have any food allergies?	
Are there certain foods that cause you problems when you eat them?	
How does your diet change with each season?	

(continued)

Cultural assessment tool	Comments
Are your food habits different on days you work versus when you are not working?	
Pregnancy and childbearing practices	
How many children do you have?	
What do you use for birth control?	
What does it mean to you and your family when you are pregnant?	
What special foods do you eat when you are pregnant?	
What foods do you avoid when you are pregnant?	
What activities do you avoid when you are pregnant?	
Do you do anything special when you are pregnant?	
Do you eat non-food substances when you are pregnant?	
Who do you want with you when you deliver your baby?	
In what position do you want to be when you deliver your baby?	
What special foods do you eat after delivery?	
What foods do you avoid after delivery?	
What activities do you avoid after you deliver?	
Do you do anything special after delivery?	
Who will help you with the baby after delivery?	
What bathing restrictions do you have after you deliver?	
Do you want to keep the placenta?	
What do you do to care for the baby's umbilical cord?	
Death rituals	
What special activities need to be performed to prepare for death?	
What special activities need to be performed after death?	
Would you want to know about your impending death?	
What is your preferred burial practice? Interment, cremation?	
How soon after death does burial occur?	
How do men grieve?	
How do women grieve?	
What does death mean to you?	
Do you believe in an afterlife?	
Are children included in death rituals?	
Spirituality	
What is your religion?	
Do you consider yourself deeply religious?	
How many times a day do you pray?	
What do you need in order to say your prayers?	
Do you practice meditation, such as TM, mindfulness meditation, etc.	
What gives strength and meaning to your life?	
In what spiritual practices do you engage for your physical and emotional health?	
Health care practices	
In what prevention activities do you engage to maintain your health?	
Who in your family takes responsibility for your health?	
Who takes care of family members when they are sick?	

Cultural assessment tool	Comments
What over-the-counter medicines do you use?	
What herbal teas and folk medicines do you use?	
For what conditions do you use herbal medicines?	
What do you usually do when you are in pain?	
How do you express your pain?	
How are people in your culture viewed or treated when they have a mental illness?	
How are people with physical disabilities treated in your culture?	
What do you do when you are sick? Stay in bed, continue your normal activities, etc.?	
What are your beliefs about rehabilitation?	
How are people with chronic illnesses viewed or treated in your culture?	
Are you averse to blood transfusions?	
Is organ donation acceptable to you?	
Are you listed as a potential organ donor?	
Would you consider having an organ transplant if needed?	
Are health-care services readily available to you?	
Do you have transportation problems accessing needed health-care services?	
Do you feel welcome when you see a health-care professional?	
What traditional health-care practices do you use? For example, mineral baths, sweating, acupuncture, acupressure, *cai gao*, coining, *moxibustion*, aromatherapy, etc.?	
What home difficulties do you have that might prevent you from receiving health care?	
Health care practitioners	
What health-care providers do you see when you are ill? Physicians, nurses?	
Do you prefer a same-sex health-care provider for routine health problems? For intimate care?	
What healers do you use beside physicians and nurses?	
For what conditions do you use healers?	

2.3   Reflective Exercise

Communication is probably the most import aspect of obtaining a good assessment. The onus of effective communication primarily lies with the health-care provider. Thus, providers must understand their own communication styles. When your personal communication practices differ from what is in the scholarly literature, posit why. See the variant cultural characteristics as a guide. Speak to each one of the following points:

1. Identify your cultural ancestry. If you have more than one cultural ancestry, chose the one with which you most closely associate.
2. Explore the willingness of individuals in your culture to share thoughts, feelings, and ideas. Identify any area of discussion that would be considered taboo?

3. Explore the practice and meaning of touch in your culture. Include information regarding touch between family members, friends, members of the opposite sex, and health-care providers.
4. Identify personal spatial and distancing strategies used when communicating with others in your culture. Discuss differences between friends and families versus strangers.
5. Discuss your culture's use of eye contact. Include information regarding practices between family members, friends, strangers, and persons of different age groups.
6. Explore the meaning of gestures and facial expressions in your culture. Do specific gestures or facial expressions have special meanings? How are emotions displayed?
7. Are there acceptable ways of standing and greeting people in your culture?
8. Discuss the prevailing temporal relation of your culture. Is the culture's worldview past, present, or future oriented? Temporality also include punctuality.
9. Discuss the impact of your culture on your nursing and/or health care. Be specific, not something that is very general.

Chapter 3
Barriers to Culturally Competent Health Care

Barriers to culturally competent health care are defined as obstacles or hindrances that make it difficult to obtain or access good health or health care. Despite efforts and goals in the United States and many other countries around the world to reduce or eliminate obstacles, significant disparities continue, especially with what the professional literature refers to as vulnerable populations. Vulnerable populations include the following:

- the economically disadvantaged
- racial and ethnic minorities (including those of mixed race or ethnicities)
- the uninsured and underinsured
- older adults
- children
- the homeless
- people with human immunodeficiency virus (HIV)
- people who are stigmatized for whatever reason, which includes substance misuse
- people with chronic health conditions and severe mental illness
- rural populations
- people with low acculturation and low levels of education
- gender disparities
- refugees and undocumented immigrants.

Overall these disparities are due to environmental, psychological, and social factors; maldistribution or shortage of human and fiscal resources; and lack of access to health-care systems. From this list, it seems clear that under the right set of circumstances, any group or individual can be vulnerable.

Barriers can be categorized in several domains that will be discussed in this chapter: language and health literacy, availability, accessibility, affordability, appropriateness, accountability, adaptability, acceptability, awareness, attitudes,

© Springer Nature Switzerland AG 2019
L. D. Purnell, E. A. Fenkl, *Handbook for Culturally Competent Care*,
https://doi.org/10.1007/978-3-030-21946-8_3

approachability, alternative practices, and additional services. It should be noted that these domains are not categorically imperative or perfectly distinct categories; they can overlap.

3.1 Language and Health Literacy

Language and health literacy barriers include the medical jargon used by health-care providers, inadequate reading level of the patient, or lack of fluency in English or in the patient's mother tongue. Several studies in the United States have identified that a lack of fluency in language is the primary barrier to receiving adequate health care, and not just for people for whom English is a second language.

Language barriers may involve both oral and written language. Interpretation refers to oral communication while translation refers to written communication. Interpretation requires more than word-for-word substitution. Professional interpreters must not only demonstrate bilingual proficiency, but they must also be knowledgeable of idioms, slang, and colloquialisms as well as medical terminology. Sign language may be a concern for some because there is no standard worldwide sign language. Besides American Sign Language, there are multiple Arabic sign languages. Language and literacy barriers may result in the following:

- Language barriers can negatively affect perceptions of patient and provider care, resulting in patients not returning for follow-up care.
- Language barriers may result in a patient leaving against medical advice. A patient who perceives that he or she is not understood by the health-care provider or who does not understand a provider who is using medical technology may simply leave.
- Limited English proficiency (LEP) can result in unsafe situations caused by poor patient comprehension of their medical condition, treatment plan, or discharge instructions; an inaccurate and incomplete medical history being obtained by the health-care provider; ineffective or improper use of medications or serious medication errors; improper preparation for tests and procedures; and poor or inadequate informed consent.
- **Be alert to comments and behaviors from patients and family members that might mean they do not understand the provider. Such cues include "Can I take these forms home to complete?" or "I forgot my glasses. I'll wait until (friend or family member) comes to help me."**
- LEP patients who do not receive professional interpretation services at admission and discharge have longer lengths of stay and higher readmission rates compared with patients who receive professional interpretation services.
- Just because someone speaks the language does not necessarily mean he or she has the skills to be successful at interpretation.
- Improper interpretation is prone to additions, substitutions, omissions, volunteer opinions, and sematic errors.

- Compared with professional medical interpreters, ad hoc interpreters such as patients' family members or house staff frequently make medical interpretation errors. These errors are significantly more likely to have clinical consequences.
- **Use only certified interpreters when at all possible.**
- **Use dialect specific interpreters whenever possible.**
- **Create a system for interpretation and translation.**
- **Compile a list of commonly used words in the dominant languages of the patients using the facility. This list does not replace the need for an interpreter but may be helpful if a professional interpreter is not available.**
- **Allow time for interpretation.**
- **Be aware that social class differences can affect interpretation.**
- **Maintain eye contact with both the patient and the interpreter.**
- Non-professional interpretation and using children or other family members can lead to a breach of confidentiality. Also, patients might not speak freely, especially regarding issues such as sexuality, drug and alcohol misuse, and domestic violence.
- **Do not place the patient in a precarious situation by asking the patient in front of the family if it is acceptable for a family member or friend to interpret.**
- **Do not use family members except in nonconfidential situations such as obtaining demographic and admitting data and teaching about medication administration and dietary requirements.**
- **Do not use individuals known to the family as interpreters due to confidentiality issues.**
- Language barriers may involve written as well as oral communication. It is therefore important to provide written materials in the patient's native language.
- **Translate satisfaction surveys into the languages of the population served.**
- **Translate treatment plans and medication requirements into the languages of the patients who come to the facility.**
- **Translate pamphlets on common illnesses and diseases into the languages of the patients who come to the facility.**
- **Ensure that translated material is at a grade 6 reading level.**
- **Create pamphlets on high-risk behaviors and occupations that are high-risk.**

3.2   Availability

Availability is defined as a health-related system offering a needed service and doing so at a time that is reasonable. Patients are routinely referred to Emergency Departments for care when health-care providers' offices and clinics are closed, adding to patient dissatisfaction and increased cost to the patient and health-care system. Pharmacy hours can be severely restricted in the evening and at night, especially in high crime unsafe neighborhoods, making their services unavailable.

- **Create fast-track systems with expanded hours of services in Emergency Departments that can be staffed by advanced practice nurses.**

- **Open satellite urgent care centers making services more readily available and taking some of the workload off the Emergency Department.**
- **Provide patients with a temporary supply of medications until the local pharmacy is open.**

3.3   Accessibility

Accessibility is defined as the ability of the patient to actually get to a service when needed. Transportation services may not be available, or challenging road conditions and rivers and mountains may make it difficult for people to obtain needed health-care services when no health-care provider is available in their immediate region. Although federal law requires public facilities to be accessible to people with disabilities, there are still private clinics where accessibility may be a problem. Extremely rural hospitals have critical access facilities receive cost-based reimbursement but access can still be a concern when the closest facility is 50+ miles away.

- **Make vans available at convenient locations and at specified times to transport patients to clinics.**
- **Provide guidance for accessing services and navigating the health-care delivery system by advertising in local newspapers, in newsletters, and on public transportation systems. Local convenience stores and restaurants are additional sites to advertising services.**
- **Provide telephone triage systems where patients and families can talk with a nurse to determine the acuity of a health concern.**

3.4   Affordability

Affordability is defined as having the financial resources to access and use health-care services. The service might be available, but if patients do not have financial resources, they may wait until their health condition is more severe, resulting in longer recovery times and an increase in cost.

- **Partner with local philanthropic agencies to provide financial services.**
- **Involve social workers to get patients funds on a temporary basis.**

3.5   Appropriateness

Appropriateness is defined as having a service that is needed and congruent with the patient's and family's cultural belief system. Maternal and child services might be available, but there might be a greater need for geriatric and psychiatric services. Perhaps the critical access hospital does not have an infectious disease specialist.

- **Partner with larger organizations that have full-service capabilities to offer hours in rural areas on a regular basis.**
- **Offer full-service health promotion and wellness services, testing, and education a few days each week on a regular basis.**

3.6 Organizational Accountability

Organizational accountability is defined as health-care providers' seeking resources for their own education and learning about the cultures of the people they serve.

- **Provide cultural competence in orientation programs.**
- **Obtain local, regional, or national organizations and personnel to provide cultural competence workshops.**
- **Update staff annually on the practices, values, and health beliefs of the populations the organization serves.**
- **Provide staff with assessment forms and guides for culturally competent assessment and interventions.**
- **Incorporate cultural data on intake assessment forms.**
- **Provide print and online resources for staff.**

3.7 Adaptability

Adaptability is defined as the health-care environment being able to change and to offer additional services when needed. For example, a mother brings her child to the clinic for an immunization. Can she get a mammogram at the same time or must she make another appointment?

- **Provide several times during the week when full-service health promotion and wellness, services, testing, and education are available. Advertise these hours in local stores, houses of worship, banks, and post offices.**

3.8 Acceptability

Acceptability is defined as being able to meet the requirements and provide satisfaction to a patient or family needing a health-care service. Are services and patient education offered in a language preferred by the patient?

- **Match sex of the patient and health-care provider whenever requested and if available.**
- **Provide ethnic concordant providers if requested whenever possible.**

3.9 Awareness

Awareness is defined as the patient and family being cognizant of needed health-care services. Is the patient aware that needed services exist in the community? The service may be available, but if patients are not aware of it, the service will not be used.

- **Advertise in local newspapers, in newsletters, and on public transportation systems services that are available in the community. Include convenient stores, grocery stores, houses of worship, and restaurants. Include advertisements in foreign languages where appropriate.**
- **Partner with local and regional radio and television stations (including foreign-language stations) that advertise the services that are available to patients in the community.**

3.10 Attitudes

Attitude is defined as a state of mind or feeling about a patient's or family's health beliefs and values. Adverse subjective beliefs and attitudes from caregivers increase the chance that the patient will not return for needed services until the condition is more severe. Do health-care providers have negative attitudes about patients' home-based traditional practices?

- **Do not tell patients that they are wrong for using home-based treatments.**
- **Do not let judgmental attitudes interfere with acceptance of patients and families.**

3.11 Approachability

Approachability is defined as the patient or family feeling it is easy to talk or deal with health-care providers. Do patients feel welcomed? Do health-care providers and receptionists greet patients in the manner in which they prefer?

- **Greet patients formally until told to do otherwise.**
- **Introduce yourself by name and position.**
- **Maintain eye contact without staring when greeting patients and families. Note that in some cultures, people do not maintain eye contact with people in higher social positions.**

3.12 Alternative Practices

Alternative practices are defined as complementary and alternative health-care practices and traditional and folk practices. While some practices are not harmful and may be helpful, some patients may overuse folk practices and folk practitioners before seeking allopathic care.

- **Incorporate nonharmful alternative practices into treatment plans.**
- **Partner with local folk healers and educate them on when referral to allopathic care is advisable.**
- **Refer patients to alternative folk practitioners when the condition (for example: evil eye) warrants it.**

3.13 Additional Services

Additional services are defined as value-added benefits that improve health-care access and adaptability. For example, are child- and adult-care services available if a parent must bring children or an aging parent with them for an appointment?

- **Network with local organizations and churches that can volunteer sitter services with professional oversight.**
- **Assist patients with navigating the intricacies of the health-care system.**
- **Assist patients and families with completing bureaucratic forms.**
- **Conduct focus groups to determine community needs.**

3.14 Reflective Exercises

1. What barriers do you see to culturally competent care in your organization? School, work, etc.
2. What language barriers to you see in your community such as English as a second language, educational levels, health literacy, etc.?
3. What affordability concerns for health care do you see in your community?
4. What complementary/alternative health-care practices do you use?
5. What complementary/alternative health-care practices are available in your community?
6. How do you want to be addressed? First name, last name with a title?
7. How do you address older people in your community? First name, last name with a title?

8. Is public transportation readily available to health-care services in your community? What might be done to improve them?
9. What does your organization do when a patient needs an interpreter?
10. In addition to English, what other languages are health care instructions provided in your organization?

Bibliography

Agency for Healthcare Quality and Research (2016) TeamSTEPPS enhancing safety for patients with limited English proficiency module. www.ahrq.gov/teamsteppstools/

Betancourt J, Green A, Carrillo JE, Ananeh-Firempong A (2003) Defining cultural competence: a practical framework for addressing racial/ethnic disparities in health and health care. Public Health Rep 118:293–302

Center for Health Care Strategies (2016) Reducing barriers to health care: practical strategies for local organizations. www.chcs.org

Flores G (2005) The impact of medical interpreter services on the quality of health care: a systematic review. Med Care Res Rev 62(3):255–299

Flores G, Laws MB, Mayo SJ, Zuckerman B, Abreu M, Medina L, Hardt EJ (2003) Errors in medical interpretation and their potential clinical consequences in pediatric encounters. Pediatrics 111(1):6–14

Giger J, Davidhizar R, Purnell L, Harden J, Phillips J, Strickland O (2007) American Academy of Nursing Expert Panel Report: developing cultural competence to eliminate health disparities in ethnic minorities and other vulnerable populations. J Transcult Nurs 18(2):95–102

Institute of Medicine of the National Academies (2016) Crossing the quality chasm. http://nationalacademies.org/hmd/reports/2001/crossing-the-quality-chasm-a-new-health-system-for-the-21st-century.aspx

Jacobs EA, Agger-Gupta N, Hm Chin A, Piorowski A, Hardt A (2016) Language barriers in healthcare settings. California Endowment. www.calendow.org

Office of Minority Health (2016) National Standards on Culturally and Linguistically Appropriate Services (CLAS). http://minorityhealth.hhs.gov/templates/browse.aspx?lvl=2&lvlID=15

Purnell L (2013) Transcultural health care: a culturally competent approach. F.A. Davis, Philadelphia

Purnell L, Davidhizar R, Fishman D, Strickland O, Allison D (2011) A guide to developing a culturally competent organization. J Transcult Nurs 22(1):5–14

The Joint Commission (2010) Advancing effective communication, cultural competence, and family centered care: a roadmap for hospitals. www.jointcommission.org/assets/1/6/aroadmapforhospitalsfinalversion727.pdf

The Joint Commission (2011) Patient-centered communication standards and EPs. http://www.imiaweb.org/uploads/pages/275.pdf

Chapter 4
People of African American Heritage

4.1 Overview and Heritage

African Americans, the second largest ethnocultural group after Hispanics in the US, are largely the descendants of Africans who were forcibly brought to this country as slaves between 1619 and 1860. A significant cultural divide may separate this group from recent immigrants from Africa and other countries. However, the U.S. Census Bureau includes all Blacks together.

Today, African Americans usually refer to themselves as black or black Americans, whereas previously the terms *Negro*, *colored*, and people of color were popular, although in some settings one can still see these terms used. The **variant cultural characteristics of culture** (see Chap. 1) describe variables that contribute to the diversity of the African American population. Note that this information on African Americans is aggregate data and does not fit every person who self-identifies as African American.

Most families place a high value on education and make great sacrifices so at least one child can go to college. However, many African Americans continue to be underrepresented in managerial and professional positions, overrepresented in "blue-collar" positions, and more likely to be employed in hazardous environments and occupations, resulting in occupation-related diseases, illnesses, and injuries.

4.1.1 Communications

- The dominant language of African Americans is English. However, many refer to an informal language known as Black or African American English (AAE), or Ebonics. For example, some may pronounce *th* as *d*. Therefore, the word *these* may be pronounced *dese*. **Do not stereotype African Americans as speaking**

© Springer Nature Switzerland AG 2019
L. D. Purnell, E. A. Fenkl, *Handbook for Culturally Competent Care*,
https://doi.org/10.1007/978-3-030-21946-8_4

only in Black English because most African Americans are articulate and competent in the formal English.

- What transpires within the family is viewed as private and not appropriate for discussion with strangers. A common phrase that reflects this perspective is, "Don't air your dirty laundry."

- Speech is dynamic and expressive with a volume that is often loud compared with some other cultural groups. Body movements are usually involved when communicating with others. Facial expressions can be very demonstrative. **This loud voice volume does not necessarily reflect anger, but merely dynamic expressions.**

- Most individuals are comfortable with a close personal space. **However, be aware that maintaining prolonged direct eye contact may be misinterpreted as aggressive behavior by some. A few older African Americans my avoid eye which is a carry-over from the past where direct eye contact was avoided.**

- In general, most individuals are more present- than past- or future-oriented. Younger and middle-aged people may be more future-oriented as indicated by the value placed on education. Some are more relaxed about time, believing that it is more important to make an appointment than to be on time for the appointment. **Flexibility in timing appointments may be necessary for those who have a circular rather than a linear sense of time.**

- Showing respect with formality in names may be a first step in gaining trust. Most prefer to be greeted formally as Mr., Mrs., Ms., or Miss with the use of their surname because the "family name" is highly respected and connotes pride in their family heritage. **Greet patients formally as Mr., Mrs., Ms., Dr., or Rev. and the last name until told to do otherwise.**

4.1.2 Family Roles and Organization

- A high percentage of families are matriarchal and live below the poverty level. A single head of household is accepted without stigma. When nuclear families are unable to provide emotional and physical support for their children, grandmothers, aunts, and extended or augmented families readily provide assistance or take responsibility for the children. **Even though many African American families are headed by single women and are economically disadvantaged, it does not equate them with broken family structures.**

- Families are usually pluralistic in nature; gender roles and child-rearing practices vary widely depending on the variant characteristics of culture (see Chap. 1). The diverse family structure extends the care of family members beyond the nuclear family to include relatives and nonrelatives. Many family tasks such as cooking, cleaning, child care, and shopping are shared, requiring flexibility and adaptability of roles.

- **The extended family structure is important for teaching health strategies and providing support. Recognize the importance of including women in decision making and disseminating health information.**
- Given a strong work and achievement orientation, most value self-reliance and education for their children. Because parents often do not expect to get full benefit from their efforts because of discrimination, many families tend to be more protective of their children and act as a buffer between their children and the outside world.
- Respect, obedience, conformity to parent-defined rules, and good behavior are stressed for children. The belief is that a firm parenting style, structure, and discipline are necessary to protect children from danger outside of the home. In violence-ridden communities, parents try to keep young children off the streets and encourage them to engage in productive activities. Adolescents are assigned household chores as part of their family responsibility, and many seek employment for pay when they are old enough, thus learning "survival skills" very early.
- Older people, especially grandmothers, are respected for their insight. The role of the grandmother is one of the most central roles in the family, providing economic support and playing a critical role in child care. **Include grandmothers when providing support and health teaching.**
- Social status is important within the community. Occupations such as medicine, dentistry, nursing, and the clergy are highly respected.
- Those who move up the socioeconomic ladder often find themselves caught between two worlds, with roots in the African American community while interacting more within the European American and other communities. Some refer to these individuals as *oreos*—a derogatory term that means "black on the outside, white on the inside."
- Teen pregnancy continues to be a problem in many communities. Poor pregnancy outcomes, such as premature and low-birth-weight infants and obstetrical complications. Premarital teenage pregnancy is not condoned; rather, it is accepted after the fact. The teenage mother is expected to assume primary responsibility for her child, with the extended family providing a strong support system. In other instances, the infant may be informally adopted, and someone other than the mother may become the primary caregiver.
- Acceptance of same-sex relationships varies between and among families. Personal disclosure to friends and family may jeopardize relationships, thereby forcing some to remain closeted. **Do not disclose same-sex relationships to others.**

4.1.3 Workforce Issues

- African Americans acculturate into mainstream society in order to successfully survive in the workforce. Survival is sometimes met with a negative workplace atmosphere motivated by prejudicial attitudes about cultural background and/or skin color. **Immediately address cultural nuances and issues that create ethnic or racial tension in the workplace.**

- Many African Americans continue to be frustrated at their lower-level positions and the absence of African American leadership in many workplaces.

4.1.4   Biocultural Ecology

- African Americans encompass a gene pool of more than 100 racial strains. Therefore, skin color among African Americans can vary from light to very dark. Lighter-skinned people appear more yellowish brown, whereas darker-skinned African Americans appear ashen.
- **Pallor in dark-skinned people can be observed by the absence of the underlying red tones that give brown and black skin its "glow" or "living color."**
- **To assess conditions such as inflammation, cyanosis, jaundice, and petechiae, palpate the skin for warmth, edema, tightness, or induration.**
- **To assess for cyanosis in dark-skinned people, observe the oral mucosa or conjunctiva.**
- **Jaundice is assessed more accurately by observing the sclera of the eyes, the palms of the hands, and the soles of the feet, which may have a yellow discoloration.**
- Keloid formation, common among dark-skinned people, is one example of the tendency toward overgrowth of connective tissue.
- African Americans have a greater bone density than European Americans, Asians, and Hispanics, resulting in a lower incidence of osteoporosis.
- Because of the darker skin color of most African Americans, the early stages of pressure ulcers are more difficult to observe.
- Pseudofolliculitis ("razor bumps") is more common among males, whereas melasma ("mask of pregnancy") is more common among darker-skinned females during pregnancy. Many experience a disproportionate amount of pigment discoloration, with vitiligo being the most common. This is an autoimmune disease that causes skin discoloration and is associated with diabetes and thyroid disorders. If left untreated, it can cause skin cancer. Some erroneously believe that dark-skinned people are not at risk for skin cancer or sunburn because of their higher concentration of melanin.
- **Dispel myths that darker-skinned people are not at risk for skin cancer and sunburn.**
- Birthmarks are more common in this population. Mongolian spots, which are found more often in newborns, disappear over time. **Reassure family that most Mongolian spots will disappear over time.**
- Some women have alopecia related to the use of chemicals to straighten/relax the hair or braiding. **Provide factual information on the causes of alopecia.**

- The pathophysiology of hypertension in African Americans is most frequently related to volume expansion, decreased renin, and increased intracellular concentration of sodium and calcium, making African Americans more genetically prone than whites to retain sodium.
- Although there has been a decline in leading causes of death such as accidents, cancer, infant mortality, and cardiovascular diseases, the adjusted death rates for African Americans are higher than for whites. Other causes of death are homicide, cirrhosis, malnutrition, chemical dependency, and diabetes.
- African Americans are at a higher risk of misdiagnosis for psychiatric disorders and, therefore, may be treated inappropriately with drugs.
- African Americans respond to or metabolize alcohol, antihypertensives, beta blockers, psychotropic drugs, and caffeine differently than European Americans. Psychiatric patients experience a higher incidence of extrapyramidal effects with haloperidol decanoate than European Americans. They are more susceptible to tricyclic antidepressant delirium, showing higher blood levels and a faster therapeutic response. As a result, they experience more toxic side effects. **Observe African American patients closely for side effects related to tricyclics and other psychotropic medications.**
- Light eyes dilate wider in response to mydriatic drugs than do dark eyes. **This difference in response to mydriatic drugs must be taken into consideration when treating African Americans.**
- Table 4.1 lists commonly occurring health conditions for African Americans in three categories: lifestyle, environment, and genetics.

Table 4.1   Health conditions common to African Americans

Health conditions	Causes
Sickle cell disease	Genetic, environment
Hypertension	Genetic, lifestyle
Systemic lupus erythematosus	Genetic with an environmental trigger
Diabetes mellitus	Genetic, lifestyle
Glaucoma	Genetic
Cardiovascular disease	Genetic, environment, lifestyle
Cerebral vascular disease	Genetic, environment, lifestyle
Lung, colon, and rectal cancer	Environment and lifestyle
Prostate cancer	Genetic, environment
Lead poisoning	Environment
Asthma	Environment and lifestyle
HIV/AIDS	Lifestyle
Hemoglobin C disease	Genetic
Hereditary persistence of hemoglobin F	Genetic
Glucose-6-phosphate dehydrogenase deficiency	Genetic
β-Thalassemia	Genetic

- If a baby is born with the amniotic sac (referred to as a "veil") over its head or face, the neonate is thought to have special powers. In addition, a child born after a set of twins, one born with a physical problem or disability, or a child who is the seventh son in a family is thought to have special powers from God.
- The postpartum period may be greatly extended due to the belief that the mother is at greater risk than the baby. She is cautioned to avoid cold air and is encouraged to get adequate rest to restore the body to normal. **Incorporate nonharmful traditional practices into postpartum care.**
- Postpartum practices care can involve the use of a bellyband or a coin placed on top of the infant's umbilical area to prevent the umbilical area from protruding outward. **Teach immediate and extended family methods for keeping equipment and objects clean that are used on the umbilicus.**

4.1.8 Death Rituals

- For most individuals, death does not end the connection among people, especially families. Relatives communicating with the deceased's spirit are one example of this endless connection. Some believe in "voodoo death," a belief that illness or death may come to an individual via a supernatural force. **Voodoo** is more commonly known as "root work," "hex," "fix," "conjuring," "tricking," "mojo," "witchcraft," "spell," "black magic," or "hoodoo."
- Some African Americans do not know about or complete advance directives. This is attributed, in part, to the fact that end-of-life decisions are usually made by the family. **A culturally relevant discussion and education in the African American community is needed regarding advance directives and durable power of attorney.**
- Most families do not rush to bury the deceased, allowing time for relatives to travel from far away to attend funeral services. Therefore, it is common for the burial service to be held 5–7 days after death.
- The body must be kept intact after death. It is common to hear "I came into this world with all my body parts, and I'll leave this world with all my body parts!" **Explain the legal requirements of autopsy.**
- One response to hearing about a death of a family member is "falling out," which is manifested by sudden collapse and paralysis and the inability to see or speak. However, the individual's hearing and understanding remain intact. **Understand that "falling out" is a cultural response to the death of a family member or severe emotional shock and not a medical condition requiring emergency intervention.**
- Some individuals are less likely to express grief openly and publicly. However, most do express their feelings openly during the funeral. Funeral services encourage emotional expression and catharsis by incorporating religious songs into the ceremony and by providing a visual display of the body. **Accept varied responses to bereavement.**

4.1.9 Spirituality

- Most African American Christians are affiliated with the Baptist and Methodist denominations. However, many other distinct religious groups are represented, including African, Episcopal, Jehovah's Witnesses, Church of God in Christ, Seventh-Day Adventist, Pentecostal, Apostolic, Presbyterian, Lutheran, Roman Catholic, Nation of Islam, and other Islamic sects. **Determine specific church affiliation at the time of the intake interview.**
- Churches play a major role in the development and survival of African Americans. There is no disjunction between the black Church and the black community; whether one is a church member or not is beside the point.
- **Determine the importance and meaning of the black Church to patients.**
- Most individuals take their religion seriously and expect to receive a message in preaching that helps them in their daily lives. Religious involvement is associated with positive mental health. Most people take an active part in religious activities. Participation may involve group singing, creating original words to songs, spontaneous testimony of a personal spiritual view, or expression of deep emotion. Singers might be encouraged with cries of "Sing it, sister!" or "That's all right."
- Most strongly believe in the use of prayer for all situations they encounter. They use prayer for the sake of others who are experiencing problems. Prayers reflect the trust and faith one has in God. **Support and allow time for praying for individuals who desire to do so. Ask if anything special is needed for praying.**
- Many believe in the "laying on of hands" while praying. Certain individuals are believed to have the power to heal the sick by placing hands on them. African Americans may pray in a language that is not understood by anyone but the person reciting the prayer. This expression of prayer is referred to as "speaking in tongues." **"Speaking in tongues" does not mean that the patient needs mental health intervention.**
- Having faith in God is a major source of inner strength. Whatever happens is "God's will." Thus, many are perceived to have a fatalistic view of life. **This does not mean that individuals do not care about their health or are not willing to practice illness prevention measures.**
- Most consider themselves spiritual beings; sickness is viewed as a separation between God and man. God is the supreme healer; thus, health-care practices center on religious and spiritual activities. **Partnering with community churches is a good strategy to promote healthy lifestyles.**
- Some associate illness or genetic/hereditary disorders in children with the sins of their parents. In this instance, they believe that healing can occur only through the prayers of a faith healer. **Recognize and support spiritual beliefs and practices that are a source of comfort, coping, and support. Provide factual information about genetic/hereditary disorders.**

4.1.10   Health-Care Practices

- Because a few tend to be suspicious of health-care providers, they may see a physician or nurse only when absolutely necessary. Some older people use the *Farmers' Almanac* to choose what are thought to be good times for medical and dental procedures.
- Many believe in natural and unnatural illnesses. Natural illness occurs in response to normal forces from which individuals have not protected themselves. Unnatural illnesses come via different people or spirits. Health is viewed as harmony with nature, whereas illness is seen as a disruption in this harmonic state due to demons, "bad spirits," or both.
- **In treating an unnatural illness, seek clergy for assistance and recognize and support the patient to pray to a supreme being.**
- Individuals commonly use home remedies, consult folk healers (root doctors), and receive treatment from allopathic health-care providers. **To render services that are effective and culturally acceptable, do a thorough cultural assessment and become partners with the community. Focus groups in churches and community groups can provide insight into health-care practices acceptable to African Americans.**
- **Form partnerships between public health and faith communities.**
- When taking prescribed medications, it is common practice to take the medications differently than prescribed. For example, in treating hypertension, some take their antihypertensive medication on an "as needed" basis. **Other helpful medical treatment modalities in treating hypertension include relaxation techniques and transcendental meditation. Explain the importance of taking medications as prescribed.**
- Needed health-care services may not be affordable for those in lower socioeconomic groups. Some services, although accessible, may not be culturally relevant. For example, a health-care provider may prescribe a strict diabetic diet to a newly diagnosed diabetic African American patient without taking into consideration the dietary habits. **Always include the patient and family when doing dietary counseling.**
- Some may have general distrust of health-care providers and the health-care system. The unequal distribution or underrepresentation of ethnic minority health-care providers can be a barrier to care for some. **When possible, encourage the support of similar ethnic minorities to promote healthy interactions.**
- Some perceive pain as a sign of illness or disease; thus, regularly prescribed medicine may not be followed if individuals are not in pain. Some believe that suffering and pain are inevitable and must be endured, thus contributing to their high tolerance level for pain. Prayers and the laying on of hands are thought to free the person from all suffering and pain, and people who still experience pain are considered to have little faith. **Dispel myths about pain and pain medications.**

- The high frequency of misdiagnosis among blacks/African Americans contributes to their reluctance to trust mental health providers. Many are more likely to report hallucinations when suffering from an affective disorder, which may lead to the misdiagnosis of schizophrenia.
- Close family and spiritual ties allow one to enter the sick role with ease. Extended and nuclear family members willingly care for sick individuals and assume their responsibilities without hesitation. Sickness and tragedy bring families together, even in the presence of family conflict. **Engage extended family members in the care of family members when self-care becomes a concern.**
- Blood transfusions are generally accepted unless the patient belongs to a religious group, such as Jehovah's Witnesses, that does not permit this practice.
- Low levels of organ donation may be related to social practices, religious beliefs, and cultural expectations. Five reasons for the low level of organ donation include lack of information about kidney transplantation, religious fears and superstitions, distrust of health-care providers, fear that donors would be declared dead prematurely, and racism (some prefer to give their organs to other African Americans). **Provide factual information at the patient's educational level about the infusion of blood and blood products and organ donation and transplantation.**

4.1.11 Health-Care Providers

- Physicians are recognized as heads of the health-care team, with nurses having lesser importance. However, as nurses are becoming more educated, African Americans are holding them in higher regard.
- Folk practitioners can be spiritual leaders, grandparents, elders of the community, voodoo doctors, or priests. Voodoo doctors are consulted for unnatural illnesses or for removing a hex. An individual may place a hex on a person because of resentment of achievement, sexual jealousy, love, or envy. A hex can be placed on the victim by using a piece of the individual's hair, fingernail, blood, or some other personal item of the victim. Victims usually seek help from a voodoo or conjure doctor to have the hex removed with magic or religious powers.
- **Include spiritual leaders and voodoo practitioners in care if the patient wishes.**
- Folk practitioners are respected and valued in the African American community and are used by all socioeconomic levels. Some perceive health-care providers as outsiders, and they may be suspicious and cautious of health-care providers they have not heard of or do not know.
- **Because interpersonal relationships are highly valued in this group, it is important initially to develop a sound, trusting relationship.**
- Whereas some individuals may prefer a health provider of the same gender for urologic and gynecologic conditions, gender is generally not a major concern in

selection of a health-care provider. Men and women can provide personal care to the opposite sex. On occasion, young men prefer that another man or older woman give personal care. Some women prefer female primary care physicians. **Respect these wishes and provide same-gender care providers when possible.**

4.2   Reflective Exercises

1. Given the heritage and diversity of the African American population in the US, what cultural and social issues do you consider important in with patients from this culture?
2. What is your knowledge of the social and historical issues that have affected African Americans?
3. Recognizing that some African Americans are more present oriented, how can you ensure that a patient follows through on preventive health-care advice?
4. How would you assess for jaundice, rashes, anemia, and sunburn in dark skinned people.
5. What resources are available for drug metabolism differences among ethnic/racial groups?
6. You are caring for an African American hypertensive patient who had a stroke. His family brings him the following foods: collard greens with bacon fat, deep-fried pork rinds, breaded fried okra, fried chicken, and ham hock with black-eyed peas. How would you go about providing dietary counseling with the family?
7. A postpartum young mother is in an exercise program. She informs you that she does not want to continue the formal exercise program because her grand-mother provides her with herbs and home remedies. How do you encourage her to consider exercises while using her home remedies?
8. Recognizing the importance of religion in the African American culture, what strategies can you use to support positive health behaviors?
9. You are working in a multi-cultural, multi-racial environment. One colleague has been overheard making derogatory comments about different racial groups. What might you do to address this employee's comments?
10. A 22-year old African American male with known hypertension visits the clinic because of a persistent headache. His blood pressure is 188/120. He is almost 200% of his ideal weight. He admits to taking his blood pressure medicine only when he has a headache or "feels bad". What some important teaching strategies for this young man?

Bibliography

Abbyad C, Robertson TR (2011) African American women's preparation for childbirth from the perspective of African American health-care providers. J Perinat Educ 20(1):45–53

Campinha-Bacote J (2013) People of African American ancestry. In: Purnell L (ed) Transcultural health care: a culturally competent approach, 3rd edn. F.A. Davis Co., Philadelphia

Jacobs EA, Rolle I, Ferrans CE, Whitaker EE, Warnecke RB (2006) Understanding African American' views of the trustworthiness of physician. J Gen Intern Med 21(6):642–647

Jardine SA, Dallaflor A (2012) Examining gender dynamics in the context of African American families. J Pedagogy Pract 4(4):1–4

Labore N, Mawn B, Dixon J, Andemariam B (2017) Exploring transitions to self-management with the culture of sickle cell disease. J Transcult Nurs 28(1):70–79

Rutherford J (2016) Funeral customs of African American Southern Baptists. http://peopleof.oureverydaylife.com/funeral-customs-african-american-southern-baptists-6189.html

Sage encyclopedia of pharmacology and society (2016) http://sk.sagepub.com/reference/download/the-sage-encyclopedia-of-pharmacology-and-society/i1498.pdf

Thorburn S, Bogart LM (2005) Conspiracy beliefs about birth control: barriers to pregnancy prevention among African Americans of reproductive age. Health Educ Behav 32(4):474–487

Townsend A, March AL, Kimball J (2017) Can faith and hospice coexist: is the African American church the key to increases hospice utilization for African Americans. J Transcult Nurs 28(1):32–40

Waters C (2000) End-of-life directives among African Americans: a need for community-centered discussion and education. J Community Health Nurs 17(1):25–37

Wicher CP, Meeker MA (2012) What influences African American end-of-life preferences? J Health Care Poor Underserved 23(1):28–58

Chapter 5
People of American Indian/Alaskan Native Heritage

5.1 Overview and Heritage

This chapter provides a general overview of American Indians and Alaskan Natives (AI/AN) followed by a more detailed description of the Navajo as one exemplar. Space does not permit an extensive discourse on each tribe. The 2010 Census reports 2.9 million people identified as American Indian and Alaska Native alone. Another 2.3 million reported being American Indian and Alaska Native in combination with one or more other races.

There are more than 567 federally recognized sovereign AI/AN **tribes**, of which 229 are Alaskan Natives and are divided into three groups: Indians, Aleuts, and Eskimos. In addition, there are over 200 non-federally recognized tribes, each with its unique customs and beliefs. Some Alaskan Natives do not mind being called Eskimos or Indians; however, some do not like to be called Native Americans or Eskimos and prefer Alaskan Native, Alaska Indian, or Inuits. The amount of Indian blood necessary to be considered a tribal member or American Indian varies with each tribe. The Cherokee requires no minimum.

The Indian Health Service (**IHS**) was designed to fulfill the U.S. government's role in providing health care to American Indians and Alaskan Natives (AI/AN); however, inadequate funding has left those who depend on the IHS with only limited health care.

- Overall, AI/ANs are extremely diverse groups of people. Some are very traditional, some are highly acculturated, and some selectively combine tradition with the European American culture. Many share similar, but not exact, traditions, beliefs, and values.
- AI/ANs are a particularly vulnerable population in terms of accessing health care.
- High poverty rates, individual high-risk health behaviors, social factors, low educational levels, and chronic conditions make it particularly important for accessing and receiving ongoing supportive and preventive care.

© Springer Nature Switzerland AG 2019
L. D. Purnell, E. A. Fenkl, *Handbook for Culturally Competent Care*,
https://doi.org/10.1007/978-3-030-21946-8_5

Table 5.1 Regions of the Indian Health Service

Region	Area(s) served
Aberdeen Area	North Dakota, South Dakota, Iowa, and Nebraska
Alaska Area	State of Alaska
Albuquerque Area	New Mexico, Colorado, and Texas
Bemidji Area	Indiana, Minnesota, Michigan, and Wisconsin
Billings Area	Montana and Wyoming
California Area	California and Hawaii
Nashville Area	Eastern US
Navajo Area	Arizona, New Mexico, and Utah
Oklahoma Area	Oklahoma, Kansas, and Texas
Phoenix Area	Arizona, California, Nevada, and Utah
Portland Area	Oregon
Tucson Area	Southern Arizona

- One-third of AI/ANs are in families with incomes below the federal poverty line.
- One in five adults has not graduated from high school.
- Almost half of uninsured adults do not have a usual source of care, making it difficult to receive preventive services and timely care for acute health problems.

The IHS is divided into 12 regions, each with hospitals, clinics, and outreach areas. The regions are listed in Table 5.1. Federally recognized tribes are considered sovereign nations, giving them independent authority over their territories with the power to rule and make laws.

- Although the IHS includes access to some specialty care and has established urban Indian health clinics, shortages in funding and a lack of facilities and health-care professionals have left many eligible AI/ANs struggling to obtain timely, high-quality health care.
- The AI/AN population is a young population with a median age of 25 years compared with 35 years for the US overall.

5.1.1 Communications

- AI/ANs are collective cultures and avoid conflict and confrontation.
- Each tribe has its own language, but some languages are similar to each other.
- Younger people speak English. Some older people may know only their traditional tribal language or perhaps Spanish in the southwestern part of the US.
- **Use professional interpreters whenever possible. Confidentiality is taken seriously. Provide interpreters unknown to the patient or family.**

- **Greeting with Mr., Mrs., or Miss shows respect and is always safe. While use of the given name is preferred by some; wait for permission to use it.**
- Asking for help with pronunciation of a surname is appreciated.
- **Greet everyone in the room and establish relationships.**
- Depending on the situation, punctuality may not be valued. Lateness is not meant as disrespect but is the result of a different sense of time. **Explain the importance of timeliness if required and be flexible with appointment times whenever possible. Be aware that many do not have reliable transportation.**
- Silence is a means of respect and thinking before speaking. **Interrupting someone while he/she is talking is considered rude.**
- Strong emotions are not usually displayed verbally or nonverbally.
- Direct questioning expecting an immediate answer is frowned upon. **Ask questions in an indirect way, using a phrase such as "Tell me about..."**
- One does not have the right to speak for someone else.
- Directly telling someone that he/she is wrong is considered rude.
- Maintaining continued eye contact with someone during a conversation is considered disrespectful. **Do not maintain direct and sustained eye contact during an assessment.**
- Singling out a specific individual for an answer is disrespectful because if the person is wrong, he/she will meet peer disapproval.
- A long pause in a conversation does not necessarily mean the person is finished talking. **Wait for a nonverbal gesture that says the person has finished.**
- **Do not use idioms, colloquialisms, or abstract ideas; those unacculturated to the dominant culture might not understand them.**
- **Greet AI/ANs with a *light* handshake, not a firm, pumping handshake.**
- **A nod of the head may denote "I hear you," not necessarily agreement with what has been said.**
- Loudness may be interpreted as aggressiveness.
- Most appreciate a space greater than 2 ft during a conversation, especially with strangers, and for some maybe up to 6 ft. **Take cues from the patient.**

5.1.2 *Family Roles and Organization*

- The extended family is the primary social unit in most AI/AN cultures.
- Most define family as members made up of fictive (blood-related) and nonfictive (non–blood–related) kin, extended family, and tribal community. One is rarely alone or without family.
- Some tribes are patriarchal and patrilocal (a social system where a married couple lives with the husband's family); others are matriarchal and matrilocal (a social system where a married couple lives with the wife's family) in structure. **Given the individuality in any family system, determine the family spokesperson ahead of time before a decision has to be made.**

5.1.6   Nutrition

- Although great variation exists, most AI/AN diets are high in fat and sugar.
- **Include a dietary assessment as part of the intake interview.**
- Availability of fresh fruits and vegetables may be severely limited, especially during winter in cold climates and on reservations.
- **Ask if any particular food is taboo, what foods are consumed for health promotion, and what foods are consumed during illness.**

5.1.7   Pregnancy and Childbearing Practices

- Birth control practices are usually an individual choice.
- Some do not discuss birth control with providers until significant trust is established.
- Each tribe has specific foods believed to be beneficial for pregnancy. **Ask about specific foods at each stage of the pregnancy.**
- Women in many tribes remain active throughout their pregnancies.
- Some believe that wearing a necklace may cause the baby to be born with the umbilical cord wrapped around its neck.
- Some believe that if a pregnant woman walks through a door backwards, she will have a breech delivery.
- Some believe that blowing up balloons may cause premature rupture of the membranes.
- Some believe that leaving projects unfinished may prolong labor.
- Stoicism during labor and delivery is valued by some. **Ask frequently if pain medicine is needed.**
- Younger, more urban generations may not adhere to any traditional birthing practices.
- For some tribes, the father is not usually present during the birth. **Do not make the father feel guilty if he does not want to participate until after the delivery.**
- Delivering in a kneeling or squatting position is preferred by some.
- In some tribes, the placenta is saved and buried, returning it to mother earth.
- Postpartum, the mother may be isolated according to the sex of her child; 1 month for a boy and 2 months for a girl. However, this is not a widespread belief.
- Many women resume their normal activities within a week and are expected to care for their babies as well as other family members.
- Women often turn to their mothers for advice after birth.
- Some tribes do not name the baby for a week or longer after the delivery and may include an English name along with a tribal name. **Do not assume that not naming the baby means the parents do not care about or do not want the baby.**
- Some tribes often name a newborn after a recently deceased respected relative.
- Both sexes are seen as gifts from nature.

5.1.8 Death Rituals

- In general, AI/AN religions do not have precise beliefs about life after death.
- Some believe in reincarnations with rebirth as a human, as a ghost, as an animal, or as a combination of these.
- Burial ceremonies vary by tribe. For example, the burial ceremony of the Zuni requires individuals to take 3 days off from work.
- Autopsy is generally acceptable. **Explain legal requirements if necessary.**
- **Expect and respect a wide variation on bereavement.**

5.1.9 Spirituality

- Some tribes have preserved their spiritual beliefs in writing or as an oral tradition.
- Some adhere to the Native North American Church. Others may be more traditional, practice mostly Christian religions, or combine the two.
- The Native North American Church practices have four requirements for good health: food, sleep, cleanliness, and good thoughts. The four divisions of nature are also a holistic part of spirituality: spirit, mind, body, and life.
- Routes of spiritual connectedness include singing, drumming, dancing, and using sweat lodges. **Warn about dehydration when using sweat lodges.**
- Table 5.2 lists the major values between AI/AN tribes and the European American culture.

5.1.10 Health-Care Practices

- Religion, medicine, and healing are inseparable among AI/AN populations.
- Diseases can be attributed to natural or unnatural forces, the scientific theories of illness and disease, or a combination of both.
- Natural remedies are preferred, although the specific herbal remedies vary significantly by tribe and region of the country and depend on environmental influences for what is available.
- **Ask patients about specific herbal and tribal remedies so they do not conflict with or potentiate allopathic medicines and treatment.**
- **Incorporate IHS programs that give general directions for health-care providers: Breastfeeding Promotion and Support, Diabetes Treatment and Prevention, Epidemiology, Elder Care Initiative, Health Weight for Life, Influenza, and HIV/AIDS (www.IHS.gov).**
- **Family conferences are appreciated with serious illness.**
- Mental illness may be seen as violation of tribal taboos or loss of harmony with the environment.
- Organ donation is not usually practiced, but it is usually an individual choice.

Table 5.2 Tribal values versus European American values

Tribal traditional cultural values	European American values
Group, clan, or tribal emphasis	Individual emphasis
Present oriented	Future oriented
Time, always with us	Time, use every minute
Age	Youth
Cooperation	Competition
Harmony with nature	Conquest of nature
Giving, sharing	Saving
Pragmatic	Theoretical
Mythological	Scientific
Patience	Impatience
Mystical	Skeptical
Shame	Guilt
Permissiveness	Social coercion
Extended family and clan	Immediate family
Nonaggressive	Aggressive
Modest	Over confident
Silence	Noise
Respect others' religion	Convert others to religion
Religion, way of life	Religion, a segment of life
Land, water, forest	Land, etc.
Belong to all	Private domains
Beneficial, reasonable	Avarice, greedy
Use of resource	Use of resource

5.1.11 Health-Care Providers

- "Shaman" is often used refer to traditional AI/AN healers.
- Spirits may be encouraged to occupy the shaman's body during public lodge ceremonies. Drum beating and chanting aid this process.
- Tribal spiritual healers may be used in combination with allopathic providers.
- **Ask for disclosure of all treatments and explain possible complications.**
- **Partner with traditional healers.**

5.2 People of Navajo Heritage

Navajo Indians claim the distinction of being the largest tribe: at least one-fourth Navajo blood is required to be considered a member of the tribe. The Navajo Indians are nomadic and wander great distances, searching for adequate grazing grounds for their sheep. Because of severe economic conditions and high unemployment rates, significant migration occurs into and out of the reservations. Commercial activities on reservations are limited to businesses owned by Navajo Indians or partnerships in which a Navajo must own the controlling interest. Many Native Americans who leave their reservations experience culture shock, and many return to their reservation on a regular basis to refresh and renew themselves through Blessingway ceremonies.

Rug-weaving and jewelry-making are the most common forms of art among the Navajo. They are also noted for sand paintings, which were traditionally used in healing ceremonies by medicine people and not originally intended for sale.

5.2.1 Communications

- The Navajo language was not reduced to writing until the 1970s; consequently, many older people speak only their native language, and few are literate in the English language. The younger populations are usually bilingual, with their native tongue spoken primarily in the home. **Be extremely careful when attempting to use the Navajo language, because minor variations in pronunciation may change the entire meaning of a word or phrase. Differences in pronunciation, particularly while speaking with an older Navajo, may cause a misunderstanding. Such misunderstandings make subsequent caring for the individual difficult. It is often safer to use an interpreter.**
- Talking loudly among Navajo Indians is considered rude. Voice tones are quiet but not monotone. Their language is full of inflections with different meanings, making the language melodious with a quiet volume.
- Most generally do not share inner thoughts and feelings with anyone outside their clan. It may take nontribal health-care providers a long time to build trust.
- Most Navajo are comfortable with long periods of silence. Interest in what an individual says is shown through attentive listening skills. One may be considered immature if answers are given quickly or one interrupts another who is forming a response. **Failing to allow adequate time for processing information may result in an inaccurate response or no response. Allow time to respond to questions.**
- Touch among the Navajo is unacceptable unless one knows the person very well. **Observation body language for determining cues related to the permissibility of touch. Ask permission and explain the reason for touch.**
- The acceptable personal space for American Indians is greater than that of American cultures. **Do not take offense if the patient stands at a greater distance from what you are accustomed.**
- Shaking hands is the traditional greeting. The handshake is light, more of a passing of the hands. **Shaking hands with the Navajo is an acceptable greeting.**
- Pointing with the finger is considered rude. Rather than pointing a finger to indicate a direction, individuals shift their lips toward the desired direction. **Do not point with your finger.**

- Sustained direct eye contact is rude and possibly confrontational. Even close friends do not maintain eye contact. **Not maintaining eye contact does not mean the patient or family is being less than truthful.**
- The time sequence is present, past, and future, in that order. Very little planning is done for the future because the Navajo view is that many things are outside of one's control and may affect or change the future. In fact, the Navajo language does not have a future-tense verb. To plan for the future is sometimes viewed as foolish. Events do not always start on time, but rather time starts when the group gathers. **To help prevent frustration in scheduling events, time factors need to be taken into consideration and the speaker made aware of these unique time perceptions. Appointments may not always be kept, especially if someone else in the clan needs help.**
- Older people are addressed as grandmother or grandfather or as mother or father by their clan. Otherwise, they are called by a nickname. **A health-care provider can call an older Navajo "grandmother" or "grandfather" as a sign of respect.**

5.2.2 Family Roles and Organization

- Whereas men are considered important, grandmothers and mothers are at the center of Navajo society. **When providing family care, note that no decision is made until the appropriate older woman is present. Find the appropriate gatekeeper; otherwise, time is lost and the problem must be addressed again later.**
- The relationship between brother and sister is often more important than the relationship between husband and wife.
- Even though children may be named at birth, their names are not revealed until their first laugh, when they are considered to officially have a soul and self-identity. This protects the children and keeps them in tune with the Holy People.
- Infants are kept in the cradleboard until they begin walking. However, hip dysplasia may be exacerbated by cradleboards. The use of diapers has decreased the incidence of hip dysplasia because diapers bind the hips in a slightly abducted position. **Encourage the use of diapers, and explain the hazard of using a cradleboard.**
- The use of formula has become popular, resulting in an increased incidence of bottle caries because many babies go to bed with a bottle of juice or soda pop. This practice causes children to lose their teeth by the age of 4 years. **Educate parents about dental caries and the hazards of sweet drinks and the importance of breast-feeding.**
- A primary social premise is that no person has the right to speak for another. Thus, children are allowed to make their own decisions; for example, children may be allowed to decide if they want to take their medicine. **Explain the importance of taking medicines as prescribed.**

- An important ceremonial ritual for teenage girls is the onset of menarche, which is celebrated with special foods that symbolize passage into adulthood, with only aunts and grandmothers participating.
- The family unit consists of the nuclear family and relatives such as sisters, aunts, and their female descendants.
- Family goals do not center on wealth or the attainment of possessions. In fact, if one person has more wealth than other relatives, the member who has more has a responsibility to assist relatives who have less.
- A sister's children are considered the same as her own children. If a mother dies or for some other reason cannot care for her children, the grandmother or sister will raise the children as her own.
- Older people are looked on with clear deference. Younger adults are faced with the responsibility of caring for relatives.
- There are few nursing homes, and hospitals are forced to keep patients until nursing home placement is found. When nursing home placement is found, it may be at a great distance from the family, making it difficult for family visits.
- Alternative lifestyles are not discussed but are quietly accepted. Special individuals exist who are not looked on with disfavor but rather are accepted as being different.

5.2.3 Workforce Issues

- Navajo are compelled to attend ceremonies and often must take time from work to do so. **Balance the needs of the individual against organizational requirements for taking time off work. Allow annual leave or allow employees to earn religious compensatory time.**
- Many Indians avoid people with whom they are in conflict. Persistence in dealing immediately with the conflict causes additional ill feelings. If one tribal member has a disagreement with another, that member does not address the second party face to face. Rather the member tells a third person, who then relays the information to the second party. **Resolve conflicts through third-party compromise.**
- One individual should not be singled out to answer a question because one's mistakes are generally not forgotten by the group. Improved success is achieved if the American Indian is allowed to observe a task several times before being asked to demonstrate it. Their first effort at completing the task in front of a group should occur without error. This is especially true when delivering care to their people. Mistakes that have been made are discussed in the community. **Allow adequate time for observation for a greater chance of success the first time a procedure is performed.**
- Educational levels may be low. **Work assignments should be made in clear, concrete terms.**
- Younger supervisors may not be respected because they are not perceived as possessing the life experiences necessary to lead. Major decisions are made by the

group with the assistance of the group leader, who is generally the senior woman. Thus, a young male manager on the reservation may face resistance when attempting to direct a work group.

- Indian Health Service is required to hire a Native American when possible. This law is referred to as the Indian Preference Law.
- English may not be their primary language. **Allow extra time for a verbal response to think about a response and translate it into English.**
- Most American Indian students are not good test takers. Although the individual may have the knowledge necessary to complete an examination, the translation of the knowledge into written form is especially difficult. When tested verbally on the same material, students tend to score higher on the examination. **To the extent possible, examinations that consist of return demonstration and that do not use abstract terminology are preferred. Provide special classes that teach multiple-choice test taking.**

5.2.4 Biocultural Ecology

- Skin color varies from light brown to very dark brown. **To assess for oxygenation in darker-skinned people, examine the mucous membranes and nail beds for capillary refill. Anemia is detected by examining the mucous membranes for pallor and the skin for a grayish hue. To assess for jaundice, examine the sclera rather than relying on skin hue. Newborns commonly have Mongolian spots on the sacral area. Do not mistake these spots for bruises and suspect child abuse.**
- The Navajo appear Asian, with epithelial folds over the eyes. They are generally taller and thinner than other American Indian tribes. The Navajo have traditionally been good runners and excel in relay races and long-distance running. **Remember that these characteristics are not seen with everyone; variations in this population do exist.**
- Table 5.3 lists commonly occurring health conditions in the Navajo in three categories: lifestyle, environment, and genetics. See Chap. 2.
- The water on the Navajo reservation is often impure and unchlorinated, making those who drink it susceptible to water-borne diseases such as shigellosis. Salmonella is common because of the lack of refrigeration and hypothermia because of frequent snowstorms and conditions that limit Navajo ability to gather wood. **Teach patients how to protect themselves from rodent-borne diseases. At every opportunity teach about the combination of environmental and lifestyle factors to decrease the incidence of parasitic and infectious diseases and illnesses and injury prevention.**
- Lidocaine reactions occur in 29% of the Navajo population as compared with 11–15% of European Americans. **Carefully observe for lidocaine reactions.**

Table 5.3 Health conditions common to Navajos

Health conditions	Causes
Shigellosis	Environment, lifestyle
Salmonella	Environment, lifestyle
Respiratory illnesses	Environment, lifestyle
Otitis media	Environment, lifestyle
Tuberculosis	Environment, lifestyle
Plague	Environment, lifestyle
Tick fever	Environment, lifestyle
Muerto Canyon Hanta virus	Environment, lifestyle
Diabetes mellitus	Genetic, environment, lifestyle
Combined immunodeficiency syndrome (SCIDS)	Genetic
Navajo neuropathy	Genetic
Albinism	Genetic
Alcohol misuse	Genetic, environment, lifestyle
Suicide	Environment, lifestyle
Homicide	Environment, lifestyle
Unintentional injuries	Environment, lifestyle
Gastrointestinal disease	Environment, lifestyle
Infant mortality	Environment, lifestyle
Heart disease	Genetic, environment, lifestyle
Cirrhosis	Environment, lifestyle
Fetal alcohol syndrome	Environment, lifestyle

5.2.5   High-Risk Health Behaviors

- Alcohol use is more prevalent than any other form of chemical abuse. Health problems related to alcoholism include motor vehicle accidents, homicide, suicide, and cirrhosis.
- Spousal abuse is common and is frequently related to alcohol use. The wife is the usual recipient of the abuse, but occasionally the husband is abused.
- Suicide is a problem among the adolescent population. Whereas space does not permit an extensive review of the American Indian/Alaskan Native National Suicide Prevention Strategic Plan, health-care providers are encouraged to access the plan at www.ihs.gov/PublicAffairs/DirCorner/2011_Letters/AIANNationalSPStrategicPlan.pdf **Educate elementary school teachers on suicide ideology and make referrals to health-care providers as necessary.**
- Noncompliance with seat-belt use is high. **Teach patients the importance of using child safety seats, helmets, and seat belts.**

5.2.6 Nutrition

- Life events and religious ceremonies are celebrated with food. Food is not generally associated with promoting health or illness. Herbs are used in the treatment of many illnesses to cleanse the body of ill spirits or poisons. **Encourage and teach about healthy food choices and preparation practices.**
- Sheep are a major source of meat, and sheep brains are a delicacy. Fry bread and mutton are cooked in lard. Access to fresh fruits or vegetables is minimal except during the fall. Squash is common at harvest time. Corn is an important staple. Corn pollen is used in the Blessingway and many other ceremonies.
- Diets may be deficient in vitamin D because many individuals suffer from lactose intolerance or do not drink milk. **Recommend nondairy foods that are high in calcium or making stews and purees using animal bones.**

5.2.7 Pregnancy and Childbearing Practices

- Traditional Navajos do not practice birth control and, thus, do not limit the size of their families. Large families are considered favorably because, in times past, many children died at an early age. **Encourage prenatal care, carefully explaining the detrimental effects on the mother and infant if prenatal care is not sought.**
- Twins are not considered favorably and are frequently believed to be the work of a witch, in which case one of the babies must die. Sometimes the mother may have difficulty caring for two infants. Twins may be readmitted to the hospital for neglect and failure to thrive. **Culturally sensitive counseling assists adoption. Encourage tribal members to adopt the children.**
- It is especially important to adhere to the many prescriptive and restrictive taboo practices related to pregnancy, which involve both husband and wife. See Table 5.4, Navajo Taboos Regarding Expectant Women, and Table 5.5, Taking Care of Yourself During Pregnancy: Navajo Rules for Expectant Couples.
- During labor, the mother wears birthing necklaces made of juniper seeds and beads to assist with a safe birth. Woven belts or sashes are used to help push the baby out. This practice is also used by Navajo midwives in caring for their patients.
- Many Navajo are reluctant to deliver their babies in hospital settings. They know that people have died in hospitals and thus perceive that pregnant women should not be around the dead or in a place where people have died.
- A taboo practice among the Navajo is purchasing clothes for an infant before birth. **Do not interpret not buying clothes for the infant as not wanting the baby.**

Table 5.4 Navajo Taboos regarding expectant women

1. Do not wear two hats at once; you will have twins (or two wives)
2. Do not hit babies in the mouth; they will be stubborn and slow to talk
3. Do not have a weaving comb (rug) with more than five points; your baby will have extra fingers
4. Do not have a baby cross its fingers; its mother will have another one right away
5. Do not swallow gum while you are pregnant; the baby will have a birthmark
6. Do not kill animals while your wife is pregnant; the baby will look like a bird
7. Do not stand in the doorway when a pregnant woman is present
8. Do not make a slingshot while you are pregnant; the baby will be crippled
9. Do not go to ceremonies while pregnant; it will have a bad effect on the baby
10. Do not eat a lot of sweet stuff while you are pregnant; the baby won't be strong
11. Do not sleep too much when you are about to have a baby; the baby will mark your face with dark spots
12. Do not look at a dead person or animal while you are pregnant; the baby will be sickly because of bad luck
13. Do not jump around if you are pregnant or ride a horse; it will induce labor
14. Do not cut gloves off at the knuckles, the baby will have short, round fingers
15. Do not cut a baby's hair when it is small; it won't think right when it gets older
16. Do not put on a Yei mask while your wife is pregnant; the baby will have a big head and look strange
17. Do not let a baby's head stay to one side in the cradleboard; it will have a wide head
18. Do not watch or look at an accident while your wife is present; it will affect the baby
19. Do not sew on a saddle while your wife is pregnant; it will ruin the baby's mouth

Ursula Wilson, July 1987. IHS inservice seminar, Tuba City Indian Health Center, Tuba City, AZ

- Immediately after birth, the placenta is buried as a symbol of the child being tied to the land. Sometimes it is burned in a fire. This is considered a safe place because fire is sacred and protects the baby against evil spirits.
- After birth, the baby is given a mixture with juniper bark to cleanse its insides and rid it of mucus. In addition, a ceremonial food of corn pollen and boiled water is given. Corn symbolizes healthy nutrients and an enduring nature.

5.2.8   Death Rituals

- One death taboo involves talking with patients concerning a fatal disease or illness. Effective discussions require that the issue be presented in the third person, as if the illness or disorder occurred with someone else. **Never suggest that the patient is dying. To do so would imply that the provider wishes the patient dead. If the patient does die, it would imply that the provider might have evil powers.**

Table 5.5 Taking care of yourself during pregnancy: Navajo rules for expectant couples

During prenatal period	
Mind/soul	
Do's	Don'ts
• Keep the peace • Keep thoughts good • Talk with "corn pollen sprinkled" words • Say morning (dawn) prayers • Have shielding prayers if you have nightmares	• Argue with partner or others • Scold children • Allow bad thoughts to occupy mind for long period of time • Talk negatively or with criticism
Body	
Do's	Don'ts
• Eat foods good for baby • Get up early and walk around • Have a Blessingway ceremony for a safe delivery	• Drink milk or eat salt or foods taken away by Navajo ceremonies • Lie around too much • Tie knots • Attend funerals or look at body of deceased person • Be with sick people for long or go to crowded place • Attend healing ceremonies for sick people like "Yei Bei Chai Dance" • Look at dead animals or taxidermic trophies • Look at eclipse of moon or sun • Make plans for baby or prepare layette sets until after birth • Lift heavy things • Kill living things or cut a sheep's throat • Weave rugs or make pottery
During labor	
Mind/soul	
Do's	Don'ts
• Think about a good delivery • Have medicine people do "Singing Out Baby" chants • Have medicine person perform "Unraveling" songs if necessary	• Let too many people observe labor; only people who are helping you in some way
Body	
Do's	Don'ts
• Loosen your hair • Drink cornmeal mush • Wear juniper seed beads • Burn cedar • Hold onto sash belt when ready to push • Have someone apply gentle fundal pressure during pushing effort • Get in squatting position for pushing • Drink herbal tea to relax if necessary • Drink herbal tea to strengthen contractions if necessary	• Braid or tie hair in a knot • Tie knots

Table 5.5 (continued)

After birth of baby (postpartum period)	
Do's	Don'ts
• Bury the placenta	• Drink cold liquids or be in cold draft
• Drink juniper/ash tea to cleanse your insides	• Smell afterbirth blood for too long
• Drink blue cornmeal mush	• Show signs of displeasure if baby soils on you or during diaper change
• Smear baby's first stool on your face	• Have sexual intercourse for 3 months after delivery
• Breast-feed your baby	
• Wrap sash belt around waist for days after delivery	

Ursula Wilson, July 1987. IHS inservice seminar, Tuba City Indian Health Center, Tuba City, AZ

- The body must go into the afterlife as whole as possible. The body is not buried for approximately 4 days after death.
- A cleansing ceremony must be performed after an individual dies, or the spirit of the dead person may try to assume control of someone else's spirit. Family members are reluctant to deal with the body because those who work with the dead must have a ceremony to protect themselves from the deceased's spirit.
- If the person dies at home, the Hogan must be abandoned, or a ceremony must be held to cleanse it.
- Before burial, a ring is placed on the index finger of each hand, and the shoes are put on the wrong feet. This allows living relatives to recognize individuals if they come back and present themselves at ceremonial dances. **Do not wear rings on your index fingers, as older people may not want to be around them.**
- Excessive displays of emotion are not considered favorably among some tribes. **Support survivors and permit family bereavement and grieving in a culturally congruent and sensitive manner that respects the beliefs of the tribe.**

5.2.9 Spirituality

- The American Indian religion predominates in many tribes. Sometimes hospital admissions are accompanied by traditional ceremonies and consultation with a pastor. Even if people are strong in their adopted beliefs, they honor their parents and families by having a traditional healing ceremony.
- Many Navajo start the day with prayer, meditation, corn pollen, and running in the direction of the sun.
- Spirituality for most is based on harmony with nature. The meaning of life is derived from being in harmony with nature. The individual's source of strength comes from the inner self and also depends on being in harmony with one's surrounding.
- Prayers ask for harmony with nature and for health and invite blessings to help the person exist in harmony with the earth and sky. Prayer helps the Navajo to attain fulfillment and inner peace with themselves and their environment.

Table 5.6 Basic concepts of traditional Indian medicine	• Indians believe in a Supreme Creator
	• Each person is a threefold being composed of mind, body, and spirit
	• All physical things, living and nonliving, are part of the spiritual world
	• The spirit existed before it came into the body, and it will exist after it leaves the body
	• Illness affects the mind and the spirit as well as the body
	• Wellness is harmony
	• Natural unwellness is caused by violation of a taboo
	• Unnatural wellness is caused by witchcraft
	• Each individual is responsible for his or her own health

- Spirituality cannot be separated from the healing process in holistic ceremonies. Illness results from not being in harmony with nature, from the spirits of evil people such as witches, or from violation of taboos. Healing ceremonies restore an individual's balance mentally, physically, and spiritually. **Assist with arrangements for healing ceremonies and practicing traditional religious beliefs and prayer.**
- The core concepts of traditional Indian medicine are shown in Table 5.6.

5.2.10   Health-Care Practices

- Many older people do not understand the theory of germs.
- Asking patients questions in order to make a diagnosis fosters mistrust. This approach is in conflict with the practice of traditional medicine men who tell people what is wrong without their having to say anything.
- The medicine man is expected to diagnose the illness and prescribe necessary treatments for regaining health. **In Western health care, the health-care provider asks the patient what he or she thinks is wrong and then prescribes a treatment. This practice is sometimes interpreted by American Indians as ignorance on the part of the white healer.**
- Often great distances must be traveled to reach hospitals or health-care facilities. Many families do not have adequate transportation and must wait for others to transport them into town. Immunizations may be missed because parents do not have transportation. **Pay close attention to the immunization status of patients on their arrival at the emergency department or clinic. If patients are not current with immunizations, scheduling an appointment may be a waste of time because they may not be able to return until a ride is found. Take time to administer the immunization on the spot or make a referral to the public health nursing office.**

- Pain control is may be ineffective because the actual intensity of the Indian's pain is not obvious to the health-care provider and because patients do not request pain medication.
- Pain is viewed as something that is to be endured. Herbal medicines may be used without the knowledge of the health-care provider. **Establishing trust will encourage patients to fully disclose herbal treatments used for pain control. Offer pain medicine, and explain that it will promote healing.**
- Mental illness results from witches or witching (placing a curse) on a person. In these instances, a healer who deals with dreams or a crystal-gazer is consulted. Individuals may wear turquoise to ward off evil; however, a person who wears too much turquoise is sometimes thought to be an evil person and, thus, someone to avoid.
- Rehabilitation as a concept is relatively new. **Explain concepts of rehabilitation.**
- Those with physical or mental handicaps are not considered different; rather, the limitation is accepted, and a role is found for them within the society.
- Cultural perceptions of the sick role for the American Indian are based on the ideal of maintaining harmony with nature and with others. Older people frequently work even when they are seriously ill and often must be encouraged to rest.
- Autopsy and organ donation are unacceptable practices to traditional American Indians. **Explain that autopsy is a legal requirement in some instances.**

5.2.11 Health-Care Provider

- Native healers are divided primarily into three categories: those working with the power of good, the power of evil, or both. Generally, these healers are divinely chosen and promote activities that encourage self-discipline, self-control, and acute body awareness. Some healers are endowed with supernatural powers, whereas others only have knowledge of herbs and specific manipulations. Acceptance of Western medicine is variable, with a blending of traditional health-care beliefs.
- Many are suspicious of American Indian physicians. Many health concerns of American Indians can be treated by both traditional and Western healers in a culturally competent manner when these health-care providers are willing to work together and respect each other's differences.
- Western health-care providers, traditional medicine men, and herbal healers receive respect on the reservation. However, not all individuals accord equal respect to these groups, and many prefer one group over the other or use all three.
- Navajo tribal health-care providers divide their knowledge into preventive measures, treatment regimens, and health maintenance. An example of a preventive measure is carrying an object or a pouch filled with objects prescribed by a medicine

Table 5.7   Traditional American Indian health-care providers

1. People who can use their power only for good can transform themselves into other forms of life and can maintain cultural integration in times of stress
2. People who can use their powers for both evil and good are expected to do evil against someone's enemies. People in this group know witchcraft, poisons, and ceremonies designed to afflict the enemy
3. The diviner diagnostician, such as a crystal-gazer, can see what caused the problem but cannot implement a treatment. Another example of this type is a hand trembler. These people, instead of using crystals, practice hand trembling over the sick person to determine the cause of an illness
4. Specialist medicine people treat the disease after it has been diagnosed and specialize in the use of herbs, massage, or midwifery
5. Those who care for the soul send guardian spirits to restore a lost soul
6. Singers, who are considered to be the most special, cure through the power of their song. These healers use laying on of hands and usually remove objects or draw disease-causing objects from the body while singing

man that wards off the evil of a witch. **Do not remove medicine pouches from patients.**

- The various types of traditional health-care providers are shown in Table 5.7.
- Male health-care providers are generally limited in the care they provide to women, especially during menses. Women are generally modest and wear several layers of slips. This practice is very common among elderly women. **Provide a same-gender health-care provider for intimate care.**

5.3   Reflective Exercises

1. How does suicide, violence, and rape in AI/AN communities compare with the national average? What can health-care providers do to help decrease these conditions.
2. What are the major health conditions of AI/ANs? On which of these can the health-care provider have a major impact.
3. What are the major high-risk health behaviors among AI/ANs? What are some major tactics to help decrease these behaviors?
4. List the major values between AI/AN tribes and the European American culture.
5. Some AI/ANs lack trust in Indian physicians. Why might this happen? What can you do to help gain their trust in Indian physicians?
6. What are the names of traditional healers? What conditions do they treat.
7. What tactics can be used to incorporate traditional healers with allopathic providers?
8. List several birthing practices of the Navajo.
9. What are the tenets of the Native American Church?
10. How might you approach explaining to a patient and family of an impending death of a Navajo?

Bibliography

Childbearing Practices of Native Alaskans (n.d.) http://www.hawaii.hawaii.edu/nursing/RNAlaskan10.html

Clark Callister L (2016) Integrating cultural beliefs and practices when caring for childbearing women and families. http://nursekey.com/integrating-cultural-beliefs-and-practices-when-caring-for-childbearing-women-and-families/

Disparities (2016) Indian Health Service. https://www.ihs.gov/newsroom/factsheets/disparities/

Hanson JD (2012) Understanding prenatal health car for American Indian women in a northern plains tribe. J Transcult Nurs 23(1):29–37

Health Care for Urban American Indian and Alaska Native Women (2015) http://www.acog.org/Resources-And-Publications/Committee-Opinions/Committee-on-Health-Care-for-Underserved-Women/Health-Care-for-Urban-American-Indian-and-Alaska-Native-Women

Improving Cross-cultural Communication: Awareness of Native American Cultures (2007) American Indian Disability Technical Assistance Center. http://ruralinstitute.umt.edu/images/tanac_issues_briefs/Issues_Brief_2007_2.pdf

Indian Preference Law (n.d.) Indian Health Service. https://www.ihs.gov/careeropps/indianpreference/

Official Native American Church of North America (n.d.) http://nacna.weebly.com/

Shaeffer Riley N (2016) The new trail of tears. Encounter Books, New York

Spirituality (2002) Alaskan Native Knowledge Network. http://www.ankn.uaf.edu/NPE/spirituality.html

Stopbullying.gov (2026) https://www.stopbullying.gov/

The Navajo Nation Government (2013) http://www.navajopeople.org

Traditional Native American Values and Behavior (n.d.) http://nwindian.evergreen.edu/curriculum/ValuesBehaviors.pdf

U.S. Census Bureau (2012) The American Indian and Alaskan Native Populations: Census 2010. http://www.census.gov/prod/cen2010/briefs/c2010br-10.pdf

U.S. Department of Health and Human Services (n.d.) Indian Health Service. http://www.ihs.gov/

U.S. Department of the Interior: Bureau of Indian Affairs (2016) https://www.bia.gov/

U.S. Department of the Interior: Bureau of Indian Education (2016) https://www.bie.edu/

Williams C (2002) INUIA. Communication Across Cultures. http://www.alaskaahec-rotations.org/images/uploads/Communication_Across_Cultures.pdf

Wilson U (1983) Nursing care of the American Indian patient. In: Orque MS, Block B, Monroy LSA (eds) Ethnic nursing care: a multicultural approach. Mosby, St. Louis

Chapter 6
The Amish

6.1 Overview and Heritage

Today's Amish live in rural areas in a band of more than 20 states, stretching westward from Pennsylvania, Ohio, and Indiana to as far west as Montana, with some scattered settlements in Florida and the province of Ontario, Canada. As affordable land has decreased, some Amish have migrated to Argentina and Bolivia. The Old Order Amish, so-called for their strict observance of traditional ways that distinguish them from other, more progressive "plain folk," are the largest and most notable group. The Amish emerged after 1693, when they parted ways with the **Anabaptist** movement that originated in Switzerland in 1525. After experiencing severe persecution and martyrdom in Europe, the Amish and related groups immigrated to North America in the seventeenth and eighteenth centuries. Some variant groups are named after their factional leaders (for example Egli and Beachy Amish); some are called conservative Amish Mennonites; and others, New Order Amish. No Amish live in Europe today.

Even though the Amish have persisted in relative social isolation, they continue to adapt and change at their own pace, accepting innovations selectively. Although most Amish homes do not have electricity and electronic labor-saving devices and appliances, that does not preclude their openness to using state-of-the-art medical technology necessary for health promotion. The Amish selectively adapt to technology when it is necessary; some use telephones for business purposes and other technology such as electrically powered ventilators for home care when required. With the advent of technology and not being able to order from such catalogues as Sears, they have been forced to order online. In many Amish communities, local stores that cater to the Amish will place electronic orders for them.

© Springer Nature Switzerland AG 2019
L. D. Purnell, E. A. Fenkl, *Handbook for Culturally Competent Care*,
https://doi.org/10.1007/978-3-030-21946-8_6

6.1.1   Communications

- English is the language of school, of written and print communications, and of contact with most non-Amish outsiders.
- At home and in the immediate Amish communities, **Deitsch/Deutsch**, or Pennsylvania German, is used. In this highly contextual culture, less overt verbal communication is required, and more reliance is placed on implicit, often unspoken, understandings. Much of what passes for "general knowledge" in the information-rich popular culture is screened, or filtered, out of Amish awareness.
- **Expect all Amish patients of school age and older to be fluently bilingual. They can readily understand spoken and written directions and answer questions presented in English, although their own terms for some symptoms and illnesses may not have exact equivalents in Deitsch and English.**
- Most Amish have severely restricted their own access to print media, permitting only a few newspapers and periodicals. Most have also rejected electronic media such as radios, television, and entertainment and information applications of film and computers. In the past where the Amish used mail order catalogues, they no longer to do so. In some communities, the Amish are able to go to community stores where personnel will order supplies and equipment for them online.
- The Amish are clearly not outwardly demonstrative or exuberant. Fondness and love of family members is held deeply but privately. **Demut**, humility, is a priority value, the effects of which may be observed in public as a modest and unassuming demeanor. **Hochmut**, pride or arrogance, is avoided because of frequent verbal warnings.
- **The expression of joy and suffering is not entirely subdued by dour or stoic silence.**
- The Amish present in an unpretentious, quiet manner, with modest outward dress in plain colors lacking any ornament, jewelry, or cosmetics.
- They are unassertive and nonaggressive and avoid confrontational speech styles and public displays of emotion.
- Amish self-perception is grounded in the present. In public, Amish avoid eye contact with non-Amish, but in one-on-one clinical contacts, patients can be expected to express openness and candor with unhesitating eye contact.
- They are generally punctual and conscientious about keeping appointments.
- Using first names with Amish people is appropriate because there are only a limited number of surnames. For example, it is preferable to use John or Mary during personal contacts rather than Mr. or Mrs. Miller. Within Amish communities, individuals are identified further by nicknames, residence, or a spouse's given name.
- **Greet Amish patients with a handshake and a smile.**
- Telephones and computers (a few Amish businesses have them) and automobiles are generally owned by nearby non-Amish neighbors and used by Amish only when it is deemed essential, such as for reaching health-care facilities.
- **Ask Amish if they have access to a telephone where messages may be left if needed. If educational videos are shown in the health-care organization, ask Amish if they wish to view them or if they prefer printed material.**

6.1.2 Family Roles and Organization

- Amish society is patriarchal, but women are accorded high status and respect. Practically speaking, husband and wife may share equally in decisions regarding the family farming business. In public, the wife may assume a retiring role, deferring to her husband, but in private they are typically partners. **Do not assume that the spokesperson for the family is the primary decision maker.**
- The Amish family pattern is the three-generational family. This kinship network includes consanguine relatives consisting of the parental unit and households of married children and their offspring.
- The highest priority for parents is child rearing, a charge given them by the church.
- Young people older than 16 years may experiment with non-Amish dress and behavior, but the expectation is that they will be baptized Amish before marriage. Unmarried children live in the parents' home until marriage. Single adults are included in the social fabric of the community.
- Families are the units that make up church districts. The size of church districts is measured by the number of families rather than by the number of church members.
- Grandparents have respected status as elders; they provide valuable advice, material support, and services that include child care to the younger generation. Family emotional and physical proximity to older adults facilitates elder care.
- **Include the extended nuclear family in health educational activities.**

6.1.3 Workforce Issues

- The Amish place a high value on hard work, with little time off for leisure or recreation.
- In the past, the Amish have worked almost exclusively in agriculture and farm-related tasks. Amish avoidance of compromising associations with "worldly" organizations, such as labor unions, restricts them to non-union work, which often pays lower hourly rates. Work settings that rely heavily on modern technology are not open to the Amish.
- Some Amish with carpentry and other businesses have landlines and or cell phones that are located in the shop or outside the building in a weather protected box. Several families may share one telephone. **Ask if there is a telephone where messages may be left.**
- Amish may avoid eye contact with non-Amish, but in one-on-one contacts in the work environment, they can be expected to express openness and candor with unhesitating eye contact. **Do not take offense at sustained eye contact.**
- Because English is the language of instruction in schools and is used with business contacts in the outside world, there is generally no language barrier for the Amish in the workplace. **Do not use slang or idiomatic expressions that the Amish may not understand.**

6.1.4 Biocultural Ecology

- The Amish are essentially a closed population, with exogamy occurring very rarely. Most are of German and Swiss descent; therefore, their physical characteristics differ, with skin variations ranging from light to olive tones. Hair and eye colors vary accordingly.
- Statistics do not exist for Amish cardiovascular and respiratory diseases and illnesses or for the incidence of cancer and diabetes mellitus. Obesity is common, especially among women.
- Table 6.1 lists commonly occurring health conditions for the Amish. A description of the common genetic disorders is in Table 6.2.

Table 6.1 Health conditions common to the Amish

Health conditions	Causes
Limb-girdle muscular dystrophy	Genetic
Ellis–van Creveld syndrome	Genetic
Dwarfism	Genetic, unknown
Polydactylism	Genetic
Cartilage hair hypoplasia	Genetic
Glutaric aciduria	Genetic
Manic-depressive disorder	Genetic
Pyruvate kinase deficiency	Genetic
Hemophilia B	Genetic
Obesity	Lifestyle

Table 6.2 Common genetic disorders

- Dwarfism
- *Ellis–van Creveld syndrome* is characterized by short stature and an extra digit on each hand, congenital heart defects, and nervous system involvement resulting in some amount of mental deficiency
- *Cartilage hair hypoplasia*, also a dwarfism syndrome, is characterized by short stature; fine, silky hair; and deficient cell-mediated immunity that increases susceptibility to viral infections
- *Pyruvate kinase anemia*, a rare blood-cell disease with jaundice and anemia
- *Hemophilia B*
- *Phenylketonuria* (PKU) results in an inability to metabolize the amino acid phenylalanine, evidenced as high blood levels of the substance and eventually severe brain damage if the disorder is untreated
- *Glutaric aciduria*, a progressive neurologic disease, which can be prevented by screening individuals at risk, restricting dietary protein, and thus limiting protein catabolism, dehydration, and acidosis during illness episodes
- Health-care providers need to plan for family and community education about genetic counseling and screening of newborns for genetic conditions

6.1.5   High-Risk Health Behaviors

- Farm and traffic accidents are major health concerns because of horse-drawn vehicles. Transportation-related injuries involving farm animals are the largest group. Falls from ladders and down hay holes result in numerous orthopedic injuries. **Encourage close monitoring of children who operate farm equipment and transportation vehicles.**
- Alcohol and recreational drug use occur at low rates among the Amish, although few statistics as to the frequency can be found. **Teach about safety factors. Ask about alcohol and recreational drug use in a nonjudgmental approach and when family members are not present.**

6.1.6   Nutrition

- Most Amish families grow their own produce. Typical meals include meat; potatoes or noodles or both; a cooked vegetable; bread; something pickled, such as red beets; cake or pudding; and coffee.
- At mealtimes, all members of the household are expected to be present.
- Snacks and meals tend to be high in fat and carbohydrates. Common snacks are large, home-baked cookies about 3 in. in diameter, ice cream, pretzels, and popcorn.
- **When asking about weight control, suggest reducing portion size, decreasing the amount of sugar used in baking, limiting fatty meats, and altering food preparation practices.**

6.1.7   Pregnancy and Childbearing Practices

- Children are considered gifts from God. The average number of live births per family is seven. Birth control is viewed as interfering with God's will and thus should be avoided. Nevertheless, some Amish women do use intrauterine devices, but this practice is uncommon. **Approaching the subject of birth control obliquely may make it possible for an Amish woman or man to sense the health-care provider's respect for Amish values and thus encourage discussion. "When you want to learn more about birth control, I would be glad to talk to you" is a suggested approach.**
- Women participate in prenatal classes, often with their husbands. Women are interested in learning about all aspects of perinatal care, but they may choose not to participate in sessions when videos are used. **Prenatal class instructors should inform them ahead of time when videos or films will be used so they can decide whether to attend.**

- Most women prefer home births and choose to use Amish or non-Amish lay midwives. The Amish have no major taboos or requirements for birthing. Men may be present, and most husbands choose to be involved; however, they are likely not to be demonstrative in showing affection verbally or physically.
- The laboring woman cooperates quietly, seldom audibly expressing discomfort. Women sometimes use herbal remedies to promote labor. **Knowledge about and respect for Amish health-care practices alert physicians and nurses to the possibility of simultaneous treatments that may or may not be harmful.**
- The postpartum mother resumes her family role managing, if not doing, all the housework, cooking, and child care within a few days after childbirth. Grandmothers or other female relatives often come to stay with the new family for several days to help with care of the infant and give support to the new mother. Older siblings are expected to help care for the younger children and to learn how to care for the newborn.
- When hospitalized, the family may want the patient to spend the least allowable time in the hospital. **Include extended nuclear family in care of the newborn. Ask the father how extensively he wants to be involved with labor and delivery.**

6.1.8 Death Rituals

- Families are expected to care for the aging and the ill in the home. However, when hospitalization is required, **make private arrangements for family members to stay overnight in the hospital.** A wake-like "sitting up" through the night is expected for the seriously ill and dying.
- The funeral ceremony is simple and unadorned, with a plain wooden coffin. Although grief and loss are keenly felt, verbal expression may seem muted as if to indicate stoic acceptance of suffering.
- **Accept varied means of grieving from Amish families. Just because grief is not overtly demonstrated does not mean that family members do not care about their loved one.**

6.1.9 Spirituality

- Amish religious and cultural values include honesty, order, personal responsibility, community welfare, humility, non-resistance or nonviolence, and obedience to parents, church, and God.
- Amish are socialized to sustain injuries, grieve, and move on without fixing blame or seeking punishment or redress from perpetrators.
- Amish settlements are subdivided into church districts similar to rural parishes, with 30–50 families in each district. Local leaders are chosen from their own

community and are generally untrained and unpaid. No regional or national church hierarchy exists to govern internal church affairs. To maintain harmony within a group, individuals often forgo their own wishes. In addition to Sunday services every other week, silent prayer is always observed at the beginning of a meal. A prayer may also end the meal.

- **When choosing among health-care options, families usually seek counsel from religious leaders, friends, and extended family, but the final decision resides with the immediate family.**

6.1.10 Health-Care Practices

- The body is considered to be the temple of God, and human beings are stewards of their bodies. Medicine and health care should always be used with the understanding that it is God who heals. Nothing in the Amish understanding of the Bible forbids them from using preventive or curative medical services. They are highly involved in the practices of health promotion and wellness; illness and disease prevention, and health restoration. Men are involved in major health-care decisions and often accompany the family to the chiropractor, physician, or hospital. Health-care decision is influenced by three factors: (1) type of health problem, (2) accessibility of health-care services, and (3) perceived cost of the service. Grandparents are frequently consulted about treatment options. **Include the extended nuclear family in health-care decision making.**
- Many Amish do not carry health insurance. Some have formalized mutual aid, such as the Amish Aid Society. **Ask Amish families if they have access to the Amish Aid Society and assist with contacting them if desired.**
- **Providers must be aware that some individuals may withhold important medical information from medical providers by neglecting to mention folk and alternative care being pursued at the same time.**
- **When the Amish use professional health-care services, they want to be partners in their health care and want to retain their right to choose from all culturally sanctioned health-care options.**
- Care is expressed in culturally encoded expectations, which the Amish best describe in their dialect as *abwaarde*, meaning "to minister to someone by being present and serving when someone is sick in bed."
- Some accept medical advice regarding the need for high-technology treatments such as transplants, ventilator support, or other high-cost interventions.
- The patient's family seeks prayers and advice from the bishop and deacons of the church, the extended family, and friends, but the decision is generally a personal family choice. Family members may also seek care from Amish healers and other alternative care practitioners, who may suggest nutritional supplements.
- Herbal remedies include those handed down by successive generations of mothers and daughters.
- Common folk illnesses are described in Table 6.3.

Table 6.3 Common Amish folk illnesses

• Brauche, sometimes referred to as sympathy curing or powwowing, is referred to as "warm hands" and includes the ability to feel when a person has a headache or a baby has colic
• **Abnemme** is a condition in which the child fails to thrive and appears puny. Specific treatments for the child may include incantations
• **Aagwachse**, or livergrown, meaning "hide-bound" or "grown together," includes crying and abdominal discomfort that is believed to be caused by jostling in rough buggy rides

- Health-care knowledge is passed from one generation to the next through women.
- **Inquire about the full range of remedies being used. For the Amish patient to be candid, the provider must develop a context of mutual trust and respect.**
- The Amish are unlikely to display pain and physical discomfort. **Remind Amish patients that medication is available for pain relief if they choose to accept it.**
- There are no cultural or religious rules or taboos prohibiting blood transfusions or organ transplantation and donation. Some may opt for organ transplantation after the family seeks advice from church officials, extended family, and friends, but the patient or immediate family generally makes the final decision.
- Children with mental or physical differences are sometimes referred to as "hard learners" and are expected to go to school and be incorporated into classes with assistance from other student "scholars" and parents. The mentally ill are generally cared for at home whenever possible.

6.1.11   Health-Care Providers

- Amish hold all health-care providers in high regard. Health is integral to their religious beliefs, and care is central to their worldview. They tend to place trust in people of authority when they fit Amish values and beliefs.
- Amish usually refer to their own healers by name rather than by title, although some say *brauch-doktor* or **braucher**. In some communities, both men and women provide these services. Amish folk healers use a combination of treatment modalities, including physical manipulation, massage, **brauche**, herbs and teas, and reflexology. Most prefer health-care providers who discuss health-care options, giving consideration to cost, need for transportation, family influences, and scientific information.
- **Because Amish are not sophisticated in their knowledge of physiology and scientific health care, providers should bear in mind that the Amish respect authority and that they may unquestioningly follow orders.**

6.2 Reflective Exercises

1. What are your concerns regarding access and quality care for an Amish family?
2. During your assessment of a 60 year old Amish woman with a breast lump, she reinforces that she does not want to be "cut upon" at her age. What health teaching and advice might you give to her?
3. How might health-care providers use the Amish values of the three-generational family and their visiting patterns in promoting health in the Amish community?
4. What elements would you include for an Amish patient wanting to lose weight?
5. A 42 year old Amish woman has just delivered her tenth baby. She tells the nurse that with ten children, she just does not want any more. What approach might the nurse take to discuss contraception with her?
6. An Amish family comes to the clinic with a chronically sick 4-year-old boy. They tell you that they have been seeing a folk healer but the child has not improved over the last year. What approach should the nurse take in determining care?
7. Name at least three common genetic disorders among the Amish. Describe conventional treatment. How might an Amish person react if you encouraged genetic counseling?
8. Distinguish the terms *aagwachse*, *agnemme*, and *abwarde*.
9. What language would you use to provide discharge instruction to an Amish family?
10. What is meant by "warm hands"?

Bibliography

Amish Family Values (2012) http://www.amishfamilyvalues.com/amish/amish-childbirth-amish-give-birth-home/

Amish Population Profile (2016) https://groups.etown.edu/amishstudies/statistics/amish-population-profile-2015/

How Religion Guides Traditions, Lifestyle, and Beliefs of the Pennsylvania Amish (2012) http://www.discoverlancaster.com/towns-and-heritage/amish-country/amish-religious-traditions.asp

Kraybill DB, Johnson-Weiner KM (2011) The Amish. Johns Hopkins University Press, Baltimore

Payne M, Rupar CA, Siu GM, Siu VM (2011) Amish, Mennonite, and Hutterite genetic disorder database. Pediatric Child Health 16(3):e23–e24

Pennsylvania Amish History and Beliefs (2016) http://www.discoverlancaster.com/towns-and-heritage/amish-country/amish-history-and-beliefs.asp

Wenger AFZ, Wenger MR (2012) The Amish. In: Purnell L (ed) Transcultural health care: a culturally competent approach, 4th edn. F.A. Davis Company, Philadelphia

Chapter 7
People of Appalachian Heritage

7.1 Overview and Heritage

The term *Appalachian* describes people born in the Appalachian mountain range and their descendants who live in or near Appalachia. Appalachia comprises 25 million people in 420 counties in 13 states—Georgia, Alabama, Mississippi, Virginia, West Virginia, North Carolina, South Carolina, Kentucky, Tennessee, Ohio, Maryland, New York, and Pennsylvania. Thus, one can see the tremendous diversity that exists is such a large region. Most of the region is rugged, mountainous terrain that is partially responsible for its residents' values and traditions. Substandard secondary and tertiary roads as well as limited public bus, rail, and airport facilities prevent easy access to much of the area. Although the region includes several large cities, most Appalachians live in small, isolated settlements that preserve their unique identity. German, Scots-Irish, Welsh, French, and English are the primary groups who settled the region between the seventeenth and nineteenth centuries.

Like many disenfranchised groups, the people of Appalachia have been described in stereotypically negative terms (e.g., "poor white trash") that in no way represent the people or the culture as a whole. Appalachians are loyal, caring, family-oriented, religious, hardy, independent, honest, patriotic, and resourceful. The concept of "home" is associated with the land and the family, not a physical structure. Although many are well educated, for some education beyond elementary levels is not considered important because it is not viewed as necessary to earning a living in their traditional occupations. High unemployment and low wages necessitate that many Appalachians migrate throughout the region for better economic opportunities. **Do not make negative comments such as "poor white trash" when working with Appalachian patients.**

- **Take great care to not stereotype Appalachians. Great diversity exists among Appalachians beyond urban and rural residence and includes other variant cultural characteristics (see Chap. 1). Remember that aggregate data, a research principle, may be true for the group but not the individual.**

© Springer Nature Switzerland AG 2019
L. D. Purnell, E. A. Fenkl, *Handbook for Culturally Competent Care*,
https://doi.org/10.1007/978-3-030-21946-8_7

7.1.1 Communications

- The dominant language among Appalachians is English, with many words derived from sixteenth-century Elizabethan English. A small population may still speak Elizabethan English in their home environment. This can cause communication difficulties with health-care providers who are not familiar with the dialect. Many drop the *g* on words ending in *ing*. For example, *writing* becomes *writin'*, *reading* becomes *readin'*, and *spelling* becomes *spellin'*. Consonants may be added, and vowels may be pronounced with a diphthong that can cause difficulty to one unfamiliar with this dialect; hence, *poosh* for *push*, *boosh* for *bush*, *warsh* for *wash*, *hiegen* for *hygiene*, *deef* for *deaf*, *welks* for *welts*, *whar* for *where*, *hit* for *it*, *hurd* for *heard*, and *your'n* for *your*. **If unfamiliar with the exact meaning of a word, ask the patient to elaborate.**
- Appalachians practice the *ethic of neutrality*, which helps shape communication styles, their worldview, and other aspects of the culture. The themes of the ethic of neutrality are described in Table 7.1.
- Some Appalachians may be less precise in describing emotions and are more concrete in conversations; many answer questions in a direct manner. **Use more open-ended questions when obtaining health information and eliciting opinions and beliefs about health-care practices.**
- Many Appalachians do not easily trust or share their thoughts and feelings with **outsiders** and are sensitive to direct questions about personal issues. **Sensitive topics are best approached with indirect questions and suggestions. Establish trust before asking questions that are sensitive in nature if possible.**
- Because of past experiences with large mining and timber companies, many dislike authority figures and institutions that attempt to control behavior. **If time permits, "sit a spell" and "chat" before getting down to the business of collecting health information. To establish trust, it is necessary to demonstrate an interest in the patient's family and other personal matters, drop hints instead of giving orders, and solicit patients' opinions and advice.**
- More traditional individuals may stand at a distance when talking with people in health-care situations. Direct eye contact from strangers may be considered as aggression or hostility.
- Most Appalachians are comfortable with silence and when talking are likely to speak without emotion, facial expression, or gestures.
- **Calling a person by his or her first name with the title Miss** (pronounced "miz," similar to "Ms.," when referring to women, whether single or married) **or Mr. (for example, Miss Lillian or Mr. Bill) denotes familiarity with respect.** Miss Lillian may or may not be married.

Table 7.1 Four dominant themes of the ethic of neutrality	• Avoid aggression and assertiveness
	• Do not interfere with others' lives unless asked to do so
	• Avoid dominance over others
	• Avoid arguments, and seek agreement

- **Some individuals may respond better to verbal instructions and education, with reinforcement from videos rather than from printed communications.**
- Based on fatalistic views, some individuals believe they have little or no control over nature and that the time of death is "predetermined by God." Thus, one frequently hears expressions such as "I'll be there, God willing, or if the crick (creek) don't rise."
- When individuals are not seen because they are late for an appointment and are asked to reschedule, they may not return because of difficult transportation or feelings of rejection. **Be flexible with appointments.**
- Appalachians are comfortable with silence. **Allow a modicum of silence before initiating conversations.**

7.1.2 Family Roles and Organization

- The traditional household continues to be patriarchal, with many families becoming more egalitarian in beliefs and practices, especially if the woman earns more money than the man.
- Large families are common, and children are usually accepted regardless of their negative behaviors in school or with authority figures. Publicly, parents impose strict conformity for fear of community censure. However, permissive behavior at home is unacceptable, and hands-on physical punishment, to an extent that some perceive as abuse, continues. **Work with parents to explain current laws of child abuse.**
- Older people and pregnant women are treated with increased status in the family, church, and community.
- Children, single or married, may return to their parents' home where they are readily accepted whenever the need arises.
- Grandparents frequently care for grandchildren, especially if both parents work. Elders usually live close to or with their children when self-care becomes a concern. Many adult children do not consider nursing home placement for their parents because it is perceived as the equivalent of a death sentence.
- **The family network can be a rich resource when health teaching and assistance with personalized care are needed. The family rather than the individual must be considered as the basic treatment unit.**
- Although alternative lifestyles are usually readily accepted in the Appalachian culture, **Do not disclose same-sex relationships to others.**

7.1.3 Workforce Issues

- Because many Appalachians value family above all else, reporting to work may become less of a priority when a family member is ill or other family obligations are pressing. For some, the preferred work pattern is to work for an extended period of time, take some time off, and then return to work. **Liberal leave poli-cies for funerals and family emergencies are a necessary part of the work environment.**

- More traditional individuals may stand at a distance when talking with people in the workforce. **Do not take offense if Appalachians stand at a greater dis-tance than to what you are accustomed.**
- A harmonious environment that fosters cooperation and agreement in decision making is valued. Health-care workers who come from outside the area may have some difficulty establishing rapport in the workplace if they lack an under-standing and appreciation for Appalachian workplace etiquette. **Familiarize yourself with the Appalachian culture of neutrality when working with Appalachian staff.** (See section on Communication.)
- Most Appalachians are comfortable with silence, and when talking are likely to speak without emotion, facial expression, or gestures. **Do not mistake silence and lack of facial expressions as not caring or not listening.**
- The Appalachian ethic of neutrality and the values of individualism and non-assertiveness, with a strong people orientation, may pose a dichotomous perception at work for outsiders who may not be familiar with the Appalachian way of life.
- Because many Appalachians align themselves more closely with horizontal rather than hierarchical relationships, they are sometimes reluctant to take on management and leadership roles. Many Appalachians prefer to work at their own pace, devising their own work rules and methods for getting the job done.

7.1.4 Biocultural Ecology

- Sometimes, the influence of American Indians can be observed in olive-toned skin.
- **Those of Scots-Irish background and others with light skin tones are at increased risk for skin cancer. Take precautions and teach patients to pro-tect themselves from the harmful effects of the sun.**
- Predominant occupations, particularly farming, textile manufacturing, mining, furniture-making, and logging, place residents at increased risk for respiratory diseases such as black lung, brown lung, and emphysema.
- Children are at greater risk for sudden infant death syndrome, congenital malfor-mations, and infections. Only 70% of children are immunized, compared with 90% for the nation as a whole. The area also has an incidence of childhood injuries due to burns, trauma, poisoning, child neglect, and abuse that is higher than the national average. Poor oral hygiene has high rates in Appalachia.
- **Work with community organizations and schools to promote immuniza-tions, oral hygiene, and prevention of childhood injuries.**
- Table 7.2 lists commonly occurring health conditions for Appalachians in three categories: lifestyle, environment, and genetics.

Table 7.2   Health conditions common to people of Appalachian heritage

Health conditions	Causes
Black lung, brown lung	Environment, lifestyle
Emphysema	Environment, lifestyle
Tuberculosis	Environment, lifestyle
Hypochromic anemia	Environment, lifestyle
Cardiovascular disease	Genetic, lifestyle
Sudden infant death syndrome	Genetic, lifestyle
Diabetes mellitus	Genetic, lifestyle
Otitis media	Environment, lifestyle
Parasitic infections	Environment, lifestyle
Cancer	Genetic, environment, lifestyle

7.1.5   High-Risk Health Behaviors

- High smoking rates continue throughout Appalachia. Underage use of alcohol is widespread among teens. A low rate of exercise and diets high in fats and refined sugars are also important risk factors.
- **Initiate anti-smoking campaigns in elementary schools and promote the benefits of not being overweight.**
- A ten-step pattern of health-seeking behaviors has been identified and is shown in Table 7.3. These ten steps may not always follow the sequence presented here: some steps may be omitted, and not all steps are always completed.

7.1.6   Nutrition

- Wealth means having plenty of food for family, friends, and social gatherings. Eating habits include high-cholesterol organ meats such as tongue, liver, heart, lungs (called lights), and brains.
- Common foods are sweet potato pie; molasses candy; apple beer; gooseberry pie; pumpkin cake; and pickled beans, fruit, corn, beets, and cabbage, usually high in sodium.
- Bone marrow is used to make sauces, and stomach, intestines (chitlins or chitterlings), pigs' feet, tail, and ribs are also commonly eaten.
- Many recipes contain lard, and meats are commonly preserved with salt. Low-fat game meat is usually breaded and fried with lard or animal fat, negating the overall gains from these low-fat meat sources.
- Most diets include sweet pre-packaged drinks, Kool-Aid with added sugar, very sweet iced tea, and soda.
- **Explain the benefits of low-fat preparation practices and the potential harmful effects of excess salt and refined sugar.**

Table 7.3 Health-seeking behaviors among Appalachians

1. At the onset of symptoms, self-care practices that are usually learned from mothers are implemented
2. When the symptoms persist, they call their mother, if she is available
3. If the mother is unavailable, they call the female in their kin network that is perceived as knowledgeable regarding health. If a nurse is available, they may seek the nurse's advice
4. If relief is not achieved, they use over-the-counter (OTC) medicine they have seen advertised on television for symptoms that most closely match their own
5. If that is ineffective, they use some of a neighbor's medicine
6. Next, they ask the local pharmacist for a recommendation; this usually marks the first encounter with a professional health-care provider. (Of course, they usually do not tell the pharmacist that they tried the neighbor's medicine.) The pharmacist may strongly suggest that they see a health-care provider; however, on their insistence, the pharmacist may recommend another OTC medication
7. When no relief is achieved, they seek a local health-care provider. The provider treats them to the best of his or her ability
8. If the condition does not resolve, the local health-care provider refers them to a specialist in the area
9. The specialist treats the condition to the best of his or her ability
10. If unsuccessful, the specialist refers them to the closest tertiary medical center. These ten steps may not always follow the sequence presented here; some steps may be skipped, and not all steps are always completed. The time around these ten steps may be several years. Often by the time they are referred for definitive treatment, compensatory reserves have been depleted, and they die at large medical centers. It is essential to inquire in a nonjudgmental way about all treatments used for an illness

- Babies from the first month may be fed grease, sugar, and coffee to promote hardiness. Many believe that the sooner a baby can take foods other than milk, the healthier it will be. **Factual information that describes health risks with early feeding of solid foods may help prevent later nutritional allergies in children.**
- Many children replace meals with snacks. The most common snacks are candy, salty foods, desserts, and carbonated beverages. Many adolescents skip breakfast and lunch entirely, preferring to eat snack foods. This pattern of snacking can result in deficiencies in vitamin A, iron, and calcium. **Educate students on healthy eating and food preparation practices beginning in elementary school.**

7.1.7 Pregnancy and Childbearing Practices

- Fertility control methods include birth control pills, condoms, and tubal ligation; abortion is an individual choice.
- A popular belief is that taking laxatives facilitates an abortion.
- A disproportionate number of teenage pregnancies occur at a younger age among Appalachians when compared with non-Appalachians. **Fertility practices and sexual activity, both sensitive topics for many teenagers, are areas in which**

Table 7.4   Beliefs about pregnancy

• Boys are carried higher and the mother's belly appears pointy, whereas girls are carried low
• Picture-taking can cause a stillbirth
• Reaching over one's head can cause the cord to strangle the baby
• Wearing an opal ring during pregnancy may harm the baby
• Being frightened by a snake or eating strawberries or citrus fruit can cause birthmarks
• If the mother experiences a tragedy, a congenital anomaly may occur
• If the mother craves a particular food during her pregnancy, then she should eat that food or the baby will have a birthmark similar to the craved food

outsiders unknown to the family may be more effective than health-care providers who are known.

- Beliefs about pregnancy are shown in Table 7.4.
- The birthing mother is expected to accept childbirth as a short, intense, natural process that will bring her closer to the Earth and must be endured. **Let the birthing mother know that pain medicine is available if she desires.**

7.1.8   Death Rituals

- When death is expected, family and friends may stay through the night. **Make accommodations for some family members to stay in the health-care facility.**
- Funeral services can last for 3 h. The amount of time for a service varies according to the age of the deceased. For example, the service for an elderly person is usually longer than for a younger person. The body is displayed for hours, either in the home or at the church, so that all those who wish to view the body may do so.
- At the end of the service, all who wish to view the body again may do so, with the closest relative being the last to view the body.
- The deceased is usually buried in his or her best clothes. A common practice is to bury the deceased with personal possessions.
- After the funeral services are completed, elaborate meals are served either in the home or at the church. Services are accompanied by singing before, during, and after the service.
- Clergy help families through the grieving process by providing counseling and support to family members. **Ask family if they wish to see a chaplain and help them with contacting one.**

7.1.9   Spirituality

- Many churches in the region stress fundamentalism in religious practices and use the King James Bible.
- Prayer for many Appalachians is a primary source of strength. Prayer is personally designed around specific church and religious beliefs and practices,

which vary widely throughout the region and between and among churches of similar faith.

- **Churches in many parts of Appalachia serve as the social centers of the community and are a good location for health teaching.**
- Many small churches have lay preachers instead of trained ministers. Most believe that to be a preacher, a person must have a divine calling.
- Many of the Baptist faiths believe that baptism must be performed in a river, pond, or lake so that the body can be submerged. Feet-washing (men wash men's feet, and women wash women's feet) demonstrates humility.
- Some free-will churches (for example, The Holiness Church) preach against attending movies, ball games, and social functions where dancing occurs. Other sects believe in handling poisonous snakes, although this is rare, the practice continues; it is believed that the snake will not bite those who have faith.
- Some ingest strychnine in small doses during religious services to increase sensory stimuli. This practice can precipitate convulsions if strychnine is ingested in large enough amounts.
- Fire-handling is still practiced by some groups, with the belief that the hot coals will not burn those who have faith.
- Within the context of ***fatalism*** comes the belief that what happens to the individual is largely a result of God's will.
- Meaning in life comes from the family and "living right," which is defined by each person and usually means living right with God and in the beliefs of a chosen church.
- **Forming partnerships between health-care providers and faith-related organizations for health promotion and wellness, and illness and disease prevention has strong potential for improving the health of Appalachians. Respect the spiritual beliefs of Appalachians without expressing negative comments.**

7.1.10   Health-Care Practices

- Because self-reliance activities and nature predominate over people, many believe that it is best to let nature heal.
- For some who do not believe in owing money, seeing a health-care provider may be postponed until the condition is severe.
- Many may not see formal biomedical health-care providers until self-medicating and folk remedies have been exhausted. When they finally seek formal health care, the condition has become severe, takes longer to treat, and has a less favorable outcome.
- Individuals may feel powerless regarding their own health and abdicate self-responsibility in favor of high expectations and unrealistic dependence on the health-care system, with the physician taking charge of their care completely.
- **Offering transportation on a regular schedule and by appointment may improve access.**

- A major health concern for many Appalachians is the state of their **blood,** which is described as being thick or thin, good or bad, and **high** or **low;** these conditions can be regulated by diet. Some individuals fear "being cut on" or "going under the knife" and feel that the hospital is a place where one goes to give birth or die.
- **Provide factual explanations and instructions in an unhurried manner.**
- When older people see a health-care provider, many expect immediate help. Physicians who dispense medications in their offices are considered helpful.
- A strong belief in folk medicine is a traditional part of the culture. Using herbal medicines, poultices, and teas is common practice among individuals of all socioeconomic levels. Although many of these home remedies are not harmful, some may have deleterious effects when used to the exclusion of, or in combination with, prescription medications.
- **Ascertain if individuals intend to use folk medicines simultaneously with prescription medications and treatment regimens so that these remedies can be incorporated into the plan of care and dialogue can be undertaken to prevent adverse effects.**
- **Integrating folk medicine into allopathic prescriptions has a greater chance of improving patients' compliance with health prescriptions and interventions.**
- Bureaucratic forms foster fear and suspicion of health-care providers. **Help patients complete bureaucratic forms if needed.**
- **Be aware that patients may be especially sensitive to criticism. If the provider uses language that the patient does not understand, the provider may be perceived as "stuck up." Decrease language barriers by decoding the jargon of the health-care environment.**
- Individuals with mental impairments or physical handicaps are readily accepted and not turned away. Those with a mental handicap are not perceived as crazy but rather as having "bad nerves" or being "quite turned" or "odd turned."
- The traditional belief is that disability is a natural and inevitable part of the aging process. **Provide factual information about disability and explain that it is not an inevitable part of the aging process.**
- For many, pain is something that is to be endured and accepted stoically. When a person becomes ill or has pain, personal space collapses inward and the person expects to be waited on and cared for by others. **Explain that self-care activities and taking pain medicine will hasten the healing process.**

7.1.11 Health-Care Providers

- Folk practitioners are primarily older women but may be men. Grannies and herb doctors are trusted and known to individuals and the community for giving more personalized care.
- **For patients to become more accepting of biomedical care, it is important for health-care providers to approach individuals in an unhurried manner and engage patients in decision making and care planning.**
- Generally, there is no problem providing care to opposite-gender patients.

7.2   Reflective Exercises

1. A patient wants to incorporate ginseng with prescriptive medication. What is your advice?
2. What is meant by high and low blood for Appalachians?
3. What evidence do you see with the "ethic of neutrality" among Appalachians or any other group or person?
4. Identify at least three health conditions common among Appalachians. What might you do to help decrease these conditions?
5. A colleague has made the derogatory comment, "poor white trash" about an Appalachian patient. How would you approach this colleague?
6. What tactics might you use to help decrease the high smoking rates among Appalachians?
7. A mother brings her 3-month-old male baby for a well-baby checkup. The mother tells you how wonderful the baby eats drinks his cereal mixed with milk. What advice do you give the mother?
8. How might you work with local community churches to teach health food practices and prevention of sexually transmitted infections?
9. How is *fatalism* displayed among Appalachians?
10. An overweight teenager is drinking 8–10 glasses of "sweet tea" on a daily basis. How might you encourage her to decrease the amount of "sweet tea" she drinks on a daily basis?

Bibliography

American Counseling Association (2010) Working with clients of Appalachian culture. http://www.counseling.org/resources/library/vistas/2010-v-online/Article_69.pdf

Appalachian Regional Commission (n.d.) https://www.arc.gov/

Billings DB, Norman G, Ledford K (1999) Back talk from Appalachia. University Kentucky Press, Lexington

Brynes N, Conlisk E, Jarvis C, Lowry N (2009) A prevalence study of folk remedy use by the middle-Appalachian elderly. 137th Meeting of the American Public Health Association. Philadelphia, PA

Dalton E, Miller L (2016) Peers, stereotypes and health communication through the lens of adolescent Appalachian mothers. Cult Health Spiritual 18(2):115–128

Diddle G, Denham SA (2010) Spirituality and its relationships with the health and illness of Appalachian people. J Transcult Nurs 21(2):175–182

Drake RB (2001) A history of Appalachia. University of Kentucky Press, Lexington

Huttlinger K (2013) People of Appalachian heritage. In: Purnell L (ed) Transcultural health care: a culturally competent approach, 4th edn. F.A. Davis Company, Philadelphia

McGarvey EL, Leon-Verdin M, Killos LF, Guterbook T, Cohn WF (2011) Health disparities between Appalachia and non-Appalachia counties in Virginia USA. J Community Health 36(3):348–356

Vance JD (2016) Hillbilly elegy: a memoir of a family and culture in crisis. Harper Collins Publishing, New York

Chapter 8
People of Arab Heritage

8.1 Overview and Heritage

Arabs trace their ancestry and traditions to the nomadic desert tribes of the Arabian Peninsula. They share a common language, Arabic. Most are united by Islam, the world's largest religion that originated in seventh-century Arabia. Despite these common bonds, great diversity exists among Arabs related to religious preference and other **variant characteristics of culture** discussed in Chap. 1. Many Arab Americans disappear in national studies because they are counted as white in census data. The September 11, 2001, al Qaeda terrorist attack on the US has increased the number and intensity of negative comments about Arabs. Health-care providers need to understand that few Arab Americans support terrorist attacks and that it is inappropriate to pigeonhole people by their cultural background.

First-wave Arab immigrants, primarily Christians, came to the US between 1887 and 1913 seeking economic opportunity. Most were male, illiterate, and unskilled mountain or rural immigrants who valued assimilation. Most post-1965 immigrants are **Muslims**. Arabism and Islam are intrinsically interwoven with some elements of Christianity so that Arabs, whether Christian or Muslim, share some basic traditions and beliefs. Consequently, knowledge of their religion is critical to understanding the Arab American patient's cultural frame of reference. Second-wave immigrants entered the US after World War II. Most are refugees from nations beset by war and political instability. This group includes a large number of professionals and individuals seeking educational degrees who have subsequently remained in the US. Most are Muslims and favor professional occupations.

© Springer Nature Switzerland AG 2019
L. D. Purnell, E. A. Fenkl, *Handbook for Culturally Competent Care*,
https://doi.org/10.1007/978-3-030-21946-8_8

8.1.1   Communications

- **Arabic** is the official language of the Arab world. Modern or classical Arabic is a universal form of Arabic used for all writing and formal situations ranging from radio newscasts to lectures.
- Although English is a common second language, language and communication can pose formidable problems in health-care settings. **Speak clearly and slowly, giving time for interpretation. Obtain an interpreter if necessary.**
- Communication is highly contextual, where unspoken expectations are more important than the actual spoken words. Conversants stand close together, maintain steady eye contact, and touch (only between members of the same sex) the other's hand or shoulder.
- Speech is loud and expressive and is characterized by repetition and gesturing, particularly when involved in serious discussions. Observers witnessing impassioned communication may incorrectly assume that Arabs are argumentative, confrontational, or aggressive.
- Privacy is valued, and many resist disclosure of personal information to strangers, especially when it relates to familial disease conditions. Among friends and relatives, Arabs express feelings freely.
- Good manners are important in evaluating a person's character. **If the health-care situation is not an emergency, inquire first about overall well-being and exchange pleasantries. Etiquette requires shaking hands (among men only if traditional) on arrival and departure. Devout Muslim men do not shake hands with women. When a man meets an Arab woman, he should wait for the woman to extend her hand.**
- Individuals are protected from bad news as long as possible. **Inform patients of bad news as gently as possible.**
- Table 8.1 describes techniques for communicating with Arab Americans.
- Punctuality is not taken seriously except in cases of business or professional meetings. **Explain the importance of punctuality in the American health-care system. Maintain flexibility with appointments when possible.**
- Titles are important and are used in combination with the person's first name (e.g., Mr. Khalil or Dr. Ali). Some may prefer to be addressed as mother (*Um*) or father (*Abu*) of the eldest son (e.g., Abu Khalil, "father of Khalil").
- **Address Arab patients formally until told to do otherwise.**

8.1.2   Family Roles and Organization

- Muslim families have a strong patrilineal tradition. Women are subordinate to men, and young people are subordinate to older people. In public, a wife's interactions with her husband are formal and respectful. However, behind the scenes she typically wields tremendous influence, particularly in matters pertaining to the home and children.

Table 8.1   Guidelines for communicating with Arab Americans

• Employ an approach that combines expertise with warmth
• Minimize status differences, and pay special attention to the person's feelings
• Take time to get acquainted before delving into business. If sincere interest in the person's home country and adjustment to American life is expressed, he or she is likely to enjoy relating such information, much of which is essential to assessing risk for a traumatic immigration experience and understanding the person's cultural frame of reference
• If the situation allows, sharing a cup of tea gives an initial visit a positive beginning
• Clarify role responsibilities regarding taking a history, performing physical examinations, and providing health information
• Perform a comprehensive assessment. Explain the relationship of the information needed for physical complaints
• A spokesperson may answer questions directed to the patient, but family members may edit some information that they feel is inappropriate
• Family members or an influential intermediary may act as the patient's advocate. They may attempt to resolve problems by taking appeals "to the top" or by seeking the help of an influential intermediary
• Convey hope and optimism. The concept of "false hope" is not meaningful to Arabs because they regard God's power to cure as infinite

- Older male figures assume the role of decision maker. Women attain power and status in advancing years, particularly when they have adult children.
- Among the traditional, gender roles are clearly defined and regarded as a complementary division of labor. Men are breadwinners, protectors, and decision makers. Women are responsible for the care and education of children and for the maintenance of a successful marriage by tending to their husbands' needs.
- Although women in more urbanized Arab countries often have professional careers, with some women advocating for women's liberation, the family and marriage remain primary commitments for most. The authority structure and division of labor within Arab families are often misinterpreted, fueling common stereotypes of the overtly dominant male and the passive and oppressed woman. **Do not be judgmental with family decision making and roles.**
- Family reputation is important; children are expected to behave in an honorable manner and not bring shame to the family. Children are dearly loved, indulged, and included in all family activities. A child's character and successes (or failures) in life are attributed to upbringing and parental influence.
- Childrearing patterns also include great respect for parents and elders. Children are raised not to question elders and to be obedient to older brothers and sisters. Discipline may include physical punishment and shaming.
- The father is the disciplinarian, whereas the mother is an ally and mediator, an unfailing source of love and kindness. Children are made to feel ashamed because others have seen them misbehave rather than to experience guilt arising from self-criticism and inward regret. **Explain child abuse laws in the US.**
- Adolescents are pressed to succeed academically. Academic failure, sexual activity, illicit drug use, and juvenile delinquency bring shame to the family. For girls in particular, chastity and decency are required.

- Women value modesty, especially devout Muslims, for whom modesty is expressed with their attire. Many Muslim women view the *hijab*, "covering the body except for one's face and hands," as offering them protection in situations in which the sexes mix. It is a recognized symbol of Muslim identity and good moral character. Ironically, many people associate the *hijab* with oppression rather than protection.
- Family members usually live nearby, sometimes intermarry (first cousins), and expect a great deal from one another regardless of practicality or ability to help.
- Loyalty to one's family takes precedence over personal needs. Cultural conflicts between American values and Arab values may cause significant conflicts for families.
- Sons are held responsible for supporting elderly parents. Elderly parents are almost always cared for within the home, typically until death. In the absence of the father, brothers are responsible for unmarried sisters.
- Although the Islamic right to marry up to four wives is sometimes exercised, particularly if the first wife is chronically ill or infertile, most marriages are monogamous and for life.
- Homosexuality is highly stigmatized. In some Arab countries, it is considered a crime, and participants may be killed. Fearing family disgrace and ostracism, gays and lesbians remain closeted. **Do not reveal sexual orientation to the family. Refer gay and lesbian patients to local gay/lesbian support groups.**

8.1.3 Workforce Issues

- Discrimination such as intimidation, being treated suspiciously, and negative comments about their religious practices has been reported as a major source of stress among Arab Americans.
- Muslim Arabs who wish to attend Friday prayer services and observe religious holidays frequently encounter job-related conflicts. **Make attempts at honoring Friday prayer services.**
- In the Arab world, position is often attained through one's family and connections. **Explain nepotism policies in the US.**
- Conversants stand close together and maintain steady eye contact. **Do not take offense if Arab employees stand closer to you than that to which you are accustomed or maintain steady eye contact. Remember these characteristics are culture bound.**
- Speech is loud and expressive and is characterized by repetition and gesturing, particularly when involved in serious discussions. **Impassioned communication may incorrectly be assumed that Arabs are argumentative, confrontational, or aggressive.**
- Criticism is often taken personally as an affront to dignity and family honor. In Arab offices, supervisors and managers are expected to praise their employees to assure them that their work is noticed and appreciated. Whereas such direct praise may be somewhat embarrassing for European Americans, Arabs expect and want praise when they feel they have earned it. **Give honest praise for the work of Arab employees.**

8.1.4 Biocultural Ecology

- Most Arabs have dark or olive-colored skin, but some have blonde or auburn hair, blue eyes, and fair complexions. **To assess pallor, cyanosis, and jaundice in dark-skinned people, examine the oral mucosa and conjunctiva.**
- Infectious diseases such as tuberculosis, malaria, trachoma, typhus, hepatitis, typhoid fever, dysentery, and parasitic infestations are common with newer immigrants. Schistosomiasis (or bilharziasis) infects about one-fifth of Egyptians and has been called Egypt's primary health problem.
- Glucose-6-phosphate dehydrogenase deficiency, sickle cell anemia, and the thalassemias are extremely common in the eastern Mediterranean.
- High consanguinity rates (roughly 30% of marriages in Iraq, Jordan, Kuwait, and Saudi Arabia) occur between first cousins and contribute to the prevalence of genetically determined disorders in Arab countries. Individuals are at increased risk to inherit familial Mediterranean fever, a disorder characterized by recurrent episodes of fever, peritonitis, or pleurisies, either alone or in some combination.
- Some individuals have difficulty metabolizing debrisoquine, antiarrhythmics, antidepressants, beta blockers, neuroleptics, and opioid agents.
- **Closely assess the effectiveness of narcotics such as codeine and morphine.**
- Table 8.2 lists commonly occurring health conditions for Arab in three categories: lifestyle, environment, and genetics.

Table 8.2 Health conditions common to Arabs

Health conditions	Causes
Familial Mediterranean fever	Genetic
Familial paroxysmal polyserositis	Genetic
Tuberculosis	Environment, lifestyle
Malaria	Genetic, environment
Trachoma	Environment, lifestyle
Typhoid fever	Environment
Glucose-6-phosphate dehydrogenase deficiency	Genetic
Sickle cell disease	Genetic, environment
Thalassemia	Genetic
Hepatitis A and B	Environment, lifestyle
Schistosomiasis (bilharzia)	Environment, lifestyle
Familiar hypercholesterolemia	Genetic, lifestyle
Dubin-Johnson syndrome	Genetic
Epilepsy	Genetic
Ichthyosis vulgaris	Genetic
Phenylketonuria	Genetic
Dyggve-Melchior-Clausen syndrome	Genetic
Familial hypercholesterolemia	Genetic
Schistosomiasis	Environment, lifestyle
Beta-Thalassemia	Genetic
Metachromatic leukodystrophy	Genetic

8.1.5   High-Risk Health Behaviors

- Nonuse of seat belts and helmets are major issues. **Encourage use of seat belts and helmets, explaining state laws.**
- Despite Islamic beliefs discouraging tobacco use, smoking remains deeply ingrained in the Arab culture: offering cigarettes is a sign of hospitality. **Partner with schools for anti-smoking campaigns at a young age.**
- Women may be at high risk for domestic violence, especially new immigrants, because of the high rates of stress, poverty, poor spiritual and social support, and isolation from family members. **Explain American abuse laws.**
- Sedentary lifestyle and high fat intake among Arab Americans place them at higher risk for cardiovascular diseases. **Provide factual information on the benefits of physical activity and the benefits of a low-fat diet.**
- The rates of breast cancer screening, mammography, and cervical Pap smears are low because of modesty. **Obtain female health-care providers for patients for whom modesty is a concern.**

8.1.6   Nutrition

- Although cooking and national dishes vary from country to country and seasonings from family to family, Arabic cooking shares many general characteristics. Spices and herbs include cinnamon, allspice, cloves, ginger, cumin, mint, parsley, bay leaves, garlic, and onions.
- Skewer cooking and slow simmering are typical modes of preparation. All countries have rice and wheat dishes, stuffed vegetables, nut-filled pastries, and fritters soaked in syrup. Dishes are garnished with raisins, pine nuts, pistachios, and almonds. Favorite fruits and vegetables include dates, figs, apricots, guava, mango, melon, papaya, bananas, citrus, carrots, tomatoes, cucumbers, parsley, mint, spinach, and grape leaves. Bread accompanies every meal and is viewed as a gift from God. Lamb and chicken are the most popular meats.
- Consumption of blood is forbidden; Muslims are required to cook meats and poultry until well done. Some Muslims refuse to eat meat that is not *halal* (slaughtered in an Islamic manner).
- **Obtain halal meat from Arabic grocery stores, through Islamic centers or mosques, or from one of the online distributors of frozen halal meals.**
- Muslims are prohibited from eating pork and pork products. Muslims are equally concerned about the ingredients and origins of mouthwashes, toothpastes, and medicines (e.g., alcohol-based syrups and elixirs) and capsules (gelatin coating) derived from pigs. However, if no substitutes are available, Muslims are permitted to use these preparations.
- Grains and legumes are often substituted for meats; fresh fruit and juices are especially popular, and olive oil is widely used.

Table 8.3 Ramadan calendar 2019–2025

Year	Fasting begins
2019	May 6
2020	May 24
2021	May 13
2022	May 3
2023	April 23
2024	April 10
2025	May 31

Note: Actual date may vary by 1 day in some countries, depending on the time zone
Retrieved from https://www.timeanddate.com/holidays/us/eid-al-fitr#tb-hol_obs

- Food is eaten with the right hand because it is regarded as clean. **When it is necessary to feed an Arab patient, use the right hand, regardless of the dominant hand.**
- Eating and drinking at the same time is viewed as unhealthy. **Serve beverages after the meal is eaten.**
- During Ramadan, the Muslim month of fasting, abstinence from eating, drinking (including water), smoking, and marital intercourse during daylight hours is required. See Table 8.3 for Ramadan dates through 2025.
- Although the sick are not required to fast, many pious Muslims insist on fasting while hospitalized. **Adjust meal times and medications, including medications given by non-oral routes. Provide appointment times after sunset during Ramadan for individuals requiring injections (for example, allergy shots).**
- Eating properly, consuming nutritious foods, and fasting are believed to cure disease. For some, illness is related to excessive eating, eating before a previously eaten meal is digested, eating nutritionally deficient food, mixing opposing types of foods (hot/cold, dry/moist), and consuming elaborately prepared foods. Gastrointestinal complaints are the most frequent reason for seeking care.
- **Lactose intolerance is common. Eating yogurt and cheese, rather than drinking milk, may reduce symptoms in sensitive people.**

8.1.7   Pregnancy and Childbearing Practices

- Fertility practices are influenced by traditional Bedouin values, which support tribal dominance and beliefs that "God decides family size." Procreation is regarded as the purpose of marriage; high fertility rates are favored.
- Sterility in a woman can lead to rejection and divorce. Approved methods for treating infertility are limited to artificial insemination using the husband's sperm and *in vitro* fertilization involving the fertilization of the wife's ovum by the husband's sperm.

- Many reversible forms of birth control are undesirable but not forbidden. They should be used when there is a threat to the mother's life, too frequent childbearing, risk of transmitting a genetic disease, or financial hardship.
- Irreversible forms of birth control such as vasectomy and tubal ligation are "absolutely unlawful" as is abortion, except when the mother's health is compromised by a pregnancy-induced disease or her life is threatened.
- Unwanted pregnancies are dealt with by hoping for a miscarriage "by an act of God" or by covertly arranging for an abortion.
- The pregnant woman is indulged and her cravings satisfied, lest she develop a birthmark in the shape of the particular food she craves. Girls are carried high; boys are carried low.
- Although pregnant women are excused from fasting during Ramadan, some Muslim women may still fast.
- Labor and delivery are women's affairs. During labor, women openly express pain through facial expressions, verbalizations, and body movements. Nurses and medical staff may mistakenly diagnose Arab women as needing medical intervention and pain medications inappropriately.
- Care for the infant includes wrapping the stomach at birth, or as soon as possible thereafter, to prevent cold or wind from entering the baby's body.
- The call to prayer is recited in the Muslim newborn's ear.
- Male offspring are preferred. Male circumcision is almost a universal practice, and for Muslims it is a religious requirement.
- Mothers may be reluctant to bathe postpartum because of beliefs that air gets into the mother and causes illness.
- Many believe washing the breasts "thins the milk." Breast-feeding is often delayed until the second or third day after birth because of beliefs that the mother requires rest, that nursing at birth causes "colic" pain for the mother, and that "colostrum makes the baby dumb." **Explain the importance of the immune properties of colostrum. Dispel myths.**
- Postpartum foods, such as lentil soup, are offered to increase milk production, and tea is encouraged to flush and cleanse the body.

8.1.8   Death Rituals

- **Death is accepted as God's will. Muslim death rituals include turning the patient's bed (or at a minimum the patient's head) to face the holy city of Mecca** and reading from the Qur'an, particularly verses stressing hope and acceptance.
- After death, the deceased is washed three times by a Muslim of the same sex. The body is then wrapped, preferably in white material, and buried as soon as possible in a brick or cement-lined grave with the head facing Mecca.
- Prayers for the deceased are recited at home, at the mosque, or at the cemetery.

- Among the very traditional, women do not ordinarily attend the burial unless the deceased is a close relative or husband. Instead, they gather at the deceased's home and read the Qur'an.
- Death rituals for Arab Christians are similar to Christian practices in the rest of the world. Extended mourning periods may be practiced if the deceased is a young man, woman, or child. Although weeping is allowed, beating the cheeks or tearing garments is prohibited.
- For women, wearing black is considered appropriate for the entire period of mourning.
- Cremation is not practiced.
- Families do not generally approve of autopsy because of respect for the dead and feelings that the body should not be mutilated. However, Islam does allow forensic autopsies and autopsies for the sake of medical research and instruction.
- Organ donation and transplantation as well as administration of blood and blood products are acceptable.

8.1.9 Spirituality

- Most Arabs are Muslims. Islam is the official religion of most Arab countries, and in Islam there is no separation of church and state; a certain amount of religious participation is obligatory.
- Islam has no priesthood. Islamic scholars or religious sheikhs, the most learned individuals in an Islamic community, assume the role of *imam*, or "leader of the prayer." The imam acts as a spiritual counselor.
- The major tenets of Islam are shown in Table 8.4.
- Many Muslims believe in combining spiritual medicine, performing daily prayers, and reading or listening to the Qur'an with conventional medical treatment. **Assist the patient and family in making accommodations for prayer.**
- Prominent Christian groups include the Copts in Egypt, the Chaldeans in Iraq (now largely refugees), and the Maronites. **Contacting the local imam may be a helpful strategy for Muslims struggling with health-care decisions.**
- School and work schedules revolve around Islamic holidays and weekly prayer. Because Muslims gather for communal prayer on Friday afternoons, the work

Table 8.4 The five major pillars, or duties, of Islam

• Faith, which is shown by the proclamation of the Unity of God by saying, "There is no God but Allah; Mohammed is the Messenger of Allah"
• Prayer, facing Mecca, is performed at dawn, noon, midafternoon, sunset, and nightfall
• Almsgiving is encouraged to assist the poor and to support religious organizations
• Fasting occurs to fulfill religious obligations, to wipe out previous sins, and to appreciate the hunger of the poor. (See Ramadan under Nutrition)
• A pilgrimage to Mecca (*hadj*) once in a lifetime is encouraged if the means are available

8.1.11   Health-Care Providers

- Many individuals find interacting with a health-care provider of the opposite sex quite embarrassing and stressful.
- Discomfort may be expressed by refusal to discuss personal information and by a reluctance to disrobe for physical assessments and hygiene. Women may refuse to be seen by male health-care providers. **Provide a same-sex caregiver whenever possible, especially with intimate care.**
- Knowledge held by a doctor is thought to convey authority and power.
- Most patients who lack English communication skills prefer an Arabic-speaking physician. **Provide an interpreter, preferably the same sex as the patient, when needed.**

8.2   Reflective Exercises

1. Few in-patient facilities have the ability for *halal* food preparation. What might you do to obtain *halal* food for those who require it?
2. How would you administer routine medications during Ramadan?
3. What is responsible for such a high rate of genetic diseases among Arabs?
4. What is the difference between being Arab and Muslim?
5. What are the five major pillars of Islam?
6. How would you assess for pallor, cyanosis, and jaundice among dark/skinned olive complexioned patients?
7. An Arab patient indicates that he does not have sufficient pain to need medication. His wife demands that the nurse give him a pain shot because he is in severe pain. How would you handle this situation?
8. What are approved methods for treating fertility among Arabs?
9. What groups are excused from fasting during Ramadan?
10. How might you break a grave diagnosis to an Arab patient or the family?

Bibliography

Abudabbah N (2005) Arab families: an overview. In: McGoldrick M, Giordano J, Garcia-Preto N (eds) Ethnicity and family therapies. Guilford Press, New York, pp 423–436

Al-Gazali L, Hamamy H, Al-Arrayad S (2002) Genetic disorders in the Arab world. Br Med J 333:831. http://www.bmj.com/content/333/7573/831

Al-Krenawi A, Graham JR (2000) Cultural sensitivity social work practice with Arab clients in mental health settings. National Association of Social Workers. http://www.socialworkers.org/pressroom/events/911/alkrenawi.asp

Islamic Beliefs (2016) http://www.religionfacts.com/islam/beliefs

Kabakian K, El-Kakl F, Shayboub E (2012) Birthing in the Arab region: translating research into practice. East Mediterr Health J 18(1):94–91

Kulwicki AD (2013) People of Arab heritage. In: Purnell L (ed) Transcultural health care: a culturally competent approach, 4th edn. F.A. Davis Company, Philadelphia

Musaiger AO, Takruri HR, Hassan AS, Abu-Tarboush H (2012) Food-based guidelines for Arab Gulf countries. J Nutr Metab. https://doi.org/10.1155/2012/905303

Whitaker B (2006) Unspeakable love: gay and lesbian life in the Middle East. Saqi Books, London

Yosef AR (2008) Health beliefs, practices, and priorities for health care of Arab Muslims in the United States. J Transcult Nurs 19(3):284–291

.

9.1.1 Communications

- Portuguese is the official language of Brazil and continues to dominate Brazilian communities. In the US, Brazilian Portuguese is different from its mother language in the meaning of certain words, accents, and dialects. Dialects vary among Brazilians. Language is frequently a barrier to accessing health-care needs.
- **Obtain an interpreter for Brazilians when necessary.**
- Many Brazilians continue to be of "proper" old-world orientation in which true feelings are not divulged for fear of hurting the receiver of the communication. Everything is said to be *tudo bom* (great), almost in a stoic sense. However, in the intimate circle of family and compatriots, sharing thoughts and feelings is common. **Establish trust before attempting to obtain sensitive information in the health history.**
- Young adult and adolescent Brazilians in the US are generally more acculturated because of their desire and need to assimilate into the new culture. Sharing thoughts and feelings is more common among intragenerational groups rather than intergenerational groups.
- Most Brazilians use touch and direct eye contact. Women kiss each other on both cheeks when they meet and when they say good-bye. At times, women and men kiss in the same manner. Men shake each other's hands and slap each other on the back with the other hand. This gesture frequently ends in an embrace. Children are kissed, and there is much touching. The kissing of a child frequently includes the combination of a "kiss and smell."
- Spatial distancing is close. Facial expressions and symbolic gestures are commonplace.
- **Do not take offense if Brazilians stand closer than you are accustomed.**
- Most Brazilians in the US are future-oriented. In general, they are not punctual and tend to arrive "a bit" late—from minutes to hours—especially for social occasions. Everyone seems to know the behavior of tardiness and plans around it. However, those in professional circles are punctual. **Carefully explain the importance of being on time for health-care appointments.**
- Brazilian names are lengthy, but the modern trend is to use only the first and last names. Traditionally, names appear as first name, mother's family name, and father's family name. "Junior" is added to a name if the son has been named after the father and "*Neto*" if the son has been named after the grandfather (third generation). When a woman marries, she may opt to drop her mother's maiden name and her father's name, or she may keep them both. At times "*de*," "*da/do*", or "*das/dos*" is added to a name to denote "of" and seems to be done out of tradition. No rigid protocol is apparent. Children who have no father are often given the mother's maiden name to which "*da Silva*" is added, denoting that the line of paternity is unclear.
- **Ask the patient his or her full name and the name that is used for legal purposes.**

- In day-to-day relationships, people are called by their first name or *Seu, Senhor* (more respectful) preceding the first name of a man, or *Dona* preceding the first name of a woman. Mothers, grandmothers, or respected strangers are referred to as *A Senhora,* and fathers, grandfathers, and respected men are called *O Senhor*. Doctors are addressed as *Doutor* or *Doutora,* and professors are addressed as *Professor* or *Professora*. The latter two are followed by the first name.
- **Address Brazilian patients formally with Mr., Mrs., Ms., or an appropriate title until told to do otherwise.**

9.1.2 Family Roles and Organization

- Brazilian society is one of *machismo,* with the middle and upper classes being patriarchal in structure. However, as women assert their equality, more egalitarian relationships are becoming evident. Lower socioeconomic households tend to be more matriarchal in nature.
- **Be sure to identify the family decision-maker for health concerns.**
- Children are important to Brazilian families. In the event that a mother and child face deportation from the US, the baby is generally accepted into the family and often raised by the maternal grandmother. Wealthier family members commonly raise the child of a poorer relative. However, these children often enter the family in a second-class capacity.
- Older people live with one of their children when self-care is a concern, and nursing home placement is rare. Older people are respected, seen as family counselors, and are always addressed as *O Senhor* or *A Senhora*. They are included in family activities and usually accompany their children's families on vacation. The extended family is very important and a *jeitinho* (knack) is always procured for employing relatives in any type of service, from the government to a bank, or for helping a relative get into a special university or school.
- **Include the elders and extended family in health-care decision making.**
- Godparents are an important family extension. Poor families frequently ask their *patron* and *patrona* (employer and spouse) to be godparents to their child. Godparent responsibilities include clothing, schooling, and caring for the children in case of the parents' death. The godmother is called **comadre** by the mother. **Compadre** is used in reference to the godfather.
- Although historically common in the lower socioeconomic classes, middle-class households with a single-female parent are becoming increasingly common among Brazilians in the US. In middle-class families, the "no father" status is obscured by the child receiving the same middle and last names as the mother.
- Social status is very important in the Brazilian society, demonstrated in the titles that people use with each other and the practice of listing both parents' surnames. Class separation is maintained discretely by literacy status.

- Brazilians, especially from the south and southeast of Brazil, have become more accepting of gay and lesbian relationships. For many, same-sex relationships carry a stigma. **Do not disclose same-sex relationships to family members or to others.**

9.1.3   Workforce Issues

- Brazilians value diplomacy over honesty even when they promise to attend to something the next day, knowing that it will be impossible.
- **Ask for specific dates and times for when work is to be completed.**
- Professional Brazilians show up for work on time. Less educated Brazilians may find it difficult to adhere to time schedules in the American workplace.
- **Explain the value of punctuality in the American workforce.**
- The English intonation and the pronunciation of certain words are particularly difficult for some Brazilians. Brazilians generally respect authority and are frequently more comfortable in employment situations where rules and job specifications are well defined.
- **Speak clearly and slowly with work directions and ask for clarification to ensure the employee understands the job requirements.**
- Many undocumented Brazilians find employment within the Brazilian community where they may never have to learn the new language. The categorization of Brazilians in the US under the general category of "Hispanics" adds to their discomfort. **Do not refer to Brazilians as Hispanics.**

9.1.4   Biocultural Ecology

- The "typical" Brazilian is a **moreno** with brown skin and eyes and black or brown hair. However, individuals particularly from the southern states of Brazil may have blond hair and blue eyes.
- Table 9.1 lists health conditions commonly occurring for Brazilians in three categories: lifestyle, environment, and genetics. **Because intestinal worms are common in Brazilian immigrants, parasitic diseases should be considered during health assessments.**
- Interviews have substantiated that the incidence of gastrointestinal diseases increases when Brazilians first move to the US. Changes in eating habits from the Brazilian long and ample midday dinner to American fast foods have left Brazilians in the US with gastric complaints. A genetic tendency toward lactose intolerance can contribute to some of these gastric problems. An increased incidence of allergies, especially in children of Brazilian immigrants, has also been reported.

Table 9.1 Health conditions
common to Brazilians

Health condition	Cause
Malaria	Environment, lifestyle
Chagas disease	Environment, lifestyle
Dengue fever	Environment, lifestyle
Meningitis	Environment, lifestyle
Rabies	Environment, lifestyle
Yellow fever	Environment, lifestyle
Schistosomiasis	Environment, lifestyle
Typhoid fever	Environment, lifestyle
Hansen's disease	Genetic
Hepatitis	Environment, lifestyle
Tuberculosis	Environment, lifestyle
Parasitic skin infections	Environment, lifestyle
Cholera	Environment, lifestyle
Cardiovascular diseases	Genetic, environment, lifestyle
Lactose intolerance	Genetic

- Brazil represents the largest number of people living with HIV in Latin America, an endemic infection following Brazilians to the US, and for which documentation is found. Since 2010, HIV is no longer considered grounds for inadmissibility to the US and since 2013, Brazil has had much success in rolling out a program of highly active antiretroviral therapy (HAART) as well as pre- and post-exposer therapy (PrEP/PEP) for its citizens. Nevertheless, due to socioeconomic constraints and lack of medical insurance in the US, many Brazilians living in the US, and particularly those who migrate between the US and Brazil, may be at greater risk for non-treatment of HIV infection or HIV treatment non-adherence.

  Refer patients as needed to social services to help them obtain needed services.

9.1.5 High-Risk Health Behaviors

- Because Brazilian immigrants frequently settle in Brazilian enclaves in large cities, they are subject to the same risk factors as any socially vulnerable urban subpopulation. The greatest risks are violence, drugs, and crime. Adolescents run the risk of resolving their adolescent identity crises by either banding together or joining gangs.
- **School officials can help decrease violence and crime among Brazilians by instituting after-school programs such as dance, music, sports, and other programs for teens.**

- Smoking is a high-risk behavior among Brazilians living in the US. Among men, drinking hard liquor is also prevalent. Accessibility and use of street drugs and an individual's desperate search for quick money are other identifiable high-risk behaviors and often involve living in crowded urban conditions where rent is inexpensive. Although many immigrants are aware of the ill effects of smoking and recreational drugs, loneliness and frustration are deterrents to stopping these habits.
- **Encourage smoking cessation and moderation in alcohol consumption. Help patients find community support groups to decrease loneliness and smoking cessation programs in elementary school.**

9.1.6 Nutrition

- Food is important in the celebration of all rites among Brazilians. Food and its counterpart, hunger, are often viewed as symbols that determine social relations. Food has symbolic content, is used as a reward or punishment, and establishes and maintains social relations.
- The mainstay of the Brazilian American's diet continues to be rice, beans, and farina. Roast beef, fresh chicken, and seafood are sought when they are not too expensive.
- *Cafe de manha* (breakfast) typically consists of bread with *cafe com leite* (half coffee and half hot milk). Sometimes *cuscus* (dry cornmeal mush) is served with milk. Fruit, fruit juices, and scrambled eggs, with or without sliced hot dogs, are common special breakfast fares among middle-class families. Sometimes sweet potatoes and yams grace the breakfast table.
- *O almoco* (dinner) is eaten at noon. This heavy meal, consisting of beans, rice, and farina, often includes *puree* (mashed potatoes) and *macarrao* (pasta). Desserts such as *pudim de leite* (custard), various cornmeal pastries, fruit, and *doce* (a sweet paste made by boiling sugar and fruit or fruit pulp) are common. A typical vegetable salad consists of finely cubed carrots, potatoes, and *shushu* (summer squash–like plant). A fruit salad with finely cubed fruits is also common.
- Brazilians in America have become vitamin- and health-food–conscious. Although this luxury is often not available to those who have immigrated for fast money, legal residents generally become health-food consumers. The preference, especially among young Brazilian women, is to rely on vitamins instead of food consumption to help them remain thin.
- Fruit juices are expensive, and special foods that are common to the Brazilian diet are hard to procure in the US. Food limitations are imposed by expense and inaccessibility of Brazilian mainstay foods. However, many Brazilian communities in the US have ethnic markets and restaurants. Large-chain supermarkets often carry a section of ethnic foods, some of which are reasonably priced.
- **Help patients identify low-cost nutritious, culturally acceptable foods.**

9.1.7   Pregnancy and Childbearing Practices

- Although Brazil is predominantly a Catholic country, birth control is taught and used. Women are encouraged by their physicians or clinic personnel to have tubal ligations to prevent unwanted pregnancies. Immigrants in the US generally practice birth control. Abortions of unwanted pregnancies occur as a matter of personal choice if needed. Thus, fertility practices among immigrant Brazilians are a matter of convenience with a traditional fatalistic overtone.
- **Help women identify acceptable methods of fertility control.**
- Herbal teas are used for bringing on late menstrual periods and for stimulating natural abortions. At times, single women try to become pregnant to facilitate their chance of remaining permanently in the US. This opportunity is greatly enhanced if the child is born here and has been able to attend school.
- Pregnant women are encouraged not to do heavy work and not to swim. Taboos also warn against having sexual relations during pregnancy.
- **Provide factual information about sexual relations during pregnancy.**
- Some foods are to be avoided, and specific foods are recommended during pregnancy. Taboos generally vary according to geographic region, socioeconomic status, and ethnic background. **Determine taboo and acceptable food choices for pregnant women on an individual basis.**
- Many mothers prefer to give their babies powdered dry milk in place of breastfeeding. Many women wish to regain their pre-pregnancy weight as soon as possible. Some women often feel that their milk is *fraca* (weak). Breastfeeding is linked to a social stigma that a mother who breastfeeds may often be thought of as abandoned or sexually unattractive.
- **Provide factual information on the nutritious effects and immune properties of breastfeeding.**
- A postpartum woman eats chicken soup to help her body return to normal. She is also advised not to eat spicy foods or *repadura* (a molasses candy) and not to drink *garapa* (sugar water) or *caldo de cana* (sugar cane juice) if she breastfeeds her infant.
- **Support postpartum women's nonharmful food choices and practices.**

9.1.8   Death Rituals

- The death of a baby or an infant, historically, has been and continues to be treated joyfully and without much sadness, for the child has died pure and is regarded as an angel.
- If financially possible, the families of Brazilians who die in the US personally accompany the body to Brazil for burial in the family vault. If family members cannot come to the US, relatives meet at the airport upon the body's arrival in Brazil.

- **Refer to social services and funeral homes if needed to assist with sending the body to Brazil.**
- Responses to death and grief depend on the family. The fatalistic expression, "It was God's will," helps the grieving process among the rich and the poor.
- Older people wear black for various amounts of time depending on their relationship with the family member. Frequently, the final portrait is hung in the family *chaper* or near the family altar, and prayers are recited. An eternal light burns.
- Relatives are honored on the anniversaries of their death, both at home and at masses. Often, the family places an obituary of remembrance with or without a picture of the deceased in the local newspaper on the anniversary of the death.
- **Support bereavement practices within their cultural context.**

9.1.9 Spirituality

- Although 90% of Brazilians are Catholic, various Protestant sects exist. A few, including Catholics, incorporate Indian animism, African cults, Afro-Catholic syncretism, and Kardecism, a spiritualist religion embracing Eastern mysticism.
- Aside from the ***curandeiros*** (folk healers), special healers exorcise and pray for the wellness of their patients. Saints are asked for help, and people wear medals or little pouches of special powders around their necks.
- **Accept a wide range of religious practices among Brazilians. Integrate non-harmful religious practices into health prescriptions.**
- The meaning of life is found in religion, economy, fatalism, and reality. For some, life is *uma luta* (a battle). For others, life has an almost hedonistic attitude.
- The greatest source of strength for Brazilians is their immediate and extended families.
- **Include nuclear and extended family in health-care decisions.**
- **Do not remove medals or pouches without asking permission.**

9.1.10 Health-Care Practices

- Most Brazilians do not talk about their illnesses unless the illnesses are very serious. Generally, illness is discussed only within the family. Many Brazilians feel that talking about an illness such as cancer negatively influences their condition.
- **Do not disclose the patient's health condition to others. Accept that some patients may not want to talk about their illness.**
- Because many Brazilians tend to shun hospitals, when they are hospitalized their families accompany them and stay around the clock. The patient is often brought

food from home. The family is the nucleus of responsibility for health care and is eager to participate in care. **Include the family in care of the patient whether in the hospital, long-term-care facility, or at home.**

- Brazilians residing in the US are legally required to have health insurance. However, many cannot afford to pay for medical care and thereby revert to self-care.
- Brazilians are known for their self-medication practices. Antibiotic, neuroleptic, antiemetic, and most other prescription drugs are easily obtained over the counter in Brazilian pharmacies. Once in the US, it becomes difficult to obtain the many drugs readily available in Brazil. Customarily, incoming Brazilians bring medicines requested by their friends and thus maintain the circulation of medications not available to Brazilians living in the US.
- **Encourage patients to fully disclose the use of all medications and treatments so they do not conflict with prescription medicines.**
- Because Brazilians tend to self-medicate, the procurement of necessary health care is often avoided or delayed. Consulting with someone who has the same condition or with friends who know someone who has a similar condition may be the first step taken. A trip to the local pharmacist may be the second. A third response may be a telephone call to Brazil asking for a particular medicine.
- The Brazilian culture is rich in folk practices and depends on geographic region, ethnic background, socioeconomic factors, and generation. Many Brazilians prefer to use homeopathic medicines and herbs. Traditional and homeopathic pharmacies are supplemented by *remedios populares* (folk medicines) and *remedios caseiros* (home medicines). **Ask about the use of herbal remedies.**
- Folk remedies and traditional health-care practices are intermeshed when a serious illness may be best treated by traditional caretakers. Some take homeopathic *bolinhas* (little white balls) prepared specifically for certain ailments.
- At times, support services for legal and undocumented Brazilians are hard to find for those who do not have language skills or the self-esteem to become assimilated into the culture of their newly found environments. Language is a major problem for these immigrants. They neglect to learn English and prefer to get by in their enclave community, which may be detrimental to health assessment. **Help patients find legal assistance through social services. Obtain an interpreter when necessary.**
- Another barrier to health care for Brazilians in the US is cost. This, combined with lack of knowledge about the health-care system and facilities, impedes both legal and illegal residents. **Refer patients in financial need to social services departments.**
- Brazilians generally do not like to talk about pain. However, once the emotional barrier is removed, they feel relieved to be able to discuss their discomfort. Many pain-relieving medicines are available without a prescription in Brazil. Frequently, a person requiring these on a regular basis can request that friends or friends of friends bring a supply from Brazil.
- **Explain the facts about pain medicine, and encourage disclosure of all pain medicine and treatments being used.**

- Most Brazilians do not work if they are seriously ill. Sickness is a neutral role and is considered socially exempt. This role is free of guilt, blame, and responsibility. Sickness is often seen as something that just happens.
- Among the lower socioeconomic groups, the term *nervios* refers to an all-incorporating illness. *Nervios* is the ever-present folk diagnosis that identifies the weakness, craziness, and anger associated principally with hunger.
- Better-educated Brazilians accept blood transfusions, organ donation, and organ transplantation. As in the US and other parts of the world, acceptance depends on religious credence and individual preference.

9.1.11 Health-Care Providers

- The folk-health field has many types of health-care providers. *Curandeiros* are divinely gifted, **rezadeiras** (praying women) help exorcise illnesses, card readers can predict fortunes, **espiritualistas** are able to summon souls and spirits, *conselheiros* are counselors or advisors, and **catimbozeiros** are sorcerers. Additionally, the *mae* or *pai de santo* are head priestesses or priests from the African-Brazilian Umbanda or Xango religion. All have the power to heal their believers.
- **Encourage patients to disclose the use of folk healers and treatments prescribed. Incorporate nonharmful practices into prescriptions.**
- Brazilians in the US tend to respect physicians and nurses. Medical education is prestigious and highly sought by aspiring university students.

9.2 Reflective Exercises

1. Explain the extensive name format for Brazilians? How would you address a woman and a man?
2. What are some common health conditions of Brazilians in the US?
3. Under what conditions can Brazilians with AIDS come to the US?
4. Identify at least three barriers to health care for Brazilians in the US.
5. Identify at least three high-risk health behaviors common among Brazilians. What might you do to help decrease these barriers?
6. Identify three Brazilian fold practitioners. What are the tenets of each?
7. Identify taboo practices for pregnant Brazilian women.
8. Describe communication practices of Brazilians with intimates. With non-intimates.
9. What are *comadres* and what are their roles?
10. What is the greatest source of strength among Brazilians?

Bibliography

Brazilian Americans (n.d.) http://www.everyculture.com/multi/A-Br/Brazilian-Americans.html

Coler MS, Coler MA (2013) People of Brazilian heritage. In: Purnell L (ed) Transcultural health care: a culturally competent approach, 4th edn. F. A. Davis, Philadelphia

HIV and AIDS in Brazil (2015) http://www.avert.org/professionals/hiv-around-world/latin-america/brazil

Marques NM, Lira PIC, Lima MC, Lacerda da Slva N, Filho MB, Hutty SRA, Ashworth A (2011) Breastfeeding and early weaning practices in Northeast Brazil: a longitudinal study. Pediatrics 108(4):E66. http://pediatrics.aappublications.org/content/108/4/e66

Roberts TE (2007) Health practices and expectations of Brazilians in the United States. J Cult Divers 14(4):192–197

Zong J, Batakova J (2016) Brazilian immigrants in the United States. Migration Information Source. http://www.migrationpolicy.org/article/brazilian-immigrants-united-states

10.1.2   Family Roles and Organization

- Kinship has traditionally been organized around the male lines. Each family maintains a recognized head who has great authority and assumes all major responsibilities for the family. In recent times, some men include housework, cooking, and cleaning as their responsibilities when their spouses work. Most believe that the family is most important and, thus, each family member assumes changes in roles to achieve this harmony.
- Children are highly valued in China because of the government's past mandate that each married couple may only have one child, although there are exceptions to this rule: e.g., if the first born is a female in rural areas or where the only child of a couple is disabled or dies in infancy. Resources are lavished on the child. Independence is not fostered. The entire family makes decisions for the child even into young adulthood.
- Children born in Western countries tend to adopt the Western culture easily while their parents and grandparents may to maintain traditional Chinese culture in varying degrees. Adolescents maintain their respect for elders even when they disagree with them. Teenagers value a strong and happy family life and seldom do things that jeopardize that unanimity. Adolescents question affairs of life and make great efforts to see at least two sides of every issue.
- Children feel pressure to succeed to help improve the future of the family; thus, most children and adolescents value studying over playing and peer relationships. Children are taught to curb their expression of feelings because individuals who do not stand out are successful. **Children are becoming more outspoken with acculturation and as they read more and watch television and movies from the Western world.**
- The perception of family is developed through the concept of relationships. Each person is identified in relation to others in the family. The individual is not lost, just defined differently from individuals in Western cultures.
- Young men and women enter the workforce immediately after high school if they are unable to continue their education. Many continue to live with their parents and contribute to the family even after marriage (in their 20s) and the birth of a child (in their 30s).
- Extended families are important. Children may live with their grandparents or aunts and uncles so individual family members can obtain a better education or reduce financial burdens.
- Teenage pregnancy is not common among the Chinese but it is increasing among Chinese in America.
- Older people are venerated and viewed as very wise. Children are expected to care for their parents when self-care becomes a concern; in China, law mandates this.
- **When Chinese immigrants need additional assistance, call on local Chinese organizations to obtain help.**

- Maintaining reputation is accomplished by adhering to the rules of society. True equality does not exist in the Chinese mind; if more than one person is in power, then consensus is important. If the person in power is not present at decision making meetings, barriers are raised, and any decisions made are negated unless the person in power agrees. **Ensure that the family spokesperson is in attendance at patient conferences or other times when health-care decisions must be made.**
- The word for "privacy" in Chinese has a negative connotation and means something underhanded, secret, and furtive. People grow up in crowded conditions; they live and work in small areas. The value of group support does not place a high value on privacy.
- **Offer to make arrangements for the family to remain with the patient 24 h a day if needed.**
- The Chinese may ask many personal questions about salary, life at home, age, and children. Refusal to answer personal questions is accepted as long as it is done with care and feeling. The one subject that is taboo is sex and anything related to sex. This may create a barrier for a Western health-care provider who is trying to assess a Chinese patient with sexuality concerns. **Approach issues of sexuality tactfully and indirectly and only after trust is obtained.**
- In China, homosexuality has been legal since 1997, yet the social stigmatization of LGBT people pervades, with gay relationships largely frowned upon by the state and continues to carry a stigma. **Do not disclose same-sex relationships.**

10.1.3   Workforce Issues

- Chinese adapt to the culture in the workplace quickly. They frequently call on other Chinese people to teach them and to discuss how to fit into the new culture more quickly. **Assign a Chinese as a mentor if available.**
- The Chinese are accustomed to giving co-workers small gifts of appreciation for helping them acculturate and adapt to the American workforce. Often, Americans seek opportunities to reciprocate with a gift, such as at a birthday party, farewell party, or other occasion. Some gifts are not appropriate. **Do not give gifts that are culturally inappropriate; for example, giving an umbrella means that you wish to have the recipient's family dispersed; giving a gift that is white in color or wrapped in white could be interpreted as meaning the giver wishes the recipient dead; and giving a clock could be interpreted as never wanting to see the person again or wishing the person's life to end. If other Chinese are available, seek their advice for gift giving.**
- Autonomy is limited and is based on functioning for the good of the group. When a situation arises that requires independent decision making, many times the Chinese know what should be done but do not take action until the leader or superior gives permission.
- **Teach the U.S. expectations for assertiveness and autonomy to Chinese employees.**

- Language may be a barrier for some Chinese. The Chinese language does not have verbs that denote tense, as in Western languages. Intonation in Chinese is in the words themselves, rather than in the sentence. Chinese people who have taken English lessons can usually read and write English competently, but they may have difficulty in understanding and speaking it. **Ask for a demonstration of the activity to ensure understanding.**
- Most Chinese speak in a moderate to low voice tone and consider Americans to be loud. **Be cautious about tone of voice when interacting with Chinese employees.**
- When asked whether they understand what was just said, the Chinese invariably answer yes to avoid loss of face. **Collect information in a manner that cannot be answered with a "yes" or "no" answer.**
- Negative queries are difficult to understand. **Place instructions in a specific order such as "First, gather your supplies." Second, explain the procedure, etc. Do not use complex sentences with "ands" and "buts." Have employees demonstrate instructions to ensure that they are understood.**

10.1.4 Biocultural Ecology

- Skin color among Chinese is varied. Many have skin color with pink undertones; some have yellow tones, and others are very dark. Hair is generally black and straight, but some have naturally curly hair. Most men do not have much facial or chest hair.
- Mongolian spots—dark bluish spots over the lower back and buttocks—are present in about 80% of infants.
- Bilirubin levels are usually higher in Chinese newborns, with the highest levels occurring on the fifth or sixth day after birth.
- The ulna is longer than the radius. Hip measurements are significantly smaller than Westerners (Seidel et al. 1994). **Provide instruction on adequate calcium consumption to prevent osteoporosis and fractures. Ensure diet is low in lactose. Assess patients for genetic conditions before medications are prescribed.**
- The Rh-negative blood group is rare.
- Poor metabolism of mephenytoin occurs in 15–20% of Chinese. Sensitivity to beta blockers, such as propranolol, is evidenced by a decrease in overall blood levels accompanied by a more profound response. Atropine sensitivity is evidenced by an increased heart rate. Increased responses to antidepressants and neuroleptics occur at lower doses. Analgesics have been found to cause increased gastrointestinal side effects, despite a decreased sensitivity to them. Chinese people generally have an increased sensitivity to the effects of alcohol (Levy 1993). **Carefully monitor Chinese patients on medications such as propranolol, atropine, antidepressants, and neuroleptics.**
- Table 10.1 lists commonly occurring health conditions for Chinese in three categories: lifestyle, environment, and genetics.

Table 10.1 Health
conditions common to people
of Chinese heritage

Health conditions	Causes
Lactose intolerance	Genetic, lifestyle
Thalassemia	Genetic
Hepatitis B	Environment, lifestyle
Tuberculosis	Environment, lifestyle
Pancreatic cancer	Environment, lifestyle
Diabetes	Genetic, environment, lifestyle
Glucose-6-phosphate dehydrogenase deficiency	Genetic
Nasopharyngeal cancer	Environment, lifestyle
Liver cancer	Environment, lifestyle
Stomach cancer	Environment, lifestyle, unknown?
Cardiovascular disease	Environment, lifestyle
Hepatitis B	Environment, lifestyle

10.1.5 High-Risk Health Behaviors

- Smoking is a high-risk behavior for many men and teenagers. **Partner with schools to teach hazards of smoking and substance misuse.**
- Most Chinese women do not smoke, but recently the numbers Chinese women who smoke is increasing among males, as much as 68% smoke. **Screen newer immigrants for smoking-related health conditions.**

10.1.6 Nutrition

- Food habits are important, and food is offered to guests at any time of the day or night. Foods served at meals have a specific order, with focus on a balance for a healthy body.
- The typical diet is difficult to describe; each region in China has its own traditional diet. Traditional Chinese medicine frequently uses food and food derivatives to prevent and cure diseases and illnesses and to increase strength in weak and older people.
- Peanuts and soybeans are popular. Common grains include wheat, sorghum, and maize (a type of corn). Rice is usually steamed but can be fried with eggs, vegetables, and meats as well. Many Chinese eat beans or noodles instead of rice. Meat choices include pork (the most common), chicken, beef, duck, shrimp, fish, scallops, and mussels. Tofu, an excellent source of protein, is a staple of the Chinese diet and is fried, boiled, or served cold like ice cream. Bean products are another source of protein. Many desserts or sweets are prepared with red beans.
- Fruits and vegetables may be peeled and eaten raw. Vegetables are lightly stir-fried in oil with salt and spice. Salt, oil, and oil products are important parts of

the Chinese diet. Their healthy selection of green vegetables limits the incidence of calcium deficiencies.

- Drinks with dinner include tea, soft drinks, juice, and beer. Foreign-born and older people may not like ice in their drinks. **Ask patients if they want ice in their drinks.**
- Foods that are considered *yin* and *yang* prevent sudden imbalances and indigestion. A balanced diet is considered essential for physical and emotional harmony. **Provide special instructions regarding risk factors associated with diets that are high in fats and salt.**
- Chopsticks should never be stuck in the food upright because that is considered bad luck.
- Heavy use of salted fish increases the risk for nasopharyngeal, esophageal, and stomach cancers. **Encourage low use of salty fish.**

10.1.7 Pregnancy and Childbearing Practices

- Because of the previous "one couple, one child" law in China, abortions are common in China. Most Chinese families see pregnancy as positive and important in the immediate and extended family. Pregnancy is seen as women's business, although men are beginning to demonstrate an active interest in pregnancy and the welfare of the mother and baby. Women are usually very modest and may insist on a female midwife or obstetrician. Some agree to use a male physician when an emergency arises. **Respect different views on involving men in pregnancy issues.**
- Pregnant women usually increase meat in their diets because their blood needs to be stronger for the fetus. Pregnant women may avoid shellfish during the first trimester because it causes allergies. Some may be unwilling to take iron because they believe that it makes the delivery more difficult. **Explain factual necessity about taking iron preparations, and dispel myths.**
- Traditional postpartum care includes 1 month of recovery, with the mother eating foods that decrease the yin (cold) energy. **Ask what foods the mother plans to eat postpartum, and dispel myths.**
- Many mothers do not expose themselves to the cold air and do not go outside or bathe for the first month postpartum because cold air can enter the body and cause health problems. Some postpartum women wear many layers of clothes and are covered from head to toe, even in the summer, to keep the air away from their bodies.
- Drinking and touching cold water are taboo for women in the postpartum period.
- Raw fruits and vegetables are avoided because they are considered "cold" foods. They must be cooked and be warm. Mothers eat 5–6 meals a day with high-nutritional ingredients including rice, soups, and 7–8 eggs. Brown sugar is commonly used because it helps rebuild blood loss. Drinking rice wine is encouraged

to increase the mother's breast-milk production. **Caution mothers that rice wine may prolong the postpartum bleeding time.**

10.1.8   Death Rituals

- Death is viewed as a part of the natural cycle of life; some believe that something good happens to them after they die. Death and bereavement traditions are centered on ancestor worship, a form of paying respect. Many believe that their spirits will never rest unless living descendants provide care for the grave and worship the memory of the deceased.
- The dead are honored by placing food, money for the person's spirit, or articles made of paper around the coffin.
- The belief that the Chinese greet death with stoicism and fatalism is a myth.
- The number 4 is considered unlucky because it is pronounced like the Chinese word for death; this is similar to the bad luck associated with the number 13 in many Western societies. The color white is associated with death and is also considered bad luck. Black is a bad-luck color. Red is the ultimate good-luck color. Mourners are recognized by black armbands on their left arm and white strips of cloth tied around their heads.
- The purchase of life insurance may be avoided because of a fear that it is inviting death.
- Traditional Chinese do not overtly express emotions to strangers. **Accept and be non-judgmental with a wide range of bereavement practices.**

10.1.9   Spirituality

- The main religions in China are Buddhism, Catholicism, Protestantism, Taoism, and Islam. Prayer is generally a source of comfort. Some Chinese people do not acknowledge a religion such as Buddhism, but if they go to a shrine, they burn incense and offer prayers. As immigration increases, many who practice Christian religions have become more visible.
- "Life forces" are sources of strength. These forces come from within the individual, the environment, and the past and the future of the individual and society.
- The individual may use meditation, exercise, massage, and prayer. Drugs, herbs, food, good air, and artistic expression may also be used. Good-luck charms are cherished and traditional and nontraditional medicines are used.
- The family is usually a source of strength. Individuals draw on family resources and are expected to provide resources to strengthen the family. Resources may be defined as financial, emotional, physical, mental, or spiritual. Calling on ancestors to provide strength as a resource requires giving back to the ancestors when necessary.

- In the US, churches play an important role in the local Chinese community by providing support to immigrants, students, scholars, and their families. **Network and partner with churches to provide additional support for Chinese patients.**

10.1.10 Health-Care Practices

- While many Chinese people have made the transition to Western medicine, others maintain their roots in traditional Chinese medicine, and still others practice both types of medicine. Younger people usually do not hesitate to seek allopathic health-care providers when necessary unless they believe that it does not work for them; then they use traditional Chinese medicine.
- Older people may try traditional Chinese medicine first and only seek Western medicine when traditional medicine does not seem to work. Even after seeking Western medical care, older people may continue to practice traditional Chinese medicine in some form. **Ask patients in a non-judgmental manner if they are using traditional Chinese medicine. Impress upon them the importance of disclosing all treatments because some may have antagonistic effects.**
- Traditional Chinese medical treatments are discussed in Table 10.2.
- The Chinese tend to describe their pain in terms of more diverse body symptoms, whereas Westerners tend to describe pain locally. The Western description includes words like "stabbing" and "localized," whereas the Chinese describe pain as "dull" and more "diffuse." They tend to use explanations of pain from the traditional Chinese influence of imbalances in the yin and yang combined with location and cause.
- **Chinese cope with pain by applying oils and massage, using warmth, sleeping on the area of pain, relaxation, and aspirin.**
- The balance between yin and yang is used to explain mental as well as physical health. Because a stigma is associated with having a family member who is mentally ill, many families initially seek the help of a folk healer. Many Chinese still view mental and physical disabilities as a part of life that should be hidden. Traditionally, an ill person is viewed as passive and accepting of illness. Illness is expected as a part of the life cycle.
- Many individuals feel uncomfortable touching their own bodies, which may be problematic when they need to provide their own health care, for example, with breast self-examinations and scrotal exams. People of the same sex may use touch if they are close friends or family. Men and women do not touch each other, and even couples who have been married for a long time do not show physical affection in public. Most women feel uncomfortable being touched by male health-care providers and tend to seek female providers. **Always ask permission, and explain the necessity for touching patients during examinations and treatments.**
- Families may be reluctant to allow autopsies because of their fear of being "cut up." Most accept blood transfusions, organ donations, and organ transplants.

Table 10.2 Traditional Chinese medical treatments

- Traditional Chinese medicine has many facets, including the five basic substances (*qi*, energy; *xue*, blood; *jing*, essence; *shen*, spirit; and *jing ye*, body fluids); the pulses and vessels for the flow of energetic forces (*mai*); the energy pathways (*jing*); the channels and collaterals, including the 14 meridians for acupuncture, moxibustion, and massage (*jing luo*); the organ systems (*zang fu*); and the tissues of the bones, tendons, flesh, blood vessels, and skin. The scope of traditional Chinese medicine is vast and should be studied carefully by professionals who provide health care to Chinese patients
- Acupuncture and moxibustion are used in many treatments. Acupuncture is the insertion of needles into precise points along the channel system of flow of the *qi* called the 14 meridians. The system has over 400 points. Many of the same points can be used in applying pressure (acupressure) and massage (acumassage) to achieve relief from imbalances in the system. The same systems approach is used to produce localized anesthesia
- Moxibustion is the application of heat from different sources to various points. For example, one source, such as garlic, is placed on the distal end of the needle after it is inserted through the skin, and the garlic is set on fire. Sometimes the substance is burned directly over the point without a needle insertion. Localized erythema occurs with the heat from the burning substance, and the medicine is absorbed through the skin
- Cupping is another common practice. A heated cup or glass jar is put on the skin creating a vacuum, which causes the skin to be drawn into the cup. The heat that is generated is used to treat joint pain
- Herbal therapy is integral to traditional Chinese medicine. Herbs fall into four categories of energy (cold, hot, warm, and cool), five categories of taste (sour, bitter, sweet, pungent, and salty), and a neutral category. Different methods are used to administer the herbs, including drinking and eating, applying topically, and wearing on the body

10.1.11 Health-Care Providers

- Traditional Chinese medicine health-care providers are shown great respect by the Chinese. In many instances, they are shown equal, if not more, respect than Western health-care providers.
- Some distrust Western health-care providers because of the pain and invasiveness of their treatments.
- Older health-care providers receive more respect than younger providers, and men usually receive more respect than women. Physicians receive the highest respect, followed closely by nurses with a university education. Other nurses with limited education are next in the hierarchy.
- If Chinese patients disagree with the health-care provider, they may not follow instructions. Moreover, they may not verbally confront the health-care provider because they fear that either they or the provider will suffer a loss of face.

10.2 Reflective Exercises

1. What are the most common health conditions among Chinese in the US?
2. What are the principle tenets of Traditional Chinese Medicine?

3. What are the primary Chinese languages? If an interpreter knows one of these languages and the patient speaks another Chinese language, what might the interpreter do to get the information?
4. Postpartum women might drink rice wine to increase lactation. What is the problem with this practice?
5. Your Chinese patient has been prescribed two medicines. One medicine is *tid* and the other is *qid*. Describe how you would teach your patient to take them?
6. What are the major religions practiced by Chinese in the US?
7. What seating arrangement would you make to interview a Chinese patient?
8. What is meant by "saving face" among the Chinese? Give an example.
9. What are the principles of *yin* and *yang*? Give an example with dietary practices.
10. Identify some medications where metabolism is different for a patient with Chinese heritage.

Bibliography

Bradford D (2013) Mental Health News Watch. Race, genetics, metabolism: drug therapy and clinical trials. http://www.miwatch.org/2008/04/race_genetics_metabolism_drug_1.html
Chinese American Outreach Guide (2009) http://www.nhpco.org/sites/default/files/public/Access/Chinese_American_Outreach_Guide.pdf
CIA (2012) World factbook: China. (2016). www.cia.gov/library/publications/the-world-factbook/geos/ch.html
Hsiu-Min T (2013) People of Chinese heritage. In: Purnell L (ed) Transcultural health care: a culturally competent approach, 4th edn. F.A. Davis Company, Philadelphia
Lau Y (2012) Traditional Chinese pregnancy restrictions, health-related quality of life and perceived stress among pregnant women in Macao, China. Asian Nurs Res 6(1):27–34
Levy RA (1993) Ethnic and racial differences in response to medicines: preserving individualized therapy in managed pharmaceutical programmes. Pharm Med 7:139–165
Migration Policy Institute (2016) Chinese immigration in the United States. http://www.migrationpolicy.org/article/chinese-immigrants-united-states
National Center for Complementary and Integrative Health (2013) Traditional Chinese medicine: in depth. https://nccih.nih.gov/health/whatiscam/chinesemed.htm
Seidel H, Ball J, Dains J, Benedict W (1994) Quick reference to cultural assessment. Mosby, St. Louis
Shek DT (2010) Oxford handbook of Chinese psychology. Oxford handbooks online. http://www.oxfordhandbooks.com/view/10.1093/oxfordhb/9780199541850.001.0001/oxfordhb-9780199541850-e-22
The Economic and Counsellors Office of the People's Republic of China in the United States of America (2014) Chinese customs, superstitions and traditions. http://us2.mofcom.gov.cn/article/aboutchina/custom/200411/20041100004548.shtml
World Bank (2016) Smoking prevalence among males. http://data.worldbank.org/indicator/SH.PRV.SMOK.MA
Yasuda SU (2008) The role of ethnicity in variability in response to drugs: focus on clinical pharmacology studies. Clin Pharmacol Ther 84(3):417–423
Yasuda SU, Zhang L, Huang SS (2008) The role of ethnicity in variability in response to drugs: focus on clinical pharmacology studies. Clin Pharmacol Ther 84(3):417–423

Chapter 11
People of Cuban Heritage

11.1 Overview and Heritage

The Republic of Cuba, located 90 miles south of Key West, Florida, is a multiracial society with people of primarily Spanish and African origins. Other groups include Chinese, Haitians, and Eastern Europeans. Spain, the US, and Russia (formerly the Soviet Union) significantly influence Cuba's history and culture. Mistrust of government has reinforced a strong personalistic tradition and sense of national identity evolving from family and interpersonal relationships. Desire for personal freedom, hope of refuge, political exile, and promise of economic opportunities prompted migration. Cubans in the US take great pride in their heritage, possess a strong ethnic identity, speak Spanish, and adhere to traditional Cuban values and practices. Their highest concentration is in Florida, although significant numbers live in New Jersey, New York, Illinois, and California.

11.1.1 Communications

- Many live and transact business in Spanish-speaking enclaves. Whereas the second generation speaks English, many converse with friends or peers in "Spanglish," a mixture of Spanish and English. The highly educated are more likely to speak English at home. **Assistance with required forms in the U.S. health-care system is needed by the less acculturated and those for whom English is a concern.**
- Many value *simpatía* and *personalismo* in their interactions with others. *Simpatía*, the need for smooth interpersonal relationships, is characterized by courtesy, respect, and the absence of criticism or confrontation. *Personalismo*, the importance of intimate interpersonal relationships, is valued over impersonal bureaucratic relationships.

© Springer Nature Switzerland AG 2019
L. D. Purnell, E. A. Fenkl, *Handbook for Culturally Competent Care*,
https://doi.org/10.1007/978-3-030-21946-8_11

- *Choteo*, a lighthearted attitude with teasing, bantering, and exaggerating, is often observed in their communications with others. Conversations are characterized by animated facial expressions, direct eye contact, hand gestures, and gesticulations. Voices tend to be loud, and the rate of speech is fast. **Do not interpret these behaviors as family discord.**
- Handshakes, touching, and hugs are acceptable among family, friends, and acquaintances and may be used to express gratitude to the caregiver. Touch is common between people of the same gender; older men and women rarely touch in public. **Explain the essential necessity and ask permission for touching private body areas during a physical examination.**
- Most tend to emphasize current issues and problems rather than future ones. *Hora cubana* (Cuban time) refers to a flexible period that stretches 1–2 h beyond designated clock time. **When setting up appointments, assess the patient's level of acculturation with respect to time. Explain the necessity of punctuality.**
- Cubans use two surnames representing the mother's and father's family names. Married women may also add the husband's name. Thus, **ask which name is used for legal purposes. When addressing patients, especially older people, formal rather than familiar forms of speech should be used unless told otherwise (i.e., Señor [Mr.], Señora [Mrs.], or Señorita [Miss or Ms.]).**

11.1.2 Family Roles and Organization

- Traditional family structure is patriarchal, characterized by a dominant and assertive male and a passive, dependent female. Traditionally, Cuban wives stay at home, manage the household, and care for children, whereas husbands are expected to work, provide financially, and make major decisions for the family. *La casa* **(the house), the province of the woman, and** *la calle* **(the street), the domain of the man, demonstrate the distinction in gender roles and should be respected. With acculturation more are women working outside the home, egalitarian decision making prevails in the US.**
- Honor is attained by fulfilling family obligations and treating others with *respeto* (respect). *Vergüenza*, a consciousness of public opinion and the judgment of the entire community, is considered more important for women than for men. Machismo dictates that men display physical strength, bravery, and virility and be the spokesperson, even though they might not make the decisions.
- *La familia* (the family, nuclear and extended, including godparents) is the most important source of emotional and physical support. Multigenerational (3–4 generations) households are common, including a high proportion of people 65 years and older who live with their relatives. **Family, the most important social unit, is involved in all decisions. Specifically, ask who makes which decisions for the family and who is the spokesperson. Verbal consent for invasive**

**procedures from family members as well as from next of kin is often
required. Extended family members are good resources for home care.**

- Compared with European American standards, Cubans tend to pamper and over-
protect their children. Children are expected to study, respect their parents, and
follow *el buen camino* (the straight and narrow). Boys are expected to learn a
trade or prepare for work and to stay away from vices. Girls are expected "to
remain honorable while single," to prepare for marriage, to avoid the opposite
sex, and not to go out without a chaperone. When a daughter reaches 15 years, a
quinceanera, or elaborate 15th birthday party, is typically held to celebrate this
rite of passage. Adolescents may undergo an identity crisis and reject their heritage; parents may feel their authority is being challenged.

- **Explain American child abuse laws, taking into consideration the culturally
complex definition of what may be considered child abuse.**

- Personal relationships, known as *compadrazgo*, are typical, with **heavier reliance for help on family and personal relationships than on government or
organizations.**

- **Traditional families seldom discuss sex openly; initiating these discussions
with health teaching may be welcome.**

- Little information is available on homosexuality. Same-sex behaviors among
men may be regarded as a sign of virility and power rather than homosexual
behavior. The gay lifestyle is contradictory to the machismo orientation of this
culture. Same-sex couples may be alienated from their families.

- **Given the stigma associated with homosexuality, a matter-of-fact, nonjudgmental approach must be used when questioning patients regarding sexual
practices. Do not disclose same-sex relationships to family members.**

11.1.3 Workforce Issues

- A source of tension is the tendency of Cubans to speak Spanish with other Cuban
or Hispanic co-workers. Speaking the same language allows them to form a common bond, relieve anxieties at work, and feel comfortable with one another.
**Explain upon employment the rules for speaking a language other than
English at work.**

- Traditional Cubans recognize supervisors as authority figures and treat them
with respect and deference. Cubans value a structure characterized by *personalismo*, one that is oriented around people rather than around concepts or ideas.
Personal relationships at work are considered an extension of family relationships. Because of the emphasis on the job or task in the American workplace,
many Cubans view this workplace as being too individualistic, businesslike, and
detached.

- **Explain the individualistic culture of the American workforce upon
employment.**

Table 11.1 Health
conditions common to
Cubans

Health conditions	Causes
Hypertension	Genetic, lifestyle
Coronary artery disease	Genetic, lifestyle
Obesity	Environment, lifestyle
Diabetes mellitus	Genetic, lifestyle
Lung cancer	Environment, lifestyle

11.1.4   Biocultural Ecology

- Most Cubans are white, and only 5% are black with physical features similar to those of African Americans. **Because of their European ancestry, most have skin, hair, and eye colors that vary from light to dark. To assess for cyanosis and jaundice in individuals with dark skin, assess the color of the sclera and conjunctiva, palms of the hands, soles of the feet, and buccal mucosa rather than relying on skin tone.**
- Cuban Americans tend to have lower incidences of diabetes mellitus, obesity, and hypertension than other Hispanic groups or whites. Because of their diet, which is high in sugar, many exhibit a high prevalence of tooth loss, filled teeth, gingival inflammations, and periodontitis. Table 11.1 lists commonly occurring health conditions among Cubans in three categories: lifestyle, environment, and genetics.
- Specific information related to drug metabolism is limited; however, in general, many require lower doses of antidepressants and experience greater side effects than non-Hispanic white populations.

11.1.5   High-Risk Health Behaviors

- Cuban Americans tend to exhibit a higher incidence of smoking than other Hispanic or European groups. **Encourage smoking cessation starting in elementary school.**
- Alcohol use is greater among males than females and among younger versus older groups. **When assessing alcohol consumption, determine the amount consumed during holidays and celebrations, not just on a daily basis.**
- Violent deaths account for high mortality rates among adolescents and young adults. Suicide rates also exceed those of the white non-Hispanic population.
- Use of preventive services depends on accessibility, which, in turn, is significantly influenced by education, annual income, and age.

11.1.6 Nutrition

1. Food allows families to reaffirm kinship ties, promotes a sense of community, and perpetuates customs and heritage. **Rejecting food that is offered may be perceived as rejecting the person who is offering the food.**
2. Staple foods include root crops like yams, yuca, malanga, and boniato; plantains; and grains. Many dishes are prepared with olive oil, garlic, tomato sauce, vinegar, wine, lime juice (*sofrito*), and spices. Meat is usually marinated in lemon, lime, sour orange, or grapefruit juice before cooking.
3. **Determine food preparation practices as well as customary foods consumed. Fats or flavorings used for cooking can add significant calories to meals.**
4. **All major food groups are well represented in the Cuban diet; however, fiber content and leafy green vegetables may be lacking. Lactose intolerance is common among most Hispanic groups.**
5. A leisurely noon meal (*almuerzo*) and a late evening dinner (*comida*), sometimes as late as 10 or 11 p.m., are often customary.
6. **Determine meal times and adjust medication schedules accordingly.**
7. The strong and bittersweet coffee called *café cubano* is a standard drink after meals and throughout the day.
8. Being overweight, according to U.S. actuarial tables, is seen as positive, healthy, and sexually attractive.
9. **Patients should be coparticipants in deciding an acceptable weight. Counsel patients about the differences between being overweight and obese and the health hazards associated with obesity. Incorporate patients' traditional and preferred food choices into nutritional health planning.**

See Table 11.2 for Popular Cuban American foods.

Table 11.2 Popular Cuban American foods

Entrees and side dishes			
Roast pork	*lechon*	Fried pork chunks	*masas de puero*
Sirloin steak	*palomilla*	Shredded beef	*ropa vieja*
Pot roast	*boliche*	Roasted chicken	*pollo asado*
Ripe plantains	*platanos maduros*	Green plantains	*platanos verdes*
Fried green plantains	*tostones* or *mariquita*		
Desserts			
Custard	*flan*	Egg pudding	*natilla*
Rice pudding	*arroz con leche*	Coconut pudding	*pudín de coco*
Bread pudding	*pudín de pan*		
Beverages			
Sugar cane juice	*guarapo*	Iced coconut milk	*coco frio*
Cuban soft drinks	*batidos*	Sangria, or beer	*Materva*
Strong coffee	*café cubano*		

11.1.7   Pregnancy and Childbearing Practices

- Cuban women's fertility rate is lower than that of other Hispanic American women. Cuba's current reproductive rate is among the lowest in the developing world. Even before the revolution, Cuba had the lowest birthrate in Latin America. This low fertility rate has been attributed to the fact that many women are in the workforce. Preterm births and neonatal and post-neonatal deaths are lower among Cuban American women than among other Hispanic American groups. Women are less likely to use oral contraceptives than other Hispanic American women.
- **Explain that condoms not only help prevent pregnancy but also decrease the incidence of HIV or *SIDA* (HIV) and other sexually transmitted infections. Determine on an individual basis the methods of acceptable birth control.**
- Prenatal care is higher than among other Hispanic and white non-Hispanics. **Stress that medical evaluations during pregnancy are necessary to help ensure the health of the mother and baby. When language barriers exist, use interpreters and videos and literature in Spanish for teaching.**
- Mothers tend to use advice about child health given by their spouses, mothers, mothers-in-law, and clerks and pharmacists.
- Childbirth is a time for celebration, with family members and friends congregating in the hospital. Traditionally, men have not attended the births of their children, but younger, more acculturated fathers are frequently present. **Respect the cultural beliefs of families if fathers are not wanted or do not wish to be in the delivery room.**
- Table 11.3 lists Cuban dietary and folk beliefs and practices about pregnancy.
- During the postpartum period, ambulation, exposure to cold, and bare feet place the mother at risk for infection. Family members and relatives often care for the mother and baby for about 4 weeks postpartum. Most women consider breast-feeding better than bottle feeding; approximately half choose to breast-feed. **Mothers tend to wean infants from the breast and introduce solid foods at a young age, but bottle weaning often occurs at a median age of 4 months,**

Table 11.3   Cuban dietary and folk beliefs and practices about pregnancy

- Women eat for two during pregnancy.
- Morning sickness is cured by eating coffee grounds.
- Eating a lot of fruit ensures that the baby will be born with a smooth complexion.
- Wearing necklaces during pregnancy causes the umbilical cord to be wrapped around the baby's neck.
- Raising one's arms over the head while pregnant may cause the umbilical cord to wrap around the infant's neck.
- (Respect nonharmful cultural beliefs when prescribing care with regard to nutrition. Assess which foods are eaten or avoided during pregnancy in order to provide a culturally appropriate nutritional plan).

fostering their beliefs that "a fat child is a healthy child" and that breast-feeding may contribute to a deformity or asymmetry of the breasts.

- Cutting the infant's hair or nails in the first 3 months is believed to cause blindness and deafness. **Do not cut the baby's hair or nails without asking permission.**

11.1.8 Death Rituals

- In death, as in life, the support of the extended family network is paramount. Bereavement is expressed openly as loud crying, with other physical manifestations of grief. Death is often seen as a part of life and some, especially men, may approach death stoically. **Recognize that the lack of an open display of emotion may not imply indifference to the dying or deceased.**
- The dying person is typically attended by a large gathering of relatives and friends. In Catholic families, individual and group prayers are held for the dying to provide a peaceful passage to the hereafter. Religious artifacts such as rosary beads, crucifixes, or *estampitas* (little statues of saints) are placed in the dying person's room. For adherents of *santería*, death rites may include animal sacrifice, chants, and ceremonial gestures.
- **Summon appropriate clergy to perform appropriate death rites. A place should be found for family members to gather during this time, preferably close to the dying person. Do not touch or remove little statues of saints without asking permission.**
- **Candles (electric candles are acceptable) are lighted after death to illuminate the path of the spirit to the afterlife.**
- A *velorio* (wake) lasts 2–3 days and is usually held at a funeral parlor or in the home where friends and relatives gather to support the bereaved family. Burial in a cemetery is common practice, although some may choose cremation.
- The deceased are customarily remembered and honored on their birthdays or death anniversaries by lighting candles, offering prayers or masses, bringing flowers to the grave, or gathering with family members at the grave site.

11.1.9 Spirituality

- Approximately 85% of Cuban Americans are Roman Catholic; the remaining 15% are Protestants, Jews, and believers in African Cuban Santería. Roman Catholicism is personalistic and characterized by devotion and intimate, confiding relationships with the Virgin Mary, Jesus, and the saints. **Religious beliefs and practices must be viewed in an open, sincere, and nonjudgmental manner. In the in-patient setting, privacy is important if patients and families need to perform certain rituals or prayers. A visit from a priest, rabbi, or** *santero* **may promote a sense of psychological support and spiritual well-being.**

- Significant religious holidays include *Noche Buena* (Christmas Eve), Christmas, *Los Tres Reyes Magos* (Three Kings Day), and the festivals of the *La Caridad del Cobre* and Santa Barbara.
- *Santería* or *Regla de Ocha*, a 300-year-old African Cuban religious system, combines Roman Catholic elements with ancient Yoruba tribal beliefs. Followers of *Santería* believe in the magical and medicinal properties of flowers, herbs, weeds, twigs, and leaves. Sweet herbs such as *manzanilla*, *verbena*, and *mejorana* are used for attracting good luck, love, money, and prosperity. Bitter herbs such as *apasote*, *zarzaparilla*, and *yerba bruja* are used to banish evil and negative energies. *Santería* is viewed as a link to the past and is used to cope with physical and emotional problems.
- **By law, as long as standards of safety and sanitation are maintained, allow space and privacy to be able to engage in specific religious ceremonies and rituals.**
- Physical complaints may be diagnosed and treated by a physician, whereas the *santero* may assist in balancing and neutralizing the various aspects of the illness.
- Belief in a higher power is evident in practices used to maintain health and well-being or cure illness, such as using magical herbs, special prayers or chants, ritual cleansing, and sacrificial offerings. Many tend to be fatalistic, believing that they lack control over circumstances influencing their lives. **Fatalistic beliefs do not mean that the patient is not willing to engage in preventive practices.**

11.1.10 Health-Care Practices

- Cuban cultural treatments for ailments are listed in Table 11.4.
- African Cubans may seek biomedical care for organic diseases but consult a *santero* for spiritual or emotional crises. Conditions such as *decensos* (fainting spells) or *barrenillos* (obsessions) may be treated solely by a *santero* or simultaneously with a physician.
- Many tend to seek help only in response to crisis situations. **Encourage visits to health providers for wellness checks, explaining the long-term benefits.**
- Many Cuban Americans rely on the family as the primary source of health advice. Older women provide traditional home remedies such as herbal teas or mixtures to relieve mild or moderate symptoms or cure common ailments. **Include the entire family in health promotion and health teaching to increase compliance with health prescriptions. Because health often means simply the absence of pain, explain the importance of preventive measures and health promotion activities.**
- Many, especially older, Cuban Americans were socialized into a strong health ideology and successful primary-care system while still in Cuba.

Table 11.4 Cuban cultural treatment for ailments

Herbal tea	Ailment
Cosimiento de anis (anise)	Stomachaches, flatulence, baby colic, anxiety
Cosimiento de limon con miel de abeja (lemon and honey)	Cough and respiratory congestion
Cosimiento de apasote (pumpkin seed)	Gastrointestinal worms
Cosimiento de canela (cinnamon)	Cough, respiratory congestion, menstrual cramps
Cosimiento de manzanilla (chamomile)	Stomachaches
Cosimiento de naranja agria (sour orange)	Cough and respiratory congestion
Cosimiento de savila (aloe vera)	Stomachaches
Cosimiento de tilo (linden leaves)	Anxiety
Cosimiento de yerba buena (spearmint leaves)	Stomachaches, anxiety
Fruit or Vegetable	*Ailment*
Chayote (vegetable)	Anxiety
Zanaoría (carrots)	Visual problems
Toronja y ajo (grapefruit and garlic)	Elevated blood pressure
Papaya y toronja, y pina (papaya, grapefruit, and pineapple)	Gastrointestinal parasites
Remolacha (beets)	Influenza, anemia, cough
Cascara de mandarina (fruit)	
Home remedies	Ailment
Agua con sal (saltwater)	Sore throat
Agua de coco (coconut water)	Kidney problems and infections
Agua raja (turpentine)	Sore muscle and joint pain
Bicarbonato, limon y agua (baking soda, lemon, and water)	Upset stomach or heartburn
Cebo de carnero (fat of lamb)	Applied directly on skin for contusions and swelling
Mantequilla (butter)	Applied directly on burns to soothe pain
Clara de huevos (egg white)	Applied directly to scalp to promote hair growth

Amulets

Azabache is a black stone placed on infants and children as a bracelet or pin to protect them from the "evil eye."

La manito de coral, symbolic of the hand of God protecting a person, may also be worn as a necklace or bracelet.

Los ojitos de Santa Lucía, or the eyes of Saint Lucy, may be hung on a bracelet or necklace for prevention of blindness and protection from the evil eye.

- Many Cuban Americans use traditional medicinal plants in the form of teas, potions, salves, or poultices. In Cuban communities, stores called **botanicas** sell herbs, ointments, oils, powders, incenses, and religious figurines to relieve maladies, bring luck, drive away evil spirits, or break curses. Santería necklaces and animals used for ritual sacrifice are often available at *botanicas*.

- **Determine which medicinal products are harmful and which will cause no harm.**
- **Incorporate nonharmful practices into the plan of care. Ascertain if patients are using over-the-counter medicine or sharing prescriptions of others. Explain the hazards of overuse of these medications as well as using medicines that were originally prescribed for other family members.** Cultural approaches used to treat common ailments are shown in Table 11.4.
- Language is the greatest barrier to health care. Additional barriers include bureaucratic paperwork and transportation problems, cost of services, and inconvenient hours. **Obtain interpreters when necessary and help them navigate the complex health-care system.**
- Many immigrants may suffer from loneliness, depression, anger, anxiety, insecurity, and health problems. **Determine physical, emotional, and financial resources available for caring for sick patients either at home or in care facilities.**
- Dependency is a culturally acceptable sick role that allows the extended family network to assume the chores and tasks of the sick person. Help is sought in response to crisis situations, pain, and discomfort and is often a signal of a physical disturbance that warrants consultation with a traditional or biomedical healer. **Pain is expressed with verbal complaints, moaning, crying, and groaning that may or may not signify a need for pain medication. Individual assessments for patients with pain are needed to determine patients' perceptions of causes, remedies, and possible coping behaviors. Explain pain medication may promote healing.**
- Blood transfusions and organ donations are usually acceptable.
- **Dispel myths related to organ donation, blood transfusions, and organ transplantation. Assistance from a priest or other clergy may be helpful.**

11.1.11 Health-Care Providers

- Both traditional and biomedical care are acceptable. Folk remedies may be used at home, but if the condition persists, folk health-care providers such as *santeros, practitioners of santeria* (and biomedical health-care providers may be used either simultaneously or successively. *Santeros* may prescribe treatment or perform rituals to enable ill people to recover by invoking supernatural deities to intervene to help make them well. Three hundred-year-old Afro-Cuban religion that syncretizes Roman Catholic elements with ancient Yoruba tribal beliefs and practices.
- **Ask patients if they are using folk practitioners, their reasons for using them, and the treatments that have been prescribed. Collaborating with folk healers and allopathic providers may increase compliance with health prescriptions.**

11.2   Reflective Exercises

1. What are the primary principles and practice of *Santeria*?
2. What is meant by *simpatia, personalismo,* and *choteo*? How are they displayed by Cubans?
3. Explain the extensive name format used by Cubans.
4. Name at least three health conditions common among Cubans.
5. Identify at least three folk beliefs about pregnancy common to Cubans.
6. What are some common amulets used by Cubans to teat ailments?
7. A female Cuban new diabetic patient wants to use herbs to control her diabetes. Would you encourage her to try herbs rather than allopathic treatments? Why or why not?
8. What is a *botanica*? Is there one in your community?
9. What is the most important source of emotional and physical support for most Cubans?
10. What is a *quinceanera?*

Bibliography

About Santeria (2016). http://www.aboutsanteria.com/what-is-santeria.html
Bryant CA (1982) The impact of kin, friend, and neighbor on infant feeding practices: Cuban, Puerto Rican, and Anglo families in Cuba. Soc Sci Med 16(20):1757–1765
Burgman C (n.d.) Santeria: race and religion in Cuba. http://www.crsp.pitt.edu/sites/default/files/Paper-Santeria-Burgman.pdf
Centers for Disease Control and Prevention (n.d.) Cubans. https://search.cdc.gov/search?query=cubans&utf8=%E2%9C%93&affiliate=cdc-main
CIA (2016) World factbook: Cuba. https://www.cia.gov/library/publications/the-world-factbook/geos/cu.html
Countries and Their Culture (n.d.) Cuba. http://www.everyculture.com/Cr-Ga/Cuba.html
Global Affairs Canada (2014) Cultural information—Cuba. https://www.international.gc.ca/cil-cai/country_insights-apercus_pays/ci-ic_cu.aspx?lang=eng
Migration Policy Institute (2015) Cuban immigrants in the United States. http://www.migrationpolicy.org/article/cuban-immigrants-united-states
Purnell L, Gil J (2013) People of Cuban heritage. In: Purnell L (ed) Transcultural health care: a culturally competent approach, 4th edn. F.A. Davis Company, Philadelphia

Chapter 12
People of European American Heritage

12.1 Overview and Heritage

The term European American culture in this chapter refers to the middle-class values of citizens of the mainland US. The US, the world's oldest constitutional democracy, comprises 3.5 million square miles with a population of almost 324 million people. The European American (American) culture evolved from those of early immigrants, primarily from Northern European countries. Today, the US includes immigrants or descendants of immigrants from almost every nation and culture of the world.

Americans, and individuals from other groups who have assimilated, value individualism, free speech, freedom of choice, independence, self-reliance, confidence, "doing" rather than "being," egalitarian relationships, nonhierarchal status of individuals, individual responsibility, achievement status over ascribed status, volunteerism, friendliness, openness, futuristic temporality, technology, and the ability to control the environment. Many emphasize material possessions and physical comfort. The extent to which people conform to this dominant culture depends on the variant characteristics of culture discussed in Chap. 1. Remember aggregate data are true for the group but not necessarily for the individual, so this profile may not be accurate for everyone who self-identifies with the European American culture.

12.1.1 Communications

- Although the unofficial language in the US is English, accents vary regionally but are usually understandable to all. Having well-developed verbal skills is considered important.
- Voice volume is loud compared with many other cultures.

© Springer Nature Switzerland AG 2019
L. D. Purnell, E. A. Fenkl, *Handbook for Culturally Competent Care*,
https://doi.org/10.1007/978-3-030-21946-8_12

- The rate of speech varies depending on location within the US, with people from the South generally speaking slower than people in the Northeast.
- Many individuals readily disclose personal information about themselves. Personal sharing is encouraged in a wide variety of topics, but not religion.
- People of the same sex (especially men) or opposite sex do not generally touch each other unless they are relatives or close friends. **The low-touch culture of the US is reinforced by sexual harassment policies.**
- Regardless of class or social standing of the conversants, Americans are expected to maintain direct eye contact without staring. **Just because someone does not maintain direct eye contact does not mean they are not listening or not telling the truth.**
- Men and women extend the right hand when greeting someone for the first time.
- Most Americans are future-oriented but try to balance the past with the present. Punctuality is valued in both business and social settings.
- The usual name format is given name, middle name (optional), and family name. People known to each other use the given name.
- **Greet patients using use Mr., Mrs., Miss, Ms., or appropriate title on first meeting someone and until told to do otherwise.**

12.1.2   Family Roles and Organization

- A value is placed on egalitarian relationships and decision making, although great variations exist within families. Women may have careers, and men often assist with child care, household chores, and cooking responsibilities.
- Great value is placed on children, and many laws help protect children.
- Autonomy is encouraged in children and teenagers. Children and teenagers are encouraged to have friends of the same and opposite sex.
- Teenagers are expected to refrain from premarital sex, smoking, recreational drug use, and alcohol use until they reach the age of 18 years or until they leave home. Most teenagers move out of their parents' home when their education is completed or when they turn age 18 years.
- Social attitudes toward homosexual activity vary widely and sometimes carry a stigma. In 2015, same-sex marriage became legal in the US and same-sex relationships have become more acceptable among many.
- **Do not to disclose patients' same-sex relationships to family members. Do not assume that married individuals are necessarily in a male-female marriage. Provide an environment that is welcoming to all individuals regardless of their sexual orientation.**

12.1.3 Workforce Issues

- Americans are expected to be punctual on their job, with formal meetings, and with appointments. If one is more than a few minutes late, an apology is expected.
- The American workforce stresses efficiency (time is money), operational procedures on how to get things done, task accomplishment, and proactive problem-solving.
- Intuitive abilities and common sense are not usually valued as much as technical abilities. The scientific method is valued and everything has to be proved. Americans want to know *why*, not *what*. Many are obsessed with collecting facts and figures before they make decisions. Pragmatism is valued.
- American values expect that the needs of individuals are subservient to the needs of the organization. Most Americans place a high value on "fairness," and rely heavily on procedures and policies in the decision-making process, a necessity in individualistic cultures in order to reach consensus.
- All professionals are expected to be assertive without being aggressive, a concept that is difficult for some non-Americans. When others speak in their native language at work, it may become a source of contention for both patients and health-care personnel.
- **Explain rules for foreign language use upon employment.**
- Four generations are in the workforce in the US: (a) Traditionalists, (b) Baby Boomers, (c) Generation X, and (d) the Millennials. Each group brings a different worldview, different perspectives, and varied strengths to the workforce.
- The Traditionalists were born before 1945 and are characterized as being loyal, patriotic, and hard working. Many from this group worked under "control and command" styles of management and may perceive changes in the health-care field as being too radical. They have a wealth of practical expertise to share. Most have also developed conflict resolution and negotiation skills.
- The Baby Boomers, born between 1945 and 1965, currently comprise almost half the workforce. Most are seen as idealistic, optimistic, competitive, and community-focused. They are also considered the "sandwich generation" because they are often caring for their children and their parents. This group also has a wealth of practical expertise to share and has developed conflict resolution and negotiation skills.
- Generation Xers, born between 1966 and 1980, are characterized as being highly independent and skeptical and have a "free agent" mentality and change jobs easily to meet personal needs. Most are looking for experience rather than job security.
- Millennials, born after 1981, like new ideas and change. Technology use is a given, and most are comfortable with diversity.
- **Capitalize upon the strengths of each generation. Generational descriptions are probably not accurate with many newer immigrant and ethnic populations.**

Table 12.1 Health
conditions common to
European Americans

Health conditions	Causes
Appendicitis	Unknown
Diverticular disease	Lifestyle, genetic
Colon cancer	Genetic, lifestyle, environment
Hemorrhoids	Lifestyle, genetic
Osteoporosis	Lifestyle, genetic
Osteoarthritis	Lifestyle, genetic
Cystic fibrosis	Genetic
Phenylketonuria	Genetic
Skin cancer	Lifestyle, environment
Cardiovascular disease	Genetic
Varicose veins	Genetic, lifestyle
Obesity	Lifestyle, environment
Multiple sclerosis	Genetic, environment

12.1.4   Biocultural Ecology

- The leading causes of death in Americans are heart disease, cancer, chronic obstructive lung disease, unintentional injury, diabetes, and HIV.
- Nearly one million people in the US are HIV-positive or have AIDS, with the highest concentrations in large urban areas.

Table 12.1 lists commonly occurring diseases and disorders for European Americans in three categories: lifestyle, environment, and genetics.

12.1.5   High-Risk Health Behaviors

- Cigarette-smoking has been declining in the US. Higher socioeconomic groups are more likely to drink but also more likely to drink without problems. **Encourage responsible drinking and smoking cessation beginning in elementary school.**
- High-risk sexual behavior continues with the incidence of STIs and HIV infections. **Encourage health screening and preventive behaviors such as condoms and safer-sex practices starting in elementary school.**
- Many traffic laws have been enacted, although they vary from state to state, requiring automobile seat belts, child restraints, and helmets for bicycle and motorcycle riders.
- **Encourage use of seat belts, child restraints, and helmets. Refer to state laws where they exist.**

12.1.6   Nutrition

- The typical diet is high in fats and cholesterol and low in fiber. **Encourage low-fat, low-cholesterol, and high-fiber diets using the U.S. Department of Agriculture food pyramids.**
- Given the varied topography of the US, there are no food shortages for a healthy diet. However, socioeconomic resources prohibit some from purchasing healthy foods. **Teach patients about healthy foods and preparation practices.**
- Meal times and food choices vary according to ethnic background, the region of the country, urban versus rural residents, and weekdays versus weekends.
- Breakfast is usually consumed between 6:00 a.m. and 9:00 a.m., depending on the person's work schedule. The noontime meal, typically consumed by 1:00 p.m., is called lunch in urban areas and dinner in rural areas. The evening meal, dinner in urban areas or supper in rural areas, is served between 6:00 p.m. and 9:00 p.m. and is usually the largest meal of the day.
- **Determine meal times, and adjust medications accordingly.**

12.1.7   Pregnancy and Childbearing Practices

- Commonly used methods of fertility control include natural ovulation methods, birth control pills, foams, Norplant, the morning-after pill, intrauterine devices, condoms, abortion, and sterilization procedures such as vasectomy and tubal ligations. Sterilization in the US is strictly voluntary. **Abortion remains a controversial issue.**
- Pregnant women are expected to seek preventive care, eat a well-balanced diet, and get adequate rest in order to have a healthy pregnancy and baby. Women are encouraged to breast-feed. **Discourage smoking, drinking alcohol, consuming large amounts of caffeine, and taking recreational drugs.**
- Prospective fathers are encouraged to take prenatal classes with expectant mothers and provide a supportive role in the delivery process. **Fathers who do not want to participate should not be made to feel guilty.**
- **Respect cultural beliefs associated with pregnancy and the birthing process.** Possible restrictive beliefs for pregnant women are listed in Table 12.2.
- **Integrate nonharmful cultural practices into preventive teaching and interventions.**

Table 12.2 Possible restrictive beliefs for pregnant women

• Pregnant women should refrain from being around loud noises for prolonged times.
• Wearing an opal ring during pregnancy can harm the baby.
• Eating strawberries or being frightened by a snake causes birthmarks.
• Congenital anomalies can occur if the mother sees or experiences a tragedy during her pregnancy.
• Nursing mothers should eat a bland diet to avoid upsetting the baby.
• The infant should wear a band around the abdomen to prevent the umbilicus from protruding and becoming herniated.
• A coin, key, or other metal object should be put on the umbilicus to flatten it.
• Cutting a baby's hair before baptism can cause blindness.
• If the pregnant woman raises her hands over her head while pregnant, it may cause the cord to wrap around the baby's neck.
• Moving heavy items can cause a pregnant woman's "insides" to fall out.
• Children born with physical or mental abnormalities are punishment from God for acts of the parents.

12.1.8 Death Rituals

- Death and responses to death are difficult topics for many Americans to verbalize.
- The dying person should not be left alone. **Make accommodations for a family member to be with the dying person at all times. Health-care personnel are expected to care for the family as much as for the patient during this time.**
- Most people are buried or cremated within 3 days of death; extenuating circumstances may lengthen this period to accommodate family and friends who must travel a long distance to attend a funeral or memorial service. The family decides if the deceased will be cremated or will have an open or closed casket.
- More people are choosing to remain at home and participate in a hospice for end-of-life care to have their comfort needs met. **One requirement for hospice care is that the patient must sign documents indicating that he or she does not want extensive life-saving measures performed.**
- Although women may be more expressive than men, most Americans demonstrate their grief conservatively. Generally, tears are shed, but loud wailing and uncontrollable sobbing rarely occur. Bereavement support strategies are listed in Table 12.3.

12.1.9 Spirituality

- Dominant religions in the US are those of the Judeo-Christian faiths and include Catholicism, Protestantism, and Judaism. However, there are many other recognized religions that are consistent with religious freedom, a reason that many people immigrated to the US. Although church and state are separate, many public events and ceremonies still open with a prayer.

Table 12.3 Bereavement support strategies

• Being physically present
• Acknowledging the patient's pain
• Encouraging a reality orientation
• Openly acknowledging the family's right to grieve
• Recognizing varied behavioral responses to grief
• Assisting the patient and family to express their feelings
• Encouraging interpersonal relationships
• Promoting interest in a new life
• Making referrals to other resources such as a priest, minister, rabbi, or pastoral care person, with the patient's or family's permission

- In times of illness, many people say prayers. **Ask patients if they need anything to say prayers. Inquire if the person wants to see a member of the clergy.**
- **Do not remove religious symbols or statues as they provide solace to the person. Removing them may increase or cause anxiety.**

12.1.10   Health-Care Practices

- The US continues to undergo a paradigm shift from curative and restorative medical practices and sophisticated technological care to health promotion and wellness; illness, disease, and injury prevention; and health restoration. Many view good health as a divine gift from God.
- Advance directives are an important part of medical care, which allow patients to specify their wishes concerning life and death decisions before entering an inpatient facility. These directives also designate the name of a family member or significant other to speak for the patient and make decisions when or if the patient is unable to do so. Some patients have a living will that outlines their wishes in terms of life-sustaining procedures in the event of a terminal illness or a serious accident. **Each inpatient facility has these forms available and will ask the patient what his or her wishes are. Patients may sign these forms at the inpatient facility or elect to bring their own forms.**
- As an adjunct to biomedical treatments, many people use acupuncture, acupressure, acumassage, herbal therapies, massage therapy, aromatherapy, and other complementary and alternative treatments. An awareness of combined practices when treating or providing health education to individuals and families helps ensure that therapies do not conflict with each other, intensify the treatment regimen, or cause an overdose.
- **Nonjudgmentally inquire about all therapies being used, such as foods, teas, herbal remedies, nonfood substances, over-the-counter medications, and medications prescribed or loaned by others.**
- **Often, traditional, folk, and magicoreligious practices are and should be incorporated into the plan of care for patients.**

- **To help prevent contradictory or exacerbated effects of prescription medication and treatment regimens, ask patients about self-medicating practices.**
- Pain is the "fifth vital sign." Patients should be made comfortable and not have to tolerate high levels of pain. **Offer and encourage pain medication, and explain that it will help the healing process.**
- The Americans with Disabilities Act protects individuals with a handicap from discrimination. Rehabilitation and occupational health services focus on returning individuals with handicaps to productive lifestyles in society as soon as possible.
- **Traditional practice calls for fully disclosing health conditions to patients, who then disclose the information to the family.**
- Most Americans favor organ donation and transplantation and blood or blood products transfusions. **Some people refuse blood or blood products for fear of contracting HIV.**
- **Jehovah's Witnesses do not receive blood or blood products. Assist patients in obtaining a religious leader to support them in making decisions regarding organ donation or transplantation if requested.** Health-Care Providers
- Most people combine biomedical health-care practices with traditional practices, folk healers, and magicoreligious healers.
- Standard practice is to assign staff to patients regardless of gender differences.
- **Recognize and respect differences in gender relationships when providing care. Respect patients' modesty by providing adequate privacy and assigning same-sex caregivers whenever the patient requests.**
- The advanced practice role of registered nurses is gaining respect as their numbers increase and research demonstrates and the public sees them as equal or preferable to physicians. The public holds nurses in high regard.

12.2 Reflective Exercises

1. What are the key attributes of the European American individualistic culture?
2. What are the four generations in the American workforce?
3. What is the official language of the US?
4. A 40-year-old male, John David Armstrong arrives to the medical office. How should he be addressed?
5. What is your acceptance of gays, lesbians, and transgendered people?
6. What are the leading causes of death among European Americans?
7. As a health-care provider, what might you do to help people decrease or eliminate smoking and substance misuse?
8. What types of birth control are acceptable to you?
9. How adherent are you to the tenets of your religion?
10. What are the bereavement practices of your family? Are they different from what is described in this chapter?

Bibliography

Carr D (2012) Death and dying in the contemporary United States: what are the psychological implications of anticipated death? Soc Person Compass 6(2):184–195

Centers for Disease Control and Prevention (2016) National center of HIV/AIDS, STD, and TB prevention. www.cdc.gov/nchhstp/default.htm

CIA World factbook online (2016). https://www.cia.gov/library/publications/the-world-factbook/geos/us.html

McCool WF, Simeone SA (2002) Birth in the United States: an overview of trends past and present. Nurs Clin N Am 37(4):735–746

Nimke A (2016) Communicating with 5 generations in the workforce. https://www.biztimes.com/2016/08/08/communicating-with-the-five-generations-in-your-workplace/

Purnell L (2013) People of European American heritage. In: Purnell L (ed) Transcultural health care: a culturally competent approach, 4th edn. F.A. Davis Company, Philadelphia

U.S. Department of Agriculture (2016). http://www.usda.gov/wps/portal/usda/usdahome

Chapter 13
People of Filipino Heritage

13.1 Overview and Heritage

Filipinos, predominantly of Malayan ancestry, have been influenced by the neighboring Chinese, Japanese, East Indian, Indonesian, Malaysian, and Islamic cultures and by influences from Spanish and American colonization. The Philippine archipelago consists of 7107 islands. The tropical climate is suitable for year-round agriculture and fishing, but it is affected by the seasonal northeast and southwest monsoons.

Filipino Americans are a diverse group whose regional variations influence the spoken dialect, food preferences, and traditions as well as the other variant characteristics of culture discussed in Chap. 1. Aggregate data in this profile may not be accurate for some more acculturated Filipinos. Most Filipinos who live in the US reside in California, Hawaii, Illinois, New Jersey, New York, Washington, and Texas. Since the passage of the 1965 Immigration Act, the Philippines have become the largest source of immigrants from Asia. Economic and educational opportunities and reunification with family members in the US are the primary motivating factors for emigration. Filipinos view educational achievement as a pathway to economic success, status, and prestige for the individual and the family.

13.1.1 Communications

- Although more than 100 dialects are spoken in the Philippines, most Filipinos speak the national language, Filipino or **Tagalog**. English is used for business and legal transactions, and in school beyond the third grade. Business and social interactions commonly use a hybrid of Tagalog and English (Tag-Lish) in the same sentence.

© Springer Nature Switzerland AG 2019
L. D. Purnell, E. A. Fenkl, *Handbook for Culturally Competent Care*,
https://doi.org/10.1007/978-3-030-21946-8_13

- Nouns are used in place of the generic and gender-neutral pronouns *siya* (singular "he/she") and *sila* (plural "they/them"). Hence, it is customary for Filipinos to use "he" and "she" interchangeably in reference to the same individual.
- Communication is highly contextual. The individual is acculturated to attend to the context of the interaction and to adopt appropriate behaviors.
- Most Filipinos are observant, displaying an intuitive feeling about the other person and the contextual environment during interactions. Meanings are embedded in nonverbal communication.
- The emphasis on maintaining smooth interpersonal relationships brings a consequent ambiguity in communication to prevent the risk of offending others.
- **Filipinos may sacrifice clear communication to avoid stressful interpersonal conflicts and confrontations. Saying "no" to a superior may be considered disrespectful, which predisposes an ambiguous positive response.**
- Focusing on action-oriented strategies may be seen as coercive.
- Saving face, a characteristic pattern of behavior employed to protect the integrity of both conversants, is a consequence of the cultural value placed on maintaining smooth interpersonal relations.
- During a teaching session, a Filipino patient's nod may have several meanings, such as comprehension or agreement ("Yes, I hear you"; "Yes, we are interacting"; "Yes, I can see the instructions") or some other message that may be difficult for the patient to disclose. "Yes" does not always mean understanding or agreement.
- **Asking for a return demonstration of a procedure can help ensure understanding. Speak clearly and slowly. Allow time for a response to questions.**
- One may not disagree, talk loudly, or look directly at a person who is older or who occupies a higher position in the social hierarchy. Direct eye contact depends on the extent of acculturation, amount of time in the US, age, and education. Some individuals may avoid prolonged eye contact with authority figures and older people as a form of respect. Older men may refrain from maintaining eye contact with young women because it may be interpreted as flirtation or a sexual advance. **Just because Filipino patients do not maintain direct eye contact does not mean they are not listening, are not being truthful, or do not care.**
- Most individuals are comfortable with silence and allow the other person to initiate verbal interaction as a sign of respect. Greater distance is observed when interacting with outsiders and people in positions of authority.
- Same-sex closeness and touching are normal behaviors. Young adults of the same sex may hold hands, put one arm over another's shoulder, or walk arm-in-arm.
- Most Filipinos have a relaxed temporal outlook. They have a healthy respect for the past, the ability to enjoy the present, and hope for the future. Past orientation is evident in their respect for elders, strong sense of gratitude, obligation to older generations, and honoring the memories of dead ancestors. Future orientation is manifested in the strong sense of family commitment to provide for the education of the young, parental participation in the care of their children and grandchildren, and a strong work ethic.

- Promptness for social events is determined situationally. "Filipino time" means arriving much later than the scheduled appointment, which can be from one to several hours.
- Children carry the surnames of their father and their mother. For example, Jose Romagos Lopez and Leticia Romagos Lopez are the children of Maria Romagos and Eduardo Lopez. The middle name or initial is the mother's maiden name. After marriage, Jose keeps the same name, whereas his sister's name becomes Leticia Lopez Lukban (her husband being Ernesto Lukban). The middle name is generally abbreviated as an initial. Nicknames symbolize affectionate regard for the person and are commonly used instead of the first name; hence, Nini, Baby, Bongbong.
- **Address Filipino adults by their title or professional affiliation such as Mr., Mrs., Miss, Ms., Dr., or attorney. Ask what name is used for legal purposes.**

13.1.2 Family Roles and Organization

- In contemporary families, although the father is the acknowledged head of the household, authority is considered egalitarian. The mother plays an equal and often major role in decisions regarding health, children, and finances.
- Traditional female roles include caring for the sick and children, maintaining positive relationships with kin, and managing the home.
- Parents and older siblings are involved in the care and discipline of younger children.
- In extended-family households, older relatives and grandparents share much authority and responsibility for the care and discipline of younger members. Institutionalization of aged parents is tantamount to abandonment of filial obligation.
- Parents traditionally expect their children to pursue a college education, have economically productive careers, and raise a family.
- Dating at an early age is discouraged for young daughters. Sex education and sex are not usually discussed openly within the family.
- Traditional families desire that their daughters remain chaste before marriage. Pregnancy out of wedlock brings shame to the whole family.
- Conditions such as mental illness, divorce, terminal illness, criminal offenses, unwanted pregnancy, and HIV/AIDS are not readily shared with outsiders until trust is established. In times of illness, the extended family provides support and assistance.
- A family visit to the hospital may take on the semblance of a family reunion.
- **Include nuclear and extended family for support in time of serious illnesses.**
- **Make arrangements for visitations by large numbers of family. Elicit help for visitor control from a respected family member.**

- Homosexuality may be recognized but is often considered an aberrant behavior. Thus, to save face and decrease stigmatization for the family, it may not be openly practiced or acknowledged. **Do not disclose same-sex relationships to family members.**

13.1.3 Workforce Issues

- Racism is a continuing theme voiced by Filipino nurses working with Americans. Filipino nurses have been recruited in large numbers where acute shortages of American-trained nurses exist. This has reinforced the cultural tendency toward collective solidarity by defining the context of interactions within the insider-outsider continuum. **Develop culturally specific staff development programs that include Filipinos and Americans.**
- The cultural concept of shared identity with other Filipinos creates a propensity among Filipino nurses to speak in their own dialect with each other to the exclusion of non-Filipino coworkers and patients. **Explain organizational rules about when and where a foreign language can be spoken in the American workforce.**
- Assertive communication is difficult for Filipinos who have been acculturated to avoid conflict. They may consider it impolite and disrespectful to confront or challenge the authority of a superior. When a problem with a manager occurs, a Filipino may communicate through a mediator, usually another Filipino, who is in the same level within the hierarchy as the manager.
- **Explain the cultural value of assertiveness in the American workforce.**
- Filipino feel comfortable performing what they perceive as caring tasks for patients that American nurses expect patients to do for themselves. Initially they may not be inclined to teach and demonstrate procedures to patients because of their traditional belief in doing the caring tasks for patients. **Explain the American value of self-care with patients and families.**
- The Filipino values of shared perception and being one with others create a cooperative, rather than competitive, outlook. An individual produces for the group and puts the group above his or her own personal gain. The businesslike and competitive perspectives of Americans, in which behavior is internally motivated by individual gain, may be interpreted as selfish and uncaring. Self-proclamations of accomplishments may be seen as offensive. The group to recognize a member's achievement, which is assessed in terms of how the action benefited the group.
- **Bicultural development of Filipino and non-Filipino staff should be the goal of orientation and training. Biculturalism requires awareness of self and others and the ability to adapt behaviors that build positive relationships with others who may be different from oneself. Role model to help Filipinos assert themselves and feel confident in problem-solving and conflict resolution.**
- Filipinos place importance on maintaining self-esteem and dignity by saving face and avoiding shame. A Filipino may say "yes" to avoid hurting other people's feelings.

- Their sensitivity and attention to other people's feelings are often exhibited as indecisiveness, which many Americans interpret as lack of assertiveness. Filipinos are taught not to show open disagreement. Less acculturated Filipinos may not understand the directness of Americans and, thus, may find it insulting. **Explain the expectation of assertiveness by all professionals in the American workforce.**
- The concept of individual accountability and responsibility in a highly litigious society, such as the US, may initially be difficult for Filipino nurses to understand. **Explain the litigious nature of the American culture and the need for defensive charting.**

13.1.4 Biocultural Ecology

- The typical Filipino may from a multiracial genetic background. The youthful features of Filipinos may make it difficult to assess their age. Common Filipino physical features may include jet black to brunette or light brown hair, dark- to light-brown pupils with eyes set in almond-shaped eyelids, deep brown to very light tan skin tones, mildly flared nostrils, and slightly low to flat nose bridges.
- The high melanin content of the skin and mucosa may pose problems when assessing signs of jaundice, cyanosis, and pallor. **Assess anemia, pallor, and jaundice in the conjunctiva and palms and soles of the feet.**
- Filipinos range in height from under 5 ft to the height of average Americans.
- Approximately 40% have blood type B and a low incidence of the Rh-negative factor.
- Table 13.1 lists commonly occurring health conditions for Filipinos in three categories: lifestyle, environment, and genetics (see Chap. 2).

Table 13.1 Health conditions common to filipinos

Health conditions	Causes
Thalassemias	Genetic
Glucose-6-phosphate dehydrogenase	Genetic
Lactose intolerance	Genetic, lifestyle
Coronary artery disease	Genetic, lifestyle
Diabetes mellitus	Genetic, lifestyle
Hypercholesterolemia	Genetic, lifestyle
Renal stones	Genetic
Hyperuricemia	Lifestyle
Gout	Lifestyle
Arthritis	Genetic?, lifestyle
Thyroid cancer	Lifestyle, environment
Breast cancer	Genetic, lifestyle, environment
Prostate cancer	Lifestyle, environment
Liver cancer	Lifestyle, environment

- Newborns may have Mongolian spots: bluish-green discolorations on their buttocks that are physiological and that eventually disappear.
- Liver cancer tends to be diagnosed in the late stages of the disease and appears to be associated with the presence of the hepatitis B virus. Silent carriers of the virus are common among Asians, and its presence is detected only when other problems are being evaluated. **Routinely screen for hepatitis B virus among recent immigrants.**
- Asians require lower doses of central nervous system depressants such as haloperidol, have a lower tolerance for alcohol, and are more sensitive to the adverse effects of alcohol.
- Because of availability of over-the-counter antibiotics and lack of adequate medical monitoring of these drugs in the Philippines, Filipino immigrants may be insensitive to the effects of some anti-infectives.
- A positive reaction to tuberculin or to the Mantoux test is observed because of the practice of giving bacille Guérin-Calmette vaccinations in childhood. **Chest X-rays and sputum cultures are recommended for screening and diagnosis of tuberculosis.**

13.1.5   High-Risk Health Behaviors

- Filipinos who are more acculturated have higher rates of alcohol misuse and increased smoking rates. **Explain the physical and psychological problems of tobacco and alcohol misuse.**
- Less educated Filipinos in the US have increased rates of HIV/AIDS compared with other Asians/Pacific Islanders. **Take every opportunity to include HIV/AIDS education in elementary school and in health teaching.**
- The more recent immigrants in the US frequently do not access health services until their illness is advanced because they are not aware of available services and may not have linguistic competence in the English language. **Educate the Filipino community about the availability of services. Provide interpreters as necessary.**

13.1.6   Nutrition

- Food and meal patterns emphasize generosity, hospitality, and thoughtfulness, which support group cohesiveness. Spanish, Chinese, and American influences are integrated into contemporary cuisine.
- Foods may be sautéed, fried, or served with a sauce. Rice is a staple food and is usually eaten at every meal, either steamed, fried, or as a dessert.

- Except for babies and young children, milk is almost absent in the Filipino diet, partly due to lactose intolerance. Milk in desserts such as egg custard and ice cream seems to be tolerated.
- Cold drinks or foods such as orange juice or fresh tomatoes are not served for breakfast in order to prevent stomach upset.
- Dietary calcium is derived from green leafy vegetables and seafood.
- Households may keep potted medicinal plants that are used for common colds, stomach upsets, urinary tract infections, and other minor ailments.
- Daily consumption of garlic to combat hypertension is common. Ginger root is boiled and served as a beverage to relieve sore throats and promote digestion.
- Bitter melon is prepared and eaten as a vegetable and believed to prevent diabetes. Greens such as *malunggay* and *ampalaya* leaves are used in stews to regain stamina for someone believed to be anemic or run-down. **Knowledge of indigenous food sources and meal patterns, nutritional content of Filipino foods and American food substitutes, and accessibility of traditional ingredients are important aspects of nutritional assessment and counseling.**

13.1.7 Pregnancy and Childbearing Practices

- The only acceptable methods of contraception are natural family planning methods. Abortion is considered a sin. Recent Filipino immigrants who come from urban areas and who are more educated may be open to alternative methods of birth control.
- Many Filipino women are embarrassed to have males perform vaginal examinations. **Obtain a female health-care provider whenever possible.**
- **The pregnant Filipino woman's network of family and community health advisers, whose opinions she respects, are important support for building trust and rapport in the patient-provider relationship.** Table 13.2 describes common beliefs about pregnancy.
- Some prefer to have their mothers rather than their husbands in the delivery room.
- Some women may prefer the squatting position for birthing.
- Common reasons for not breast-feeding are insufficient milk, mother working, nipple and breast problems, and mother's poor health.
- Supplementing breast-feeding with other liquids and foods occurs as early as 2 months. **Educate mothers and family members on the importance of breast-feeding.**
- Beliefs about the postpartum period are included in Table 13.3.

Table 13.2 Prenatal beliefs about pregnancy

• Some women refuse to take vitamins because they are afraid that vitamins could deform the fetus.
• Some individuals believe that when pregnant women crave certain foods, the craving should be satisfied to avoid harm to the baby. They believe that if the mother craves dark-skinned fruit or dark-colored food, the infant's skin will be dark.
• Sudden fright or stress may harm the developing fetus.
• Eating blackberries will make the baby have black spots.
• Eating black plums will give the baby dark skin.
• Eating twin bananas will result in twins being born.
• Eating apples will give the baby red lips.
• When a woman's stomach is not round or the mother's face is blemished, the baby will be a boy.
• Going outside during a lunar eclipse is harmful to the baby.
• Going out in the morning dew is bad for the baby because evil spirits are present.
• Funerals are avoided because the spirit of the dead person may affect the baby.
• Wearing necklaces may cause the umbilical cord to wrap around the baby's neck.
• Sitting by a doorway will make the delivery difficult.
• Sitting by a window when it is dark may let evil spirits come to the pregnant woman.
• Sweeping at night may sweep away the good spirits.
• Knitting might tangle the baby's intestines at birth.
• Naming the baby before it is born or after a dead person is bad luck.

Table 13.3 Postpartum practices and beliefs

• Mothers should eat plenty of hot soups (chicken with papaya) to promote milk production.
• Postpartum mothers should avoid exposure to cold and should use warm water to drink and bathe for a month.
• Showers are prohibited because they may cause arthritis. However, the woman's mother may give her a sponge bath with aromatic oils and herbs, or a *hilot*, traditional healer, may give an aromatic herbal steam bath followed by full body massage, including the abdominal muscles, to stimulate a physiological reaction that has both physical and psychological benefits.
• Eating sour or ice-cold foods may cause abdominal cramps for both the baby and mother.
• The mother and baby should not go out for a month except to visit a doctor.
• One should give money to charity or to the needy when a baby comes to one's house for the first time.
• The baby's abdomen is wrapped with a cloth until the umbilical cord falls off to prevent an umbilical hernia.
• Garlic, salt, or a rosary is placed near the baby's crib to ward off evil spirits.
• Hanging the baby's placenta in a tree will make the baby a good climber.

13.1.8 Death Rituals

- Family members generally wish to provide intimate caregiving for the patient regardless of the setting. **Make arrangements for the family to participate in caregiving.**
- Illness and death may be attributed to supernatural and magic or religious causes such as punishment from God, angry spirits, or sorcery.

- Religiosity and fatalism contribute to stoicism in the face of pain or distress as a way of accepting one's fate.
- Before the decision is made to inform the patient about his or her terminal condition, a discussion among family members should occur. They may request the doctor to not divulge the truth to protect the patient.
- The ethical principles of beneficence and malfeasance take precedence over patient autonomy.
- Planning for one's death is taboo and may be considered tempting fate. Hence, many traditional Filipinos are averse to discussing advance directives or living wills.
- A priority for the family is to gather around the dying person and during the immediate period after death to pray for the soul of the departed. **When death is imminent, contact a priest if the family is Catholic. Do not remove religious medallions, rosary beads, a scapular, or religious figures from the patient or from the bedside.**
- After death, a wake may last from 3 days to 1 week (to wait for the arrival of kin from other states or countries).
- Women generally show emotions openly by crying, fainting, or wailing. Men are expected to be more stoic and grieve silently.
- Cremation is acceptable to avoid the spread of disease.
- On the first-year anniversary of death, family and friends are reunited in prayer to commemorate this event.
- Many women wear black clothing for months or up to a year after the death. The 1-year anniversary ends the ritual mourning.

13.1.9 Spirituality

- The dominant religious affiliations are Catholicism (the majority), Protestantism, Islam, and Buddhism. Most seek medical care; they believe that part of the efficacy of a cure is in God's hands or by some mystical power.
- Novenas and prayers are often said on behalf of the sick person. Performance of religious obligations and sacraments and daily prayers are some of the ways health and peaceful death are achieved. **Provide for spiritual needs of patients by making accommodations to their practicing beliefs.**
- Strength comes from an intimate relationship with God, family, friends, neighbors, and nature. The concept of self is formed from the relationship with a divine being.

13.1.10 Health-Care Practices

- Some Filipinos are fatalistic, tending to accept fate easily, especially when they feel they cannot change a situation.

- Decisions about when, where, and from whom to seek help are largely influenced by the intimate circle of family.
- **Include nuclear and extended family in health-care decision making when the patient requests.**
- Many accept and adhere to medical recommendations while using alternative sources of care suggested by trusted friends and family members, which include indigenous medical practices. Often, major decisions are delegated to the physician rather than to the patient or family taking an active collaborative role in decision making.
- **Stress that medications need to be taken as prescribed, medications are ordered specifically for each ailment, unused drugs should be discarded.**
- A high value is placed on personal cleanliness. Keeping oneself clean and free of unpleasant body odors is viewed as good for one's health and face-saving. To be slovenly and disorderly is to be shamelessly irresponsible. Aromatic baths are taken both for pleasure and to restore balance.
- Illness in infancy and childhood may be attributed to the "evil eye." Healing rituals may involve religion (prayers and exorcism), sacrifices to appease the spirits, use of herbs, and massage.
- Balance and moderation are embedded in the hot and cold theory of healing. Change should be introduced gradually. Sudden changes from hot to cold, from activity to inactivity, from fasting to overeating, and so forth, introduce undue bodily stresses that can cause illness.
- Some believe that after strenuous physical activity, a rest should precede a shower; otherwise the person could develop arthritis. Exposure to sudden cold drafts may induce colds, fever, rheumatism, pneumonia, or other respiratory ailments. Some avoid handwashing with cold water after ironing or heavy labor. Exposure to cold such as showers is avoided during menstruation and during the postpartum period.
- Many accept minor ailments stoically and consider them natural imbalances that will run their normal course and disappear.
- Patients often do not complain of pain despite physiological indicators. Many view pain as part of living an honorable life and view this as an opportunity to reach a fuller spiritual life or to atone for past transgressions. **Offer and, in fact, encourage pain relief interventions.**
- Mental illness may be attributed to an external cause such as witchcraft, soul loss, or spirit intrusion.
- Minimal expressions of psychological and emotional discomfort may be observed. The discomfort in discussing negative emotions with outsiders may be manifested by somatic complaints or ritualistic behaviors, such as praying.
- **Explore the underlying meaning of somatization (loss of appetite, inability to sleep) and observe patient interactions with others for valuable information.**
- Many believe that mental illness is hereditary and that it carries a certain stigma.
- Family members tend to take care of emotional problems to minimize exposing the problems to outsiders. **Involving a trusted family member or friends, initiating contact with a Filipino mental health worker, especially a Filipino**

physician, may increase the likelihood of getting the person into a treatment program.

- The birth of a child with a developmental disability may be viewed as God's gift, an opportunity to become a better person or family, a curse from some unknown "angry spirit," negligence while pregnant, or a family matter that should be kept private; institutionalization may not be readily accepted.
- Organ donation is not an option, except perhaps in cases in which a close family member is involved. Many Filipinos who follow Catholic traditions believe that keeping the body intact as much as possible until death is preparation for the afterlife. **Provide factual information on organ donation and transplantation.**

13.1.11 Health-Care Providers

- Western medicine is familiar and acceptable to most. Some accept the efficacy of folk medicine and may consult both allopathic and indigenous healers.
- Folk healers are less common in the US, with the exceptions of the West Coast and Hawaii. When available, they contribute by facilitating cultural rapport between health-care providers and patients.
- The *hilot* (traditional healer) is often willing to be included in the counseling session and provide support for the patient's compliance with medical treatment. The *hilot* may provide a special prayer to be incorporated into the medically prescribed treatment plan to increase the patient's sense that all available resources are being used.
- Linguistically and ethnically congruent health-care providers are preferred.
- A health-care provider of the same gender and the same culture may encourage more Filipinos to take advantage of disease prevention services.
- **Include a Filipino primary care provider whenever possible. A bilingual person is helpful to improve communication with older Filipinos.**

13.2 Reflective Exercises

1. What is the most common languages spoken by Filipinos? What is *Taglish*?
2. What is meant by "Filipino time?"
3. Describe the name format for Filipinos.
4. Describe Filipino communication practices.
5. What are the most common health conditions of Filipinos?
6. What are acceptable birth control practices for *traditional* Filipinos?
7. Describe at least three prenatal beliefs about pregnancy.
8. What is a *hilot*?
9. What healing rituals are used for a child with the "evil eye"?
10. To what might Filipinos attribute mental illness?

Bibliography

CIA (2016) World factbook: the Philippines (2013). https://www.cia.gov/library/publications/the-world-factbook/geos/rp.html

De Castro AB, Gee GC, Takeuchi DT (2009) Job-related stress and chronic health conditions among Filipino immigrants. J Immigr Minor Health 10(6):551–558

Galanti GA (2000) Filipino attitudes toward pain medication: a lesson in cross-cultural care. J Immigr Minor Health 10(6):551–558

Johnson-Kozlow M, Matt GE, Rock CL, de la Rosa R, Conway TL, Romera RA (2011) Assessment of dietary intakes of Filipino-Americans: implications for food frequency questionnaire design. J Nutr Edu Behav 43(6):505–510

Manielleas: 30 Superstitions Filipinos Practice During Funerals (2024). http://www.manillenials.com/filipino-superstitions-funerals-pamahiin-sa-patay/

Munoz CC (2013) People of filipino heritage. In: Purnell L (ed) Transcultural health care: a culturally competent approach, 4th edn. F.A. Davis Company, Philadelphia

Philippines Guide: A Look at Filipino Language, Culture, and Etiquette (2016). http://www.commisceo-global.com/country-guides/phillippines-guide

Siojo R (n.d.) Filipino beliefs on pregnancy. https://philippines-events-culture.knoji.com/philippine-beliefs-on-pregnancy/

Chapter 14
People of German Heritage

14.1 Overview and Heritage

The Federal Republic of Germany is one of the most densely populated countries in Europe, with a population of over 80 million people. Sixty million Americans are of German heritage. The first wave of German immigrants came to the US for religious freedom and settled along the eastern seaboard. The second wave arrived between 1840 and 1860 and was fleeing political persecution, poverty, and starvation. Many worked as indentured servants. The 1930s and 1940s saw a third wave because of the rise of fascism in Germany.

Germans have a deep respect for education. Credibility, social status, and level of employment are based on educational achievement. Germans receive a stronger education than Americans, with the German undergraduate degree being equal to the American master's degree. Great diversity exists among German Americans. Based on individuality and the variant characteristics of culture as described in Chap. 1, the aggregate data in this profile might not fit some German Americans. Because many German immigrants were discriminated against in the 1940s, and after World War II, many Germans changed their names.

14.1.1 Communications

- German is the official language of Germany; however, many German dialects exist, making it difficult for some to understand each other. German is a low-contextual language, with a greater emphasis on verbal than nonverbal communication. A high degree of social approval is shown to people whose verbal skill in expressing ideas and feelings is precise, explicit, and straightforward.

© Springer Nature Switzerland AG 2019   155
L. D. Purnell, E. A. Fenkl, *Handbook for Culturally Competent Care*,
https://doi.org/10.1007/978-3-030-21946-8_14

- Feelings are considered private and are often difficult to share. Sharing one's feelings with others often creates a sense of vulnerability or is looked on as a weakness.
- Expressing fear, concern, happiness, or sorrow allows others a view of the personal and private self, creating a sense of discomfort and uneasiness. Therefore, philosophical discussions, hopes, and dreams are only shared with family members and close friends.
- Emotions are experienced intensely but are not always expressed. "Being in control" includes harnessing one's emotions and not revealing them to others. Newer generations are more demonstrative in sharing their thoughts, ideas, and feelings.
- **Do not assume that just because a person may not express or display emotions that the person does not care about his or her health or the situation.**
- In traditional families where the father plays a dominant role, little touching occurs between the father and children. This relationship may become more demonstrative as parents and children age. Affection between a mother and her children is more evident. In other families there is an outward expression of love from parents, grandparents, and extended family members; hugs and kisses are a "reaffirmation of love."
- Whereas close friends are often extended warmth through handshakes, brief embraces, and sometimes kisses, strangers are kept at arm's length and greeted formally. Germans generally are careful not to touch people who are not family or close friends.
- Most individuals place a high value on privacy.
- Doors are used to protect privacy. A closed door requires a knock and an invitation to enter, regardless of whether the door is encountered in the home, business, or in-patient facility. Even looking into a room from the outside is considered a visual intrusion.
- **Do not enter a patient's room without knocking on the door, announcing yourself, and requesting permission to enter.**
- Eye contact is maintained during conversations. Staring at strangers is considered rude.
- To focus on the present is to ensure the future. The past, however, is equally important, and Germans often begin their discussions with background information.
- Most pride themselves on their punctuality. There are rarely good excuses for tardiness or delays that disturbs the "schedule" of events. **If the health-care provider is late with an appointment or treatment, provide a thorough explanation.**
- Traditionally, Germans keep social relations on a formal basis. Those in authority, older people, and subordinates are always addressed formally. Only family members and close friends address each other by their first names. Younger generations or the more acculturated may be less formal in their interactions. Health-care providers will find that many patients vary widely in their observance of these rules of etiquette.
- **Greet patients formally unless told to do otherwise. Ask patients how they would like to be addressed.**

14.1.2   Family Roles and Organization

- Traditional families view the father as head of the household. In the US, the husband and wife are more likely to make decisions mutually and share household duties.
- Older people are sought for their advice and counsel, although the advice may not always be followed. They are admired for maintaining their independence and their continued contributions to society. Many live alone or with aging spouses. Helping parents or grandparents to remain in their own home is important. For those who grow dependent, moving in with children or in a retirement nursing home is a viable choice.
- **Include the extended family in decision making. Consider using family members to assist with personal care and treatments when the need arises.**
- Prescriptive behaviors for children include using good table manners, being polite, doing what they are told, respecting their elders, sharing, paying attention in school, and doing their chores.
- Prescriptive behaviors for adolescents include staying away from bad influences, obeying the rules of the home, sitting like a lady, and wearing a robe over pajamas.
- Restrictive and taboo behaviors for children include talking back to adults, talking to strangers, touching another person's possessions, and getting into trouble. Restrictive and taboo behaviors for adolescents include smoking, using drugs, chewing gum in public, having guests when parents are not at home, and having run-ins with the law.
- Family and lifelong friendships are highly valued. Concern for one's reputation is a strong value. One's family reputation is considered part of a person's identity and serves to preserve one's social position.
- Pregnancy outside marriage results in overt or subtle disapproval. Because families are concerned about their reputations in the community, the presence of an unwed mother taints their reputation and may result in the family being ostracized by others. If marriage follows the pregnancy, less sanctioning occurs, but the fact that pregnancy existed before marriage creates a stigma for the woman, and sometimes for the child, that may last for the rest of their lives.
- Although same-sex marriage is not legal in Germany, registered partnerships for same-sex couples have been legal since 2001, which provides most of the same rights as opposite-sex married couples receive. However, many older gays and lesbians may fear exposure because of the extreme discrimination homosexuals experienced in Nazi Germany. **When encountering gays and lesbians who need support, a referral to one of the gay and lesbian religious or community groups may be helpful. Do not disclose same-sex relationships to others.**

14.1.3 Workforce Issues

- For Germans being on time is important. Business communication should remain formal: shaking hands daily, using the person's title with the last name, and keeping niceties to a minimum.
- For the German manager, credentials and education confirm their credibility and lead to power. Germans, who have no translation for "fair play," are assertive by putting other people in their place. Nuances and jargon can frustrate the individual who does not understand idioms and colloquial expressions.
- **Help other employees understand the formality of German employees. Explain the culture of the American workforce and legalities of "fairness."**

14.1.4 Biocultural Ecology

- Germans range from tall, blond, and blue-eyed to short, stocky, dark-haired, and brown-eyed. Because many Germans have fair complexions, skin color changes and disease manifestations can easily be observed. For those with fair skin, prolonged exposure to the sun increases the risk for skin cancer. **Counsel patients on exposure to sun and recommend protection.**
- Table 14.1 lists commonly occurring health conditions for German Americans in three categories: lifestyle, environment, and genetics.

14.1.5 High-Risk Health Behaviors

- Smoking and excessive alcohol consumption remain high-risk behaviors for many Germans. **Encourage smoking cessation and limiting alcohol consumption.**
- Most individuals enjoy the outdoors, fresh air, and exercise. Sports are played for exercise and the pleasure of participating in group activities. Encourage

Table 14.1 Health conditions common to German

Health conditions	Causes
Cardiovascular disease	Genetic, lifestyle, environment
Stomach cancer	Genetic, lifestyle, environment
Myotonic muscular dystrophy	Genetic
Hereditary hemochromatosis	Genetic
Sarcoidosis	Genetic
Dupuytren's disease	Genetic
Peyronie's disease	Genetic
Cystic fibrosis	Genetic
Hemophilia	Genetic
Cholelithiasis	Genetic

water sports and other activities among older people, people with disabilities, mothers, and small children. **Encourage health club memberships for exercise.**

14.1.6 Nutrition

- Food is a symbol of celebration for Germans and is often equated with love. Children are rewarded for good behavior with food. Their infatuation with food can lead to overeating, resulting in obesity.
- Real cream and butter are used. Gravies, sauces, fried foods, rich pastries, and sausages are only a few of the culinary favorites that are high in fat content; meats are stewed, roasted, and marinated and are often served with gravies. Vegetables (fresh is preferred) are often served in a butter sauce. Foods are also fried in butter, bacon fat, lard, or margarine. Traditional food preparation methods use high-fat ingredients that add to nutritional risks. Garlic and onions may be eaten daily to prevent heart disease.
- One-pot meals such as string beans and potatoes, cabbage and potatoes, pork and sauerkraut, stews, and soups are served as family meals. Casseroles are also popular. Foods prepared with vinegar and sugar as flavorings are also favorites. Potato salad, cucumber salad, coleslaw, pickled eggs, pickled cucumbers, cauliflower, tongue, and herring are common examples of favored foods. **Encourage reducing portion size, overcoming harmful food rituals, and reducing fat intake.**
- Those who are ill receive egg custards, ginger ale, or tomato soup (without cream) to settle their stomach. Prune juice is given to relieve constipation. Soup from fresh tomato juice is used to treat a migraine headache.
- Ginger ale or 7-Up relieves indigestion and settles an upset stomach. After gastrointestinal illnesses, a recuperative diet is administered to the sick family member, beginning with sips of ginger ale over ice. If this is retained, hot tea and toast are offered. The last step is coddled eggs, a variation of scrambled eggs prepared with margarine and a little milk. If these foods are tolerated, the sick person returns to the normal diet.

14.1.7 Pregnancy and Childbearing Practices

- All usual methods of contraception (*Verhütungsmittel*) are easily available in Germany. Heterologous artificial insemination, use of contraceptive pills, and unnatural contraception are forbidden among strict Catholic Germans. Therapeutic or direct abortion is forbidden as the unjust taking of innocent life. **Assist patients in identifying acceptable fertility practices within their religious practices.**

- Prescriptive practices during pregnancy include getting plenty of exercise and increasing the quantity of food to provide for the fetus. Restrictive practices during pregnancy include not stretching and not raising the arms above the head to minimize the risk of the cord wrapping around the baby's neck.
- A child born with the membrane (the amniotic sac, also knows as a "veil") over its head is believed to be a special child, a belief shared by many cultures.
- Prescriptive practices for the postpartum period include getting plenty of exercise and fresh air for the baby. If the mother is breast-feeding, she should eat foods that enhance the production of breast milk, such as malt beverages. **Assist mothers in identifying nonharmful foods and beverages that improve milk production.**

14.1.8   Death Rituals

- Death is a transition to life with God. Because illness is sometimes perceived as a punishment, the duration and intensity of the dying process may be seen as a result of the quality of the life led by the person. Death is considered part of the life cycle, a natural conclusion to life. Individuals who embrace a set of religious beliefs may look forward to a life after death, often a better life.
- Careful selection of the clothes to be worn by the deceased and the flowers that represent the immediate family is important. Traditionally a 3-day period of mourning is reserved after the death of a family member.
- The body of the deceased is prepared and "laid out" in the home, where support from family and friends is readily available. Neighbors come to do the chores and to sit with the family of the deceased until the burial. A short service is held in the home before the body is taken to the church, where family and friends can attend a funeral service. After the church services, the body is taken to the cemetery for burial. After a short graveside service, the minister invites everyone at the graveside service to go to the home of the deceased for food.
- Wearing black or dark clothing when attending a viewing or a funeral may be expected of both family and friends. Crying in public is permissible among some families, but in others the display of grief is private. **Accept varied expressions of grief.**

14.1.9   Spirituality

- Major religions include Roman Catholicism, Methodism, and Lutheranism. Other religions, such as Judaism, Islam, and Buddhism, have substantial membership. **Recognize that individuals' decisions may vary from the formal position of their religious groups.**

- Prayer is used to ask for healing, for effectiveness of treatments, for strength to deal with the symptoms of the illness, and for acceptance of the outcome of the course of the illness. Prayers are often recited at the bedside with all who are present joining hands, bowing their heads, and receiving the blessing from the clergy.
- Most families have a family bible, which is passed down through the generations. It serves as spiritual comfort and as a reservoir of family historical data such as the dates of births, marriages, and deaths. Individual sources of strength are their beliefs in God and in nature, although they may not attend church on a regular basis.
- Family loyalty, duty, and honor to the family are strong values. **Assist patients with obtaining clergy as needed.**

14.1.10 Health-Care Practices

- In traditional families, the mother usually ensures that children receive check-ups, get immunizations, and take vitamins. Women in the family often administer folk/home remedies and treatments. German Americans use a variety of over-the-counter drugs, believing that individuals are responsible for their own health.
- Common, natural folk medicines include roots, herbs, soups, poultices, and medicinal agents such as camphor, peppermint, and spirits of ammonia. Folk medicine includes "powwowing," use of special words, and wearing charms.
- **Ascertain if over-the-counter and folk remedies are being used so as to determine if there are contraindications with prescription medications.**
- Even when experiencing pain, many individuals continue to carry out their family and work roles. Many value being stoic when experiencing pain. **Health-care providers may not be able to identify verbal or nonverbal clues about pain among Germans. Careful interviewing and astute observation must be used to accurately assess the level of pain they experience. Explain that pain medication will help the healing process.**
- Mental illness may be viewed as a flaw, resulting in this group being slow to seek help because of the lack of acceptance and the stigma attached to needing help. Physical disabilities caused by injury are more acceptable than those caused by genetic problems.
- People's discomfort with expressing personal feelings to strangers may impede the counseling process and influence the counseling methods used. Most need to discuss the past without expressing their feelings. **Recognize that patients may not readily discuss feelings, especially during the initial counseling session.**
- **Rehabilitation is consistent with Germans' health beliefs because it helps return them to productive lives.**
- Blood transfusions, organ donation, and organ transplants are acceptable medical interventions.

14.1.11   Health-Care Providers

- Health-care providers hold a relatively high status among Germans. This admiration stems from the love of education and respect for authority.
- Most individuals accept care from either gender. Some younger and older, more traditional women prefer intimate care from a same-sex health-care provider.

14.2   Reflective Exercises

1. Distinguish German communication practices between family and outsiders.
2. What protocol does Germans expect with a "closed door"?
3. Describe taboo behaviors for German adolescents.
4. Name and describe at least three genetic diseases common among Germans.
5. What are the primary foods a German might use while recuperating from and illness?
6. Describe some prescriptive and restrictive German postpartum practices.
7. What might you do to encourage a stoic German to accept pain medication?
8. How is mental illness perceived by Germans?
9. Identify at least three common folk remedies commonly used by Germans.
10. Describe a traditional German funeral.

Bibliography

CIA (2016) World factbook: Germany. https://www.cia.gov/library/publications/the-world-factbook/geos/gm.html

Ho G, Cardamone M, Farrar M (2015) Congenital and childhood myotonic dystrophy: current aspects of disease and future directions. World J Clin Pediatr 8(4):66–80

Kulig JC, Fan H (2011) Death and dying beliefs and practices among low-German speaking Mennonites: application to practice. http://www.mennonitehealth.com/sites/mennonitehealth.com/files/ULeth_DD_Final.pdf

Lehmann-Willenbrock N, Allen JA, Meinecke AL (2013) Observing culture: differences in U.S.-American and German team meeting behaviors. Psychology Faculty. http://digitalcommons.unomaha.edu/cgi/viewcontent.cgi?article=1102&context=psychfacpub

Steckler J (2013) People of German heritage. In: Purnell L (ed) Transcultural health care: a culturally competent approach, 4th edn. F.A. Davis Company, Philadelphia

Chapter 15
People of Guatemalan Heritage

15.1 Overview and Heritage

People of Guatemalan heritage are among the Hispanic/Latino population in the US. Whereas Guatemalans may share a common Spanish language with other Hispanic groups, they are, nonetheless, a unique cultural group. Health-care providers must be knowledgeable regarding variant cultural characteristics (see Chap. 1) to provide culturally competent care for patients of Guatemalan heritage, remembering that this aggregate profile will not fit every Guatemalan.

Guatemala, referred to as "eternal spring," has an estimated population of over 15 million. Guatemala is inhabited by Mestizo (mixed Amerindian Spanish—in local Spanish called *Ladino*), K'iche, Kaqchikel, Mam, Q'eqchi and other Mayan, and indigenous non-Mayan.

Civil war during the 1960s, 1970s, and 1980s created widespread death, loss of land rights, economic instability, and disruption of the way of life for most Guatemalans. As a result, as many as one million fled to Mexico and the US seeking safety and political asylum. Despite the many years since the civil war, Guatemala continues to be plagued by "abject" poverty, with over half of the population living below the poverty level.

15.1.1 Communications

- The major languages in Guatemala include the official language, Spanish, which is spoken by 60% of the population, and Amerindian languages, spoken by the remaining 40%. Officially, 23 Amerindian languages are recognized.
- Some Mayan men do not have a formal education but are able to speak Spanish because of frequent interactions with Spanish speakers. **Obtain an interpreter**

© Springer Nature Switzerland AG 2019
L. D. Purnell, E. A. Fenkl, *Handbook for Culturally Competent Care*,
https://doi.org/10.1007/978-3-030-21946-8_15

when needed. Note: Interpreters are not available for many of the Amerindian languages.

- Although topics such as income and investments are taboo among many Guatemalans, they generally like to express their inner beliefs, feelings, and emotions once they get to know and trust a person. Meaningful conversations often become loud and seem disorganized. To the outsider, the situation may seem stressful or hostile, but this intense emotion means the conversants are enjoying each other's company. **Do not assume that a loud voice volume means anger.**

- **Within the context of respect (*respeto—goes on the glossary as well*), encourage open communication and sharing. Develop the patient's sense of trust by inquiring about family members before proceeding with the usual business if the patient's condition permits.**

- Guatemalans place great value on closeness and togetherness, including when they are in an in-patient facility. They frequently touch and embrace and like to see relatives and significant others. Touch between men and women, between men, and between women is acceptable, including in the health-care environment. **Explain the necessity of and ask permission before touching during a physical examination.**

- Traditional Guatemalans consider sustained eye contact when speaking directly to an older person to be rude. Direct eye contact with superiors may be interpreted as insolence. **Do not assume that lack of eye contact means the patient or family is being less than truthful or not listening. Avoiding direct eye contact with superiors is a sign of respect.**

- Guatemalans tend to value the past and live in the present, being more concerned with today than the future that is uncertain for many.

- Because work is a priority in the life of many Guatemalans, many seek health care only when their illness has progressed to the point of preventing them from working or carrying out their duties or roles within the family. Taking time to go to the doctor means time lost from work and loss of pay.

- Punctuality is difficult for many because of limited transportation and unexpected family needs. **Carefully explain the necessity of timeliness in keeping appointments in the U.S. health-care system.**

- Guatemalans who have a Hispanic heritage use the Spanish format for names. At birth, a child is given a first name (Ovidio) followed by the surname of his father (Garcia), and then the surname of his mother (Salvador), resulting in Ovidio Garcia Salvador. In referring to him as Mr. (*Señor*) it would be appropriate to use Señor Garcia. Men's names remain the same through their lifetime. However, when a woman named Jovita Garcia Salvador marries Francisco Vasquez Gutierrez, she then becomes Jovita Garcia de Vasquez or simply Jovita Garcia Vasquez. She then becomes Mrs. (*Señora*) Vasquez. **Ask what the legal name is for medical record keeping.**

- **To convey respect, address the Guatemalan in a formal manner unless otherwise requested by the patient.**

- Many Guatemalans shake hands softly. To give a firm handshake indicates aggressive behavior.

15.1.2 Family Roles and Organization

- Traditionally, the man has been the head of household and is the primary decision maker, "breadwinner," and provider for the family.
- Women's roles have traditionally involved raising the children. **More Guatemalan women are entering the workforce outside of the home, resulting in more egalitarian male/female roles.**
- Guatemalans place high value on the family. Most families are nuclear—comprised of a father, mother, and children. Extended family is important to Guatemalans and may include grandparents, aunts, uncles, and cousins. **Determine the spokesperson in the family.**
- Children are a gift from God and are highly valued in Guatemalan society. Sons are usually more valued than daughters. Children are taught to be obedient and demonstrate respect for older people.
- In Mayan communities, family members and other adults take an active part in raising a child. Values include being humble, content, and respectful of others; working hard; avoiding arguments; and placing the needs of the family before one's own individual needs.
- Disobedience among children in Guatemala may be handled with physical punishment. **Explain child abuse laws in the US.**
- Many Guatemalans do not believe in homosexuality, sexual activity among the unmarried, or infidelity. Persons involved in these activities must do so in secrecy. **Do not disclose same-sex relationships to the family.**
- Larger cities in the US offer organizations such as *Ellas*, a support group for Latina lesbians; El Hotline of Hola Gay, an organization with information and referrals in Spanish; and Dignity, a gay Catholic support organization. **Refer patients to gay support groups when requested.**

15.1.3 Workforce Issues

- During the civil war, residents were permitted to migrate to the US and apply for political asylum, allowing them to stay permanently in the US.
- Guatemalans who migrate to the US today often find employment in the agricultural industry, including hand-harvesting crops and working on egg or chicken farms, citrus processing plants, and plant and flower nurseries.
- **Assess patients for diseases related to farming, including pesticides.**
- Some find positions in housekeeping or maintenance in businesses or schools. Because punctuality is not valued in Guatemala, Guatemalan employees may arrive for work late. They might not wear a timepiece, be able to tell time, or understand the importance of punctuality in the US.
- **Stress the necessity of and importance that the US places on punctuality.**

- Those of lower socioeconomic status or with formal education and English language skills usually acquire positions with responsibility but little authority. They prefer to get along well with others and not criticize or voice complaints when treated poorly. Moreover, the Guatemalan is likely to remain in a position equal to his or her peers rather than seek a promotion.

15.1.4   Biocultural Ecology

- Most Guatemalans are a mixture of Spanish and Mayan Indian heritage.
- There is a small population of black Guatemalans with ancestry from the Caribbean and Africa. This accounts for variations in skin color, facial features, hair, body structure, and other biological variations. Black Guatemalans tend to have black hair, black skin, and dark eyes and be of average or taller height with a medium to large build.
- Guatemalans who are predominantly Spanish have the appearance of Caucasians. They may have blonde or brown hair, fair (white) complexion, and blue eyes and be of average or taller height with a medium to large build.
- Guatemalans with predominantly Mayan Indian ancestry tend to have black hair, brown skin, and dark eye color and are of short height with a petite build.
- Jaundice, cyanosis, and petechiae are more difficult to detect among dark-skinned Guatemalans. **Exam the sclera and buccal mucosa for, tongue, lips, nailbeds, and palms of hands and soles of feet.**
- The leading causes of mortality in Guatemala are pneumonia, influenza, perinatal conditions, intestinal infectious diseases, and nutritional deficiencies. **Assess newer immigrants for health conditions common in their home country.**
- Other common health conditions include malaria or dengue fever, gastritis, intestinal worms, pulmonary tuberculosis, and diarrhea resulting from shigella, amoebas, and giardia. Table 15.1 lists commonly occurring health conditions for Guatemalans in three categories: lifestyle, environment, and genetics.
- Eye disorders include dacrocytosis and trachoma. Cataracts (even in children), congenital toxoplasmosis, toxic optic neuropathy (usually related to TB treatment), vitamin A deficiency, and actinic allergy. **Assess Guatemalans in the US for health conditions common in their home country.**

Table 15.1 Health conditions common to Guatemalans	Health conditions	Causes
	Lactase deficiency	Genetic
	Gastritis	Environment, lifestyle
	Malaria	Environment, lifestyle
	Tuberculosis	Environment, lifestyle
	Eye disorders	Lifestyle

15.1.5 High-Risk Behaviors

- Alcohol is readily available and widely abused. Men who immigrate to the US Guatemala may find themselves drinking alcohol excessively, even if they did not prior to migration. This may be due to such factors as (a) the stress of living in another country illegally; (b) being away from family, friends, and support systems; (c) fears of inadequate work and deportation; and (d) illness and being victims of violence and injury.
- Guatemalans of low socioeconomic status have received little health education, have had limited access to health care, and often believe illness is punishment from God.
- Guatemalan families readily participate in immunization programs for their children yet do not participate themselves. **Encourage immunizations as part of routine care.**
- Most women do not participate in routine screening for breast and cervical cancer. Few men participate in routine screening for testicular or colon cancer.
- **Educate Guatemalans in a family context for disease, illness, and injury prevention; assist with low-cost referrals.**

15.1.6 Nutrition

- Food to Guatemalans signifies physical, spiritual, and cultural wellness.
- Foods vary among Guatemalans based on cultural traditions and accessibility. Gari'funa cuisine reflects the Caribbean coast and includes recipes from African ancestors. Common foods include sea bass, flounder, red snapper, tarpon, shrimp, ceviche, coconut milk, tomato, onion, garlic, lime, and lemon.
- Corn is highly valued in the Mayan culture and is the chief crop and the basis for many food products and meals. The Mayan diet primarily consists of maize, black beans, rice, chicken, squash, tomatoes, carrots, chilies, beets, cauliflower, lettuce, cabbage, chard, leek, onion, and garlic. These foods are used to make tortillas, *atole* (liquid corn drink), *pinol* (chicken-flavored corn gruel), *pepi'an* (chicken stew with squash seeds, hot chilies, tomatoes, and tomatillos [small green tomato]), and *caldos* (soups made of chicken stock and vegetables). Green leafy plants are often added to soups and stews. Preferred beverages are Hibiscus tea and coffee.
- Ladino foods reflect their Spanish ancestry and include maize, rice, black or red beans, beef, chicken, pork, milk, cheese, plantains, carrots, peppers, tomatoes, squash, avocado, cilantro, chilies, onion, garlic, lemon, lime, and chocolate. Common dishes include *arroz con pollo* (chicken with rice), *chile rellenos* (peppers stuffed with pork or beef with carrot, onion, and tomato and fried in egg batter), and *tamales* (chicken or pork in a corn paste steamed in banana leaves or corn husks). Again, green leafy plants are often added to soups and stews. Foods are seasoned with toasted squash seeds ground to a powder. Guatemalan food is not served spicy.
- **Complete a nutritional assessment as part of the intake interview.**

15.1.7   Pregnancy and Childbearing Practices

- For religious reasons, most do not believe in contraception or abortion. Women who desire contraception will not actually seek it because of lack of support from their spouse, family, and church. When the Guatemalan patient does express a desire to learn more about family planning, she may request the session include her husband.
- **Arrangements should be made to accommodate the patient's request.**
- Mayan women do not believe in lying down to give birth or delivering in a hospital. The idea that a Guatemalan woman in the US might need to deliver in the hospital may be perceived as foreign and frightening. **If possible, provide a tour of the labor and delivery unit well before her due date.**
- If someone eats in front of the pregnant woman without offering her food, she believes she will have a miscarriage.
- Following delivery, the placenta has to be burned, not buried, because it is disrespectful to the earth to do so.
- Guatemalan women may continue breast-feeding until the child reaches the age of 5 years. They may be breast-feeding a new baby while continuing to breast-feed a toddler.
- A bag with garlic, lime, salt, and tobacco is hung around the baby's neck, and a red thread is used to tie the umbilical cord to protect the baby, provide strength, and denote respect for the ancestors. **Do not remove amulets or other items that are seen as protective for the baby.**

15.1.8   Death Rituals

- Some Guatemalans relate their illness to "punishment" or impending death to "God's will" and refuse an intervention or heroic measures.
- When a Guatemalan dies in the US, the family may request repatriation. It is important for the final resting place to be the home country. Often, immediate and extended family pool their resources to send the body home.
- **Refer families to Social Services and funeral homes to assist with transporting the body to Guatemala.**
- Guatemalans believe in burial; they do not practice cremation.
- Yellow is the color of mourning. Yellow flowers are placed at the grave. Food is placed at the head for the spirit of the departed.
- **Ladinos mourn the dead by wearing black. Maya do not believe in this practice.**

15.1.9 Spirituality

- Most Guatemalans are Roman Catholic. When the Spanish brought Roman Catholicism to Guatemala, some Maya converted to Christianity. Others continued to practice their Mayan religion. Still other Guatemalans combined beliefs and practices of the two. In some cases, Guatemalans integrated aspects of Catholicism into their lives while continuing to believe in the spirituality of their ancestors in private.
- Two practices influenced by the Spanish are *guachibal* and *cofradia*. *Guachibal* involves the practice of keeping an image of a Christian saint in the home and celebrating on the particular saint's day. *Cofradia* refers to a "religious brotherhood" that serves to maintain the "cult" of a particular saint. *Cofradias* consist of dues-paying members and elected leaders. In addition to religious activities, they often serve needy persons in the community by visiting the sick and paying for funerals.
- Family provides Guatemalans with meaning in their lives. Life revolves around the nuclear and extended family. Spirituality helps to explain life and the circumstances faced by Guatemalans.
- Whether Catholic, Protestant, or traditional Mayan, many believe that life's events happen for a reason. The reason may be attributed to favor from God or gods when positive experiences occur and to punishment or disfavor from God or gods when negative events occur. Some feel nothing can be done to change the outcome of these experiences. This belief is referred to as *fatalism*.
- When illness occurs, many Guatemalans turn to their faith for strength, wisdom, and hope. **Do not remove statues or pictures of patron saints. Contact chaplaincy for spiritual care services for the patient and family through their church or hospital resources.**

15.1.10 Health-Care Practices

- The preferred mode of treatment among Ladinos is medication administered by hypodermic injection. If an infant has a cold, Ladinos believe an injection is necessary to treat it effectively. If someone has the flu, an IV infusion is preferred. **Provide factual information about allopathic medical practice.**
- Health-care seeking among Guatemalans generally occurs by first seeking advice from a mother, grandmother, or other respected elder. If this approach is unsuccessful, then the family usually seeks health care from folk healers.
- Several medications, including antibiotics, purchased in the US only with a prescription are available without a prescription in Guatemala. **Ascertain if**

Guatemalan patients are receiving medicines from family or friends from their home country.

- Modern medical care may be the last resort. Many are fearful of hospitals. In Guatemala, when hospital care is necessary, patients are often seriously ill, resulting in death, which perpetuates the belief that "hospitals are places where patients go to die."
- Most Guatemalans engage in folk medicine practices and use a variety of prayers, herbal teas, and poultices to treat illnesses. Many of these practices are regionally specific and vary among families. Many of these practices are harmless, but some may contradict or potentiate therapeutic interventions. **Specifically ask patients whether they are using folk medicine.**
- Many diseases are caused by a disruption in the hot-and-cold balance of the body. Thus, eating foods of the opposite variety may either cure or prevent specific hot-and-cold illnesses and conditions.
- Physical or mental illness may be attributed to an imbalance between the person and the environment. Influences include emotional, spiritual, and social state, as well as physical factors such as humoral imbalance expressed as either too much hot or cold.
- **Nonjudgmentally ask patients and families about the use of folk and traditional therapies.**
- Folk practitioners are consulted for several notable conditions. *Mal de ojo* (evil eye) is a folk illness that occurs when one person (usually older) looks at another (usually a child) in an admiring fashion. Another example of *mal de ojo* is if a person admires something about a baby or child, such as beautiful eyes or hair. Such eye contact can be either voluntary or involuntary. Symptoms are numerous, ranging from fever, anorexia, and vomiting to irritability. The spell can be broken if the person doing the admiring touches the person admired while it is happening. Children are more susceptible to this condition than women, and women are more susceptible than men.
- Guatemalans often delay seeking health care until they are incapacitated by illness, disease, or injury. Many times, they are unaware of the dangers associated with working in agriculture in the US. They may be exposed to pesticides and dangerous equipment without proper training. Although the government has specific laws in place to protect farm workers, enforcement is limited. Companies may not tell the workers the dangers or the workers may not understand owing to language differences. **Be aware of the chemical exposures, laws, and resources available in order to adequately provide care for and advocate for the patient.**
- Barriers to health care in the US include *availability, accessibility/reliability, affordability,* a*ppropriateness, accountability,* a*daptability, acceptability, awareness,* a*ttitudes, approachability, alternative practices and practitioners, and additional services* (see Chap. 3).
- Guatemalans tend to view health and illness in relation to their ability to perform duties associated with their roles. As long as women are functioning in their role

of caring for the home and family, and men are functioning in their job, then they feel "healthy."
- When an illness prevents normal functioning required for their roles, then Guatemalans view it seriously. The cause of debilitating illness or disease may be viewed as punishment from God rather than lack of prevention or early detection.
- When symptoms persist, fear may keep the Guatemalan patient from seeking medical attention.
- Family members would rather care for their loved one at home if at all possible. **Demonstrate respect for the cultural values by providing community resources that enable them to adequately care for a family member at home if they so desire.**
- Blood transfusions and organ donation are not common in Guatemala, and with many living outside of the realm of modern medicine, residents may be uneducated regarding the medical situations in which they are indicated, benefits, and risks. Questions related to organ donation may be puzzling and elicit fear and anxiety.
- Some Guatemalans fear venipuncture because taking blood leaves the body without enough blood to keep them strong and healthy. **Assess the level of education and understanding of the Guatemalan patient regarding blood and blood products and use an interpreter who is competent in the language and culture of the patient in order to promote successful communication.**

15.1.11 Health-Care Providers

- *Modern medicine* refers to health care provided by educated physicians and nurses. *Ladino folk medicine* is provided by Ladino pharmacists, spiritualists, and lay healers (***curanderos***).
- *Mayan Indians* seek medical care from Mayan *shaman*, *bonesetters*, and *herbalists*.
- When Ladinos and Mayan Indians have access to modern medicine, the utilization of these services increases.
- In the US, Guatemalans use many Mexican traditional healers (see Chap. 24, on People of Mexican Heritage).
- Guatemalans have great respect and admiration for health-care providers, who are viewed as authority figures with clinical expertise.
- Guatemalans expect their health-care provider to have the appearance and manners of a professional. When this is not the case, Guatemalans lose confidence in the provider.
- Many fear disclosure which may result in deportation or rejection. Patients also fear confidentiality will not be maintained in the health-care setting.
- Guatemalan women are usually very modest. They may refuse to discuss personal issues or receive an examination by a male health-care provider. Likewise,

a male Guatemalan patient may refuse a female health-care provider. **Provide a same-sex provider when possible, especially for intimate care.**

- Because Guatemalans dislike conflict, they may not actually refuse care; yet they withhold personal information owing to discomfort with the health-care provider.

15.2   Reflective Exercises

1. Explain the name format for Guatemalans.
2. What languages besides Spanish are spoken by Guatemalans and Mayans?
3. What could one do to break a spell of the "evil eye"?
4. What are common infectious diseases common among Guatemalans?
5. What are common eye problems among Guatemalans? What are some reasons for these eye problems?
6. What folk illnesses are common among Guatemalans and Mayans?
7. What routes do Guatemalans prefer for medication administration?
8. If a Guatemalan dies in the US and the family wants to have the body returned to Guatemala, what resources are available to assist the family?
9. How is *fatalism* displayed among Guatemalans?
10. How would you assess jaundice, cyanosis, and petechiae in dark-skinned Guatemalans?

Bibliography

Bhatt S (2012) Health care issues facing the Maya people of the Guatemalan highlands: the current state of care and recommendations for improvement. J Glob Health Perspect 1:1–7. http://jglo-balhealth.org/article/health-care-issues-facing-the-maya-people-of-the-guatemalan-highlands-the-current-state-of-care-and-recommendations-for-improvement-2/

Callister LC, Vega R (1998) Giving birth: Guatemalan women's voices. J Obstet Gynecol Neonatal Nurs 27(3):289–295

CIA world factbook: Guatemala (2016). https://www.cia.gov/library/publications/the-world-factbook/geos/gt.html

Ellis T, Purnell L (2013) People of Guatemalan heritage. In: Purnell L (ed) Transcultural health care: a culturally competent approach. F.A. Davis, Philadelphia

Weller SC, Ruebush TK, Klein RR (1991) An epidemiological description of folk illnesses: a study of *empacho* in Guatemala. Med Anthropol 13(1–2):19–31

Chapter 16
People of Haitian Heritage

16.1 Overview and Heritage

Haiti, with a population of 10.4 million, is located in the Caribbean on the island of Hispaniola, which it shares with the Dominican Republic. The per capita annual income is $817. Only 13% of the people have access to potable water. Haiti defines itself as a black, 95% of the population, nation. In Haiti, the concept of color differs from the concept of race. The Haitian system has been described as one in which there are no tight racial categories but in which skin color and other phenotypic demarcations are significant variables. When Duvalier was elected president-for-life in 1964, many Haitians covertly immigrated to the US in small sailboats, resulting in their being labeled "**boat people**," a term that is associated with extreme poverty.

The Haitian population in the US is not well documented because many are undocumented immigrants. Even though Haitians value education, only 15% are privileged enough to attain a formal education. The illiteracy rate of 60% continues to be a major concern in Haiti. Like other ethnic groups, Haitians are very diverse according to the variant characteristics of culture as described in Chap. 1. The catastrophic 7.0 earthquake in January 2010 killed over 300,000 people, injured another 300,000, and left over one million homeless.

16.1.1 Communications

- Haiti has two official languages: French and Creole. French is the dominant language of the educated and the elite, whereas Creole is the language of the lower socioeconomic classes. Creole, a rich, expressive language, is spoken by everyone. **Develop video programs, audiocassettes, and picture brochures in Creole for providing health education.**

© Springer Nature Switzerland AG 2019
L. D. Purnell, E. A. Fenkl, *Handbook for Culturally Competent Care*,
https://doi.org/10.1007/978-3-030-21946-8_16

- Most are very expressive with their emotions, including loud, animated speech. Pain and sorrow are very obvious in facial expressions.
- Most Haitians are affectionate, polite, and shy. The under-educated generally hide their lack of knowledge by keeping to themselves, avoiding conflict, and sometimes projecting a timid air or attitude. They smile frequently and often respond in this manner when they do not understand what is being said.
- Many individuals pretend to understand by nodding; this sign of approval is given to hide their limitations. **Because Haitians are very private, especially in health matters, it is inappropriate to share information with friends.**
- Many patients may prefer to use professional interpreters, who will give an accurate interpretation of their concerns. Most important, the interpreter should be someone with whom they have no relationship and will likely never see again. **Use simple and clear instructions. Ask family members to assist with interpretation only if an interpreter is not available.**
- When conversations are really animated, the conversants speak in close proximity and ignore territorial space. Sometimes the conversation is at such a high pitch and speed that, to an outsider, the conversation may appear disorganized or angry. **Do not interpret loud conversations as anger.**
- Traditionally, Haitians do not maintain eye contact when speaking with those in a position of authority. **Do not interpret lack of eye contact as not listening or not caring.**
- Haitians touch frequently when speaking with friends. They may touch you to make you aware that they are speaking to you. **Do not take offense from a casual touch on the arm or shoulder.**
- Haitians greet each other by kissing and embracing in informal situations. In formal encounters, they shake hands and appear composed and stern. Men usually do not kiss women unless they are old friends or relatives. Children greet everyone by kissing them on the cheek. Children refer to adult friends as uncle or auntie.
- Temporal orientation is a balance among the past, present, and future. The past is important because it lays the historical foundation from which one must learn. The present is cherished and savored. The future is predetermined, and God is the only Supreme Being who can redirect it.
- Haitians have a fatalistic but serene view of life. Some believe that destiny or spiritual forces are in control of life events such as health and death.
- **To achieve acceptance, build trust, and ensure compliance, be clear, honest, and open when assessing individuals' perceptions and how they perceive the forces that have an influence over life, health, and illness.**
- Most do not respect clock time; flexibility with time is the norm, and punctuality is not valued. Arriving late, even for medical appointments, is not considered impolite. **Make reminder calls for appointments, and encourage the patient about the importance of timeliness if it is important.**
- Most Haitians have a first, middle, and last name; sometimes the first two names are hyphenated as in Marie-Maude. The family name is very important in middle-

and upper-class society, promoting and communicating tradition and prestige. Families usually have an affectionate name or nickname for individuals.

- When a woman marries, she takes her husband's full name. She loses her name except on paper. Most names are of French origin, although many Arabic names are also heard. Most are formal and respectful. **Address patients by their title: Mr., Mrs., Miss, Ms., Doctor, or other title until given permission to do otherwise.**

16.1.2 Family Roles and Organization

- Traditionally, the head of the household has been the man, but today most families are matriarchal. The man is generally considered the primary income provider for the family, and daily decision-making is considered his province. Women are expected to be faithful, honest, and respectable.
- Children are valued among Haitians because they are key to the family's progeny, cultural beliefs, and values. Children are expected to be high achievers. Children are expected to be obedient and respectful to parents and older people. They are not allowed to express anger to older people.
- Physical punishment, often used as a way of disciplining children, is sometimes considered child abuse by U.S. standards. Fear of having their children taken away from them because of their methods of discipline can cause parents to withdraw or not follow through on health-care appointments if such abuse is evident (e.g., bruises or belt marks). **Educate parents about American methods of discipline and laws so they can learn new ways of disciplining their children without compromising their beliefs or violating U.S. laws.**
- In the summer, parents engage their children in health promotion activities such as giving them *lok* (a laxative), a mixture of bitter tea leaves, juice, sugarcane syrup, and oil. Children are also given *lavman* (enemas) to ensure cleanliness. This rids the bowel of impurities and refreshes it, prevents acne, and rejuvenates the body.
- Boys are given more freedom than girls and are even expected to receive outside initiation in social and sexual life. Girls are educated toward marriage and respectability. Even when they are 16 or 17 years of age, girls cannot go out alone because any mishap can be a threat to her future and bring shame to her family.
- **Assist children and family members to work through cultural differences related to proper age for dating while conveying respect for family and cultural beliefs.**
- The expression "blood is thicker than water" reflects family connectedness. An important unit for decision-making is the family council, which is composed of influential members of the family. Any action taken by one family member has repercussions for the entire family; consequently, all share prestige and shame. **Include the extended nuclear family in health-care decision-making.**

- The family system includes the nuclear, consanguine, and affinal relatives, some or all of whom may live under the same roof. **Include family members in the care of loved ones to achieve trusting relationships and compliance with treatment regimens.**
- When family members are ill and in the hospital, there is an obligation to be there for them; all family members try to visit. **Suggest alternative locations for family members to gather. Ask a family spokesperson to help regulate the number of visitors at any one time.**
- When grandparents are no longer able to function independently, they move in with their children. Older people are highly respected and are often addressed by an affectionate title such as aunt, uncle, grandma, or grandpa, even if they are not related. Their children are expected to care for and provide for them when self-care becomes a concern. Older people are family advisers, babysitters, historians, and consultants.
- Most Haitians are reluctant to place family members in nursing homes.
- Homosexuality is taboo; gay and lesbian individuals usually remain closeted. If a family member discloses homosexuality, everyone keeps it quiet; there is total denial. Gay and lesbian relationships are not talked about; they remain buried. **Do not disclose lesbian and gay relationships to family members.**
- Although divorce is common, family members, friends, the church, and elders try to counsel the couple before the divorce becomes final.
- Single parenting, widespread in Haiti, is well accepted. It is a well-accepted practice for men to have both a wife and a mistress. Both women bear children. The mistress usually raises her children alone and with minimal support from the father. Haitian women in general know that their husbands are involved in extramarital relationships but pretend not to know. **Approach health education, birth control, and safe-sex issues with sensitivity and acceptance within cultural boundaries. Provide educational brochures in French, Creole, and English.**

16.1.3   Workforce Issues

- Many Haitians in the US work two jobs to provide for their family while sending money to Haiti for those left behind. Most readily conform to the rules and regulations of the workplace. Haitians tend to settle in clusters with other Haitians. When people live and work primarily in an ethnic enclave, the native culture becomes a barrier to assimilation and acculturation into the dominant society. **Encourage Haitian employees to socialize with other cultural groups.**
- Sometimes Haitians in the workplace greet each other in their native tongue because it is easier to articulate ideas and feelings and to express support in their native language. This may be irritating to non-Haitians who consider it rude. **Explain rules for speaking a foreign language when at work.**

- Many individuals pretend to understand by nodding; this sign of approval is given to hide their limitations. **Because Haitians are very private, especially in health matters, it is inappropriate to share any information with friends.**
- When conversations are really animated, the conversants speak in close proximity and ignore territorial space. Sometimes the conversation is at such a high pitch and speed that, to an outsider, the conversation may appear disorganized or angry. **Do not interpret loud conversations as anger or take offense at spatial distancing closer than what you are accustomed.**
- Traditionally, Haitians generally do not maintain eye contact when speaking with those in a position of authority. Maintaining direct eye contact can be considered rude and insolent, especially when speaking with superiors. **Do not interpret lack of eye contact as not listening or not caring.**
- Haitians touch frequently when speaking with friends. They may touch you to make you aware that they are speaking to you. **Do not take offense from a casual touch on the arm or shoulder.**

16.1.4 Biocultural Ecology

- Different assessment techniques are required when assessing dark-skinned people for anemia and jaundice. **Examine the sclera, oral mucosa, conjunctiva, lips, nail beds, palms of the hands, and soles of the feet when assessing for cyanosis and low blood hemoglobin levels. To assess for jaundice, examine the conjunctiva and oral mucosa for patches of bilirubin pigment because dark skin has natural underlying tones of red and yellow.**
- Table 16.1 lists commonly occurring health conditions for Haitians in three categories: lifestyle, environment, and genetics.
- **Assess newer immigrants for signs of malaria, hepatitis, and infectious diseases that are common in Haiti.**

Table 16.1 Health conditions common to Haitians

Health conditions	Causes
Cholera	Lifestyle, environment
Malaria	Lifestyle, environment
Tuberculosis	Lifestyle, environment
Hepatitis	Lifestyle, environment
Parasitosis	Lifestyle, environment
Diabetes mellitus	Genetic, lifestyle, environment
Cancer	Genetic, lifestyle, environment
Cardiovascular diseases	Genetic, lifestyle, environment
Cerebrovascular disease	Genetic, lifestyle, environment

16.1.5 High-Risk Health Behaviors

- Behaviors that may be considered high-risk in American society are generally viewed as recreational or unimportant among Haitians.
- Alcohol plays an important part in society and is culturally approved for men. Women drink socially and in moderation. Cigarette smoking is practiced by men, whereas Haitian women have a very low rate of tobacco use. **Encourage smoking cessation and start educational programs in elementary school.**
- Other high-risk behaviors include non-use of seat belts and helmets when driving or using a motorcycle or bicycle because these safety measures are not normally used in Haiti. **Educate patients seat belt use, car seats for youngsters, and the need for helmets on bicycles.**
- **Use graphic videos or skits when instructing patients about these safety practices.**
- **Use Haitian radio stations when they are available for educational programs.**
- **Promote behavioral changes using church and community group activities.**

16.1.6 Nutrition

- For many, food means survival; however, food is also relished as a cultural treasure, and most Haitians retain their food habits and practices after emigrating. They prefer eating at home, take pride in promoting their culture through their food choices to their children, and discourage fast food. When hospitalized, many individuals would rather fast than eat non-Haitian food. They do not eat yogurt, cottage cheese, or "runny" egg yolks. They drink a lot of water, homemade fruit juices, and cold fruity sodas.
- **Encourage family members to bring food from home for the hospitalized patient if necessary.**
- The typical breakfast consists of bread, butter, bananas, and coffee. Children are allowed to drink coffee, which is not as strong as that consumed by adults.
- The largest meal is eaten at lunch, which includes rice and beans, boiled plantains, a salad made of watercress and tomatoes, and stewed vegetables and beef or cornmeal cooked as polenta.
- "Hot and cold," "acid and non-acid," and "heavy and light" are the major categories of contrast when discussing food. Illness is caused when the body is exposed to an imbalance of *fret* (cold) or *cho* (hot) factors. For example, soursop, a large green prickly fruit with a white pulp that is used in juice and ice cream, is considered a cold food and is avoided when a woman is menstruating; eating white beans after childbirth is believed to induce hemorrhage.
- Foods that are considered heavy, such as plantain, cornmeal mush, rice, and meat, are to be eaten during the day because they provide energy. Light foods,

such as hot chocolate milk, bread, and soup, are eaten for dinner because they are more easily digested.

- To treat a person in the hot-and-cold system, a potent drink or herbal medicine of the class opposite to the disease is administered. For example, cough medicines are considered to be in the hot category, and laxatives are in the cold category.
- Teenagers are advised to avoid drinking citrus fruit juices such as lemonade to prevent the development of acne.
- After performing strenuous activities or any activity that causes the body to become hot, one should not eat cold food because that will create an imbalance, causing a condition called *chofret*. A woman who has just straightened her hair by using a hot comb and then opens a refrigerator may become a victim of *chofret*. This means she may catch a cold and/or possibly develop pneumonia.
- When ill, pumpkin soup, bouillon, or a special soup made with green vegetables, meat, plantains, dumplings, and yams is consumed.
- Eating right entails eating sufficient food to feel full and maintain a constant body weight, which is often higher than weight standards medically recommended in the US. Men like to see "plump" women. **Negotiate a desirable weight with patients.**
- Weight loss is considered one of the most important signs of illness. A healthy diet includes tonics to stimulate the appetite and the use of high-calorie supplements such as *Akasan*, which is either prepared plain or made as a special drink with cream of cornmeal, evaporated milk, cinnamon, vanilla extract, sugar, and a pinch of salt. **Ask patients about food rituals when designing individualized dietary plans to facilitate compliance with dietary regimens that promote a healthier lifestyle. Include family in the counseling process.**
- Many women and children who come from rural areas have significant protein deficiencies. A cultural factor that contributes to this problem is the uneven distribution of protein among family members. The problem is not one of net protein deficiency in the community but rather the unwise distribution of the available protein among family members. Whenever meat is served, the major portion goes to the men on the assumption that they must be well fed to provide for the household. **Being aware that men receive the largest protein servings enables health-care providers to prepare nutritional plans that meet patients' dietary needs.**

16.1.7 Pregnancy and Childbearing Practices

- Pregnancy and fertility practices are not readily used. Most Haitians are Catholic and are unwilling to overtly engage in conversation about birth control or abortion.
- Abortion is viewed as a woman's issue and is left to her and her significant other to decide.

- **Be cautious in assessing and gathering information related to fertility control.**
- Because pregnancy is not a disease or illness, many women do not seek prenatal care. Pregnancy does not relieve a woman from her work.
- Pregnant women are restricted from eating spices that may irritate the fetus. They eat vegetables and red fruits because these are believed to improve the fetus's blood. They are encouraged to eat large quantities of food because they are eating for two.
- Pregnant women who experience increased salivation may rid themselves of the excess at places that may seem inappropriate in the US. They may even carry a "spit" cup in order to rid themselves of the excess saliva. They are not embarrassed by this behavior because they feel it is perfectly normal.
- During labor, the woman may walk, squat, pace, sit, or rub her belly. Generally, women practice natural childbirth and do not ask for analgesia. Some may scream or cry and become hysterical, whereas others are stoic, only moaning and grunting. **Women in labor need support and reassurance. Applying a cold compress on the woman's forehead demonstrates caring and sensitivity.**
- Cesarean birth is feared because it is abdominal surgery.
- Fathers do not generally participate in labor and delivery, believing that this is a private event that is best handled by women. The woman is not coached; female members of the family give assistance as needed. **Do not make men feel guilty if they do not want to participate in the delivery.**
- Postpartum, the woman takes an active role in her own care. She dresses warmly after birth as a way to become healthy and clean.
- Many believe that the bones are "open" after birth and that a woman should stay in bed during the first 2–3 days postpartum to allow the bones to close. Wearing an abdominal binder is another way to facilitate closing the bones.
- The postpartum woman also engages in "the three baths." For the first 3 days, the mother bathes in hot water boiled with special leaves that are either bought or picked from the field. She also drinks tea boiled from these leaves. For the next 3 days, the mother bathes in water prepared with leaves that are warmed by the sun. At this point, the mother takes only water or tea warmed by the sun. Another practice is to take a vapor bath with boiled orange leaves to enhance cleanliness and tighten the internal muscles. At the end of the third to fourth week, she takes the third bath, which is the cold bath. A cathartic may be administered to cleanse her intestinal tract. When the process is completed, she may drink cold water again and resume her normal activities.
- In the postpartum period, women avoid white foods such as lima beans, okra, mushrooms, and tomatoes because they are believed to increase vaginal discharge. Strength foods include porridge, rice and red beans, plantains boiled or grated with the skins and prepared as porridge (the skin is high in iron), carrot juice, and carrot juice mixed with red beet juice.
- Breast-feeding is encouraged for up to 9 months postpartum. Breast milk can become detrimental to both mother and child if it becomes too thick or too thin. If it is too thin, it is believed that the milk has "turned," and it may cause diarrhea

and headaches in the child and, possibly, postpartum depression in the mother. If milk is too "thick," it is believed to cause *bouton* (impetigo).

- **Provide factual information and dispel myths related to breast-feeding.**
- Breast-feeding and bottle feeding are accepted practices. If the child develops diarrhea, breast-feeding is immediately discontinued.
- *Lok*, a laxative similar to the one administered to the older children in the summer, is given to infants to hasten the expulsion of meconium. **Stress the risks associated with lok and the need to prevent dehydration.**

16.1.8 Death Rituals

- Generally, Haitians prefer to die at home rather than in the hospital.
- When death is imminent, the family may pray and cry uncontrollably, sometimes even hysterically. **Meet the person's spiritual needs by encouraging family members to bring religious medallions, pictures of saints, or fetishes.**
- When the person dies, all family members try to be at the bedside and have a prayer service. **If possible and if it is not too disturbing to other patients, encourage a family member to assist with postmortem care.**
- The preburial activity is called *veye*, a gathering of family and friends who come to the house of the deceased to cry, tell stories about the deceased's life, and laugh. Food, tea, coffee, and rum are in abundant supply. The intent is to show support and to join the family in sharing this painful loss.
- Another religious ritual is the *dernie priye*, a prayer service consisting of 7 consecutive days of prayer, which facilitates the passage of the soul from this world to the next.
- On the seventh day, there is a mass called *prise de deuil*, which officially begins the mourning process. After each of these prayers, there is a reception/celebration in memory of the deceased.
- Haitians have a very strong belief in resurrection and paradise; thus, cremation is not usually an option.
- Most Haitians are very cautious about autopsies. If foul play is suspected, they may request an autopsy to ensure the patient is really dead. This alleviates their fear that their loved one is being zombified. According to the belief, this can occur when the person appears to have died of natural causes but is still alive. **Carefully explain the legalities of autopsy.**

16.1.9 Spirituality

- Catholic religious beliefs are combined with voodooism. Voodooism involves communication by trance between the believer and ancestors, saints, or animistic deities. Participants gather to worship the *loa*, deities or spirits, who are believed to have received their powers from God, and who are capable of expressing themselves through possession of a chosen believer.

- Believers in voodooism attribute their ailments or medical problems to the doings of evil spirits. In such cases, they prefer to confirm their suspicions through the *loa* before accepting natural causes as the problem, which would lead to seeking Western medical care. Belief in the power of the supernatural can have a great influence on the psychological and medical concerns of patients.
- **Recognizing and accepting patients' beliefs alleviates barriers and may make patients feel more at ease to discuss their beliefs and needs.**

16.1.10   Health-Care Practices

- Good health is the ability to achieve internal equilibrium between *cho* and *fret*. To become balanced, one must eat well, give attention to personal hygiene, pray, and have good spiritual habits. To promote good health, one must be strong, have good color, be plump, and be free of pain. To maintain this state, one must eat right, sleep right, keep warm, exercise, and keep clean.
- Illness is perceived as punishment, considered an assault on the body, and may have two different causes: natural illnesses, known as *maladi Bondye* (disease of the Lord), and supernatural illnesses.
- Natural illnesses are of short duration and are caused by environmental factors such as food, air, cold, heat, and gas.
- Supernatural illnesses are caused by angry spirits. To placate these spirits, patients must offer feasts called *manger morts*. If individuals do not partake in these rituals, misfortunes are likely to befall them. Illnesses of supernatural origin are fundamentally a breach in rapport between the individual and his or her protector. Health can be recovered if the patient takes the first step in determining the nature of the illness. This can be accomplished by eliciting the help of a voodoo priest and following the advice given by the spirit itself. **To accurately prescribe treatment options, differentiate between natural and supernatural causes of illness and disease.**
- Most individuals believe that gas may provoke pain and anemia. Gas can occur in the head, where it enters through the ears; in the stomach, where it enters through the mouth; and in the shoulders, back, legs, or appendix, where it travels from the stomach. When gas is in the stomach, the patient is said to suffer *kolik* (stomach pain). Gas in the head is called *van nan tet* or *van nan zorey*, which literally means "gas in one's ears," and is believed to be a cause of headaches.
- Gas moving from one part of the body to another produces pain. Thus, the movement of gas from the stomach to the legs produces rheumatism, to the back causes back pain, and to the shoulder causes shoulder pain. **Ask patients what they think is causing their pain.**

- Foods that help dispel gas include tea made from garlic, cloves, and mint; plantain; and corn. To deter the entry of gas into the body, one must be careful about eating leftovers, especially beans.
- Most Haitians engage in self-treatment and consider these activities as a way of preventing disease or promoting health, trying home remedies as a first resort for treating illness. If they know someone who had a particular illness, they may take the prescribed medicine from that person.
- Haitians tend to keep numerous topical and oral medicines on hand. In Haiti, many medications can be purchased without a prescription, a potentially dangerous practice.
- **Admonishing patients may cause them to withdraw and not listen to instructions. Be very discreet in assessing, teaching, and guiding the patient toward safer health practices. Inquire if the patient has been taking medication that was prescribed for someone else.**
- Many Haitians in America ask friends or relatives to send medications from Haiti. Such medications may consist of roots, leaves, and European manufactured products that are more familiar to them. **Ascertain what the patient is taking at home to avoid serious complications.**
- A primary respiratory ailment is *oppression*, a term used to describe asthma. However, the term really describes a state of anxiety and hyperventilation rather than the condition. *Oppression* is considered a "cold" state, as are many respiratory conditions. A home remedy for oppression is to take a dry coconut and cut it open, fill it half with sugarcane syrup and half honey, grate one full nutmeg and add it to the syrup mix, reseal the coconut, and then bury it in the ground for a month. The coconut is reopened, the content is stirred and mixed, and one tablespoon is administered twice a day until it is finished. By the end of this treatment, the child is supposed to be cured of the respiratory problem.
- Many Haitians have low-paying jobs that do not provide health insurance, and they cannot afford to purchase it themselves. Thus, economics acts as a barrier to health promotion. Those who do not speak English well have difficulty accessing the health-care system, explaining their needs fully, or understanding prescriptions and treatments. **Obtain an interpreter when necessary.**
- Diabetes is considered a natural illness; however, most do not seek immediate medical assistance when they detect the symptoms of polyuria, excessive thirst, and weight loss. Instead, they attempt symptom management by making dietary changes on their own by drinking potions or herbal remedies. When the person finally seeks medical attention, he or she may be very sick. **Dispel myths and explain the medical regimen, diet, and medications for diabetes. Reinforce at every opportunity.**
- Pain is commonly referred to as *doule*. Many Haitians have a low pain threshold. They are verbal about the cause of their pain and sometimes moan. They are vague about the location of the pain because they believe that it is not important; they believe that the whole body is affected because disease travels, making it very difficult to assess pain accurately. Injections are the preferred method for medication administration, followed by elixirs, tablets, and capsules.

- Chest pain is *doule nan ke mwen*, abdominal pain is *doule nan vent*, and stomach pain is *doule nan ke mwen* or *doule nan lestomak mwen.*
- Nausea is expressed as *lestomak/mwen ap roule, M santi m anwi vomi, lestomak/ mwen chaje*, or *ke mwen tounin*. Because of modesty, they may discard vomitus immediately so as not to upset others. **Specific instructions should be given regarding keeping the specimen until the health-care provider has had a chance to see it.**
- Oxygen should be offered only when absolutely necessary because the use of oxygen is perceived as an indicator of the seriousness of the illness.
- Fatigue, physical weakness is interpreted as a sign of anemia or insufficient blood. Various external and internal environmental factors are believed to cause *sezisman (fright)*, thereby disrupting the normal blood flow. *Sezisman* may occur when someone receives bad news, is involved in a frightful situation, or suffers from indignation after being treated unjustly.
- The stigma attached to mental illness is strong; most do not readily admit to being depressed. Depression can be viewed as a hex placed by a jealous or envious individual.
- **Ask patients what they think is causing their depression. Negotiate treatment accordingly.**
- In an unnatural illness, the person's poor health is attributed to magical causes such as a hex, a curse, or a spell, which has been cast by someone as a result of family or interpersonal disagreement. Symptoms include burning skin, rashes, pruritus, nausea, vomiting, and headaches and often coincide with psychological problems
- Most are extremely afraid of diseases associated with blood irregularities. Patients and their families become emotional about blood transfusions. Thus, they are received with much apprehension. Additionally, blood transfusions are feared because of the potential for HIV transmission.
- **Factually explain the need for a blood transfusion, and carefully explain the procedure along with the involved risks. Involve patients and their families in the care as much as possible. Explain precautionary measures that have been taken to prevent blood contamination.**
- The body must remain intact for burial. Thus, organ donation and transplantation are not generally discussed or practiced. **Assess patients' beliefs about organ donation, and involve a religious leader to provide support and help facilitate a decision regarding organ donation or transplantation.**

16.1.11 Health-Care Providers

- Most Haitians resort to symptom management with self-care first and then spiritual care. They commonly use traditional and Western health-care providers simultaneously.
- Physicians and nurses are well respected. Physicians are men, and nurses are women. Nurses are referred to as "Miss." **Explain and re-explain relevant points to compensate for patients' knowledge deficit or language limitations.**

16.2 Reflective Exercises

1. In what languages would you prepare educational brochures for general health education for Haitians?
2. What are the primary roles of traditional Haitian men and Women?
3. Explain the principles of *fret* and *cho*. How are these principles applied in dietary practices? In health conditions?
4. What are the primary health conditions common with Haitians?
5. Explaining breast feeding practices common with Haitian mothers. Which practices are potentially dangerous and how would you address them with the mother?
6. Explain the tenets of *voodism*. How is *voodism* seen in health-care practices?
7. Distinguish natural versus supernatural illness among Haitians.
8. What is *oppression*? What education would you include in a patient who has *oppression?*
9. What are infectious diseases would you look for in recent immigrants from Haiti?
10. What are traditional Haitian verbal and nonverbal communication practices?

Bibliography

Background on Haiti and Haitian Health Culture (2010) Cook Ross. http://www.in.gov/isdh/files/cultural_primer_on_Haiti.pdf

CIA (2016) World factbook: Haiti. https://www.cia.gov/library/publications/the-world-factbook/geos/ha.html

Colin J, Paperwalla G (2013) People of Haitian heritage. In: Purnell L (ed) Transcultural health care: a culturally competent approach, 4th edn. F.A. Davis Company, Philadelphia

Cultural Competency and Haitian Immigrants: Death Rituals (2003). https://www.salisbury.edu/nursing/haitiancultcomp/death_rituals.htm

Cultural Competency and Haitian Immigrants: Family Roles and Organization (2003). https://www.salisbury.edu/nursing/haitiancultcomp/family_roles_and_organ.htm

Cultural Competency and Haitian Immigrants: Pregnancy and Childbearing Practices (2003). https://www.salisbury.edu/nursing/haitiancultcomp/preg_and_childbear_prac.htm

Death and Funeral Rituals in the Haitian Culture (2013) Haiti Observer. http://www.haitiobserver.com/blog/death-and-funeral-rituals-in-haitian-culture.html

Haiti: Global Health (2016) USAID. https://www.usaid.gov/haiti/global-health

Haiti: Poorest Country in the Western Hemisphere (2012) Haitian Ministries. http://www.haiti-healthministries.org/haiti/

Chapter 17
People of Hindu Heritage

17.1 Overview and Heritage

More than 1.2 billion people inhabit India. Over 80% of the population are Hindus, followers of Hinduism. Other significant religious groups include Moslems (13.4%), Christians (2.3%), Sikhs (1.9%), and other (1.8%). Although different religious sectors share many common cultural beliefs and practices, they differ according to the variant characteristics of culture as presented in Chap. 1. India is divided into north and south, based primarily on Dravidian and Aryan cultural variations.

Immigrants to the US come predominantly from urban areas, including all major Indian states, and they have come in two waves. The first wave began in the early twentieth century and continued to the mid-1920s. Conditions such as racial discrimination and lack of access to economic advancement made it difficult for the first wave of Asian Indians to sustain themselves or their culture in America. The second wave of immigration began after 1965 and still continues. Most individuals from this wave are highly educated. Most come to the US to attain a higher standard of living, better working conditions, and job opportunities. Secondary reasons include opportunities for additional education as well as Indian perceptions of America as a country of opportunity and freedom. Trafficking, mostly internal, of men, women, and children for the purposes of forced or bonded labor and commercial sex exploitation continues to be a major problem for Indians.

17.1.1 Communications

- Many Indian state borders are reorganized in accordance with language limitations. Languages fall into two main groups: Indo-Aryan in the north and Dravidian in the south. Hindi with numerous dialects is spoken by 41% of the population and is the national language, along with English. In addition, there

© Springer Nature Switzerland AG 2019
L. D. Purnell, E. A. Fenkl, *Handbook for Culturally Competent Care*,
https://doi.org/10.1007/978-3-030-21946-8_17

are 14 other official languages: Bengali, Telugu, Marathi, Tamil, Urdu, Gujarati, Malayalam, Kannada, Oriya, Punjabi, Assamese, Kashmiri, Sindhi, and Sanskrit. English is the language among the educated. **Provide an interpreter for those who may not speak English well.**

- Women often speak in a soft voice, making it difficult to understand or decipher what they say. Men may become intense and loud when they converse with other family members. To an onlooker, it might seem disruptive but, in general, this form of communication can be construed as meaningful when it is conducted with close friends.
- Traditional women avoid direct eye contact with men. Direct eye contact with older people and authority figures may be considered a sign of disrespect. **Do not misconstrue a lack of eye contact as not caring or not listening.**
- Touching and embracing are not acceptable for displaying affection. Even between spouses, a public display of affection such as hugging or kissing is frowned upon, being considered strictly a private matter.
- Temporality is past-, present-, and future-oriented. Time is conceived in cycles of four ages, which start with the "age of perfection" and end with the "age of degeneration."
- Punctuality in keeping scheduled appointments may not be considered important. **Do not misconstrue being late for appointments as a sign of irresponsibility or not valuing health. Reinforce the necessity of punctuality when needed.**
- Older family members are usually not addressed by name but as elder brother, sister, aunt, or uncle. A woman never addresses a man by name because the woman is not considered an equal or superior.
- Strangers are greeted with folded hands and a head bow.

17.1.2   Family Roles and Organization

- No institution in India is more important than the family. The hierarchical structure of authority is the patriarchal joint family, based on the principle of superiority of men over women.
- The male head of the family is legitimized and considered sacred by caste and religion, which delineate relationships. Central relationships in this system are based on continuation and expansion of the male lineage through inheritance and ancestor worship, related to the father-son and brother-brother relationships.
- A matrilineal system exists in a few areas in the southwestern and northeastern regions of the country; however, power rests with the men in the woman's family.
- A submissive and acquiescent role is expected of women in the first few years of married life, with little or no participation in decision-making. Strict norms govern contact and communication with the men of the family, including a woman's husband.

- Parents want their children to be successful and strongly encourage and emphasize scholastic achievement in fields that promise good employment and a high social status. Parents in America want their children to maintain ties with their families and the Indian community.
- Although many parents expect and accept the Westernization of their children, the question of marriage is still a concern for parents who have opinions about how their children should be married, whether "arranged" or partly arranged. Hindu parents or Indians from all religious traditions want their children to marry other Indians.
- Arranged marriages at a young age are considered most desirable for women. The practice of an arranged marriage continues in the US in order to minimize the stress associated with differences in castes, lifestyles, and expectations between the male and female hierarchy.
- The two major types of transfer of material wealth accompanying marriage are bride price and a dowry. Bride price is payment in cash and other materials to the bride's father in exchange for authority over the woman, which passes from her kin group to the bridegroom's kin group.
- In the joint family structure, Hindu women are considered "outsiders" and are socialized and incorporated in such a way that "jointness" and residence are not broken up; therefore, a close relationship between the husband and wife is disapproved because it induces favoring the nuclear family and dissolving the joint family. Therefore, a marriage is regarded as indissoluble.
- Older family members are held in reverence and cared for by their children when self-care becomes a concern.
- **Understand the various types of families (joint, extended, or nuclear) and determine which individual has control within the hierarchy.**
- Single-parent, blended, and communal families are not well accepted by Hindus.
- Homosexuality may cause a social stigma. **Refer lesbian, gay, or bisexual Hindu Americans to national support groups such as TriKone, a nonprofit group for lesbian, gay, and bisexual South Asians located in the San Francisco Bay area; the National AIDS hotline; or Asians Together in Washington, D.C. Do not disclose same-gender relationships to other family members.**

17.1.3 Workforce Issues

- At work, Hindus adopt American practices and cultural habits. Hierarchies of age, gender, and caste prescribe transactions among Hindus. At work, relationships are a reproduction of the authority-dependence characteristic of family and social relationships.
- In seeking to establish a personal and benevolent relationship, Hindus may be seen as too eager to please, ingratiating, or docile, all antithetical to the task of

assertion and independence. **Explain the value of assertiveness expected of all health-care providers in the American workforce.**
- Punctuality may not be considered important. **Explain the necessity of timeliness in the American workforce.**

17.1.4 Biocultural Ecology

- Indian diversity of physical types can be divided into three general groups according to the color of their skin: white in the north and northwest, yellow in areas bordering Tibet and Assam, and black in the south.
- *Indids* (whites) have a light-brown skin color, wavy black hair, dark- or light-brown eyes; are tall or of medium height; and are either dolichocephalic (long-headed) or brachycephalic (short-headed).
- *Melanids*, often referred to as the Dravidians, are the population of southern India and have dark skin ranging from light brown to black, elongated heads, broad noses, thick lips, and black, wavy hair. They are usually shorter than 5 ft 6 in. tall.
- **Pallor in brown-skinned patients presents as a yellowish-brown tinge to the skin. Pallor in dark-skinned individuals is characterized by the absence of the underlying red tones in the skin. Jaundice may be observed in the sclera. The oral mucosa of dark-skinned individuals may have a normal freckling or pigmentation. Inspection of the nail beds, lips, palpebral conjunctiva, and palms of the hands and soles of the feet shows evidence of cyanosis.**
- Table 17.1 lists commonly occurring health conditions for Hindus in three categories: lifestyle, environment, and genetics.
- **Be alert to possible signs and symptoms and risk factors associated with diseases linked to migration from India.**
- Many individuals require lower doses of lithium, antidepressants, and neuroleptics, and they may experience side effects even with the lower doses. They also are more sensitive to the adverse effects of alcohol, resulting in marked facial flushing, palpitations, and tachycardia. **Question therapeutic regimens that do not consider racial or ethnic differences.**

17.1.5 High-Risk Health Behaviors

- Alcoholism and cigarette smoking among Hindu Americans, especially among men, cause significant health problems. **Refer to "stop smoking" clinics, and encourage low alcohol consumption. Partner with elementary school teachers to introduce anti-smoking at a young age.**
- Beriberi is found in people coming from rice-growing areas, pellagra in maize-millet areas, and lathyrism in khesari-growing areas of Central India. Thiamine deficiency is common among people who are mostly dependent on rice. Thorough

Table 17.1 Health conditions common to Hindus

Health conditions	Causes
Malaria	Environment, lifestyle
Filariasis	Environment, lifestyle
Tuberculosis	Environment, lifestyle
Pneumonia	Environment, lifestyle
Cardiovascular diseases	Genetic, environment, lifestyle
Rheumatic heart disease	Genetic, environment, lifestyle
Sickle cell anemia	Genetic, environment, lifestyle
Dental diseases	Environment, lifestyle
Lactose intolerance	Genetic, lifestyle
Cancer of the cheek, nose, and mouth	Lifestyle
Breast cancer	Genetic, environment, lifestyle
Ichthyosis vulgaris	Genetic
Beriberi	Lifestyle
Thiamine deficiency	Lifestyle
Lathyrism	Lifestyle
Goiter	Environment, lifestyle
Osteomalacia	Environment, lifestyle
Endemic dropsy	Lifestyle
Stomach cancer	Environment, lifestyle
Fluorosis	Environment, lifestyle

milling of rice, washing rice before cooking, and allowing the cooked rice to remain overnight before consumption the following day results in the loss of thiamine.
- Lathyrism is a crippling disease-causing paralysis of leg muscles; it occurs mostly in adults who consume large quantities of seeds of the pulse khesari, *lathyrus sativus*, over a long period.
- Goiters are common along the sub-Himalayan tracts.
- Osteomalacia is prevalent in northwest India.
- Endemic dropsy is prevalent in west Bengal as a result of the use of mustard oil for cooking.
- Cancers of the mouth and lip are common because of chewing *pan* and tobacco.
- Fluorosis occurs in parts of Punjab, Haryana, Andhra Pradesh, and Karnataka, resulting from drinking water with large amounts of fluoride.

17.1.6   Nutrition

- Dietary habits are complex and regionally varied. Most believe that food was created by the Supreme Being for the benefit of man. The influence of religion is pervasive in food selection, customs, and preparation methods.
- Classification of regional food habits can be twofold, based on the types of cereals and fresh foods consumed. In the first category are rice and bread eaters; in

the second category are vegetarians and nonvegetarians. Vegetarianism is firmly rooted in culture; the term *nonvegetarian* is used to describe anyone who eats meat, eggs, poultry, fish, and sometimes cheese. Many Brahmins in North India consider eating meat to be religiously sanctioned. In some parts of India, eating fish is acceptable to Brahmins, whereas in other parts eating meat is sacrilegious. Assess cultural food choices upon admission to the health-care facility.

- Dietary staples include rice, wheat, *jowar, bajra*, jute, oilseeds, peanuts, millet, maize, peas, sugarcane, coconut, and mustard. Cereals supply 70–90% of the total caloric requirements. A variety of pulses or lentils, cooked vegetables, meat, fish, eggs, and dairy products are also consumed. Heavily spiced (curry) dishes with vegetables, meat, fish, or eggs are favored, and hot pickles and condiments are common. Spice choices include garlic, ginger, turmeric, tamarind, cumin, coriander, and mustard seed. Vegetable choices include onions, tomatoes, potatoes, green leaves, okra, green beans, and root vegetables.
- In North India wheat is the staple food. Other cereals are *jowar, bajra*, and *ragi*, consumed in porridges, gruels, and *rotis* (baked pancakes).
- *Bajra*, a staple food in Maratha families, is not considered favorably in Uttar Pradesh. People from Punjab do not favor fish, and people from the south generally dislike the idea of meat of any kind. In Saurashtra in the south, fish, fowl, flesh, and eggs are taboo practically everywhere.
- Women generally serve the food but may eat separately from men. Women are not allowed to cook during their menstrual periods or have contact with other members of the family.
- **Assess food rituals practiced by Hindus in relation to meal times and food selections before attempting dietary counseling.**
- Foremost among the perceptions of Hindus is the belief that certain foods are "hot" and others are "cold" and therefore should only be eaten during certain seasons and not in combination. The geographic differences in the hot and cold perceptions are dramatic; many foods considered hot in the north are considered cold in the south. Such perceptions and distinctions are based on how specific foods are thought to affect body functions. The belief is that failure to observe rules related to the hot and cold theory of diseases results in illness. A more detailed description of the hot and cold theory of diseases is provided under Health-Care Practices. **Given the diversity of Hindus in America, individually assess dietary practices and nutritional deficiencies of patients according to their ethnic origins and area of residence.**

17.1.7 Pregnancy and Childbearing Practices

- Birth control methods include intrauterine devices, condoms, and rhythm and withdrawal methods. **Women may desire education in family planning from a same-sex health-care provider as well as assistance with delivery from female physicians, midwives, or female nurse practitioners.**

Table 17.2 Practices during pregnancy

• Based on the hot and cold theory of disease, certain "hot" foods like eggs, jiggery (a course brown sugar), coconut, groundnut, maize, mango, papaya, fruit, and meat are avoided during pregnancy because of a fear of abortion caused by heating the body or inducing uterine hemorrhage.
• Pregnancy is a time of increased body heat; hence "cold" foods such as milk, yogurt, and fruits are considered good. Buttermilk and green leafy vegetables are avoided because of the belief that these foods cause joint pain, body aches, and flatulence. Minor swelling of the hands and feet is thought to result from increased heat and is not of much concern.
• Morning sickness is caused by an increase in body heat. Burning sensations during urination, scanty urine, or a white vaginal discharge are considered serious signs of significant overheating.
• Overeating and consumption of high-protein foods, including milk, are avoided because such foods result in an exaggerated growth of the baby that may lead to a difficult delivery.
• Anemia caused by iron deficiency is one of the nutritional disorders affecting women of childbearing age. This condition may be aggravated because of the practice of reducing the consumption of leafy vegetables to avoid producing a dark-skinned baby.

- Grandmothers, mothers, and mothers-in-law are considered to have expert knowledge in the use of home remedies during pregnancy and the postpartum period. Many older women frequently travel to the US to assist new mothers in antenatal and postnatal care that is consistent with traditional customs.
- The birth of a son is a blessing because the son carries the family name and takes care of the parents in their old age. The birth of a daughter is cause for worry and concern because of the traditions associated with dowry, a ritual that can impoverish the lives of those who are less affluent.
- Table 17.2 describes various cultural practices related to pregnancy.
- No taboo against the father being in the delivery room exists, but men are usually not present during birthing. Instead, they tend to wait outside the delivery room and allow female relatives to support the pregnant mother during labor and delivery. **Report labor progress to fathers who prefer to stay in the waiting room.**
- Because self-control is valued, women suppress their feelings and emotions during labor and delivery. **Closely observe nonverbal communication, such as a change in body posture, restlessness, and facial expressions during labor and provide assistance as necessary.**
- Beliefs and practices used during the postpartum period are discussed in Table 17.3.

17.1.8   Death Rituals

- A tenet of Hinduism is that the soul survives the death; death is a rebirth. Therefore, by performing a ritual bath, sprinkling holy river water over the body, covering the body with new clothes, daubing parts of the body with ghee, and

Table 17.3 Postpartum beliefs and practices

- During the postpartum period, the mother remains in a warm room and often keeps the windows closed to protect her against cold drafts. Exposure to air conditioners and fans, even in warm weather, may be considered dangerous. Provide warm clothing and additional blankets for the mother and baby to keep them warm.
- After the birth, both the mother and the baby undergo purification rites on the 11th day. The postpartum mother is considered to be impure and is confined to a room. The pollution lasts for 10 days. This period of necessitated and mandatory confinement assists in bonding between the mother and the newborn and provides the mother with adequate rest and time to tend to the baby's needs. The baby is officially named on the 11th day during the "cradle ceremony," and several rituals are performed to protect the baby from evil spirits and to ensure longevity.
- A sponge bath for the newborn is recommended until the umbilical cord falls off.
- Soft massage to the extremities is recommended before bathing the infant.
- Washing the infant's hair daily is believed to improve the quality of the hair.
- During the postpartum period, hot foods such as brinjals, drumsticks, dried fish, dhal, and greens are good for lactation. Cold foods are thought to produce diarrhea and indigestion in the infant. Cold foods such as buttermilk and curds, gourds, squashes, tomatoes, and potatoes are restricted because they produce gas. Such abstentions are primarily practiced for the baby's health because harmful influences might be transmitted through the mother's breast milk.
- Some believe that colostrum is unsuited for infants. Most women think that the milk does not "descend to the breast" until their ritual bath on the third day and, as a result, newborns are fed sugar water or milk expressed from a lactating woman.
- Breast milk is commonly supplemented with cow's milk and diluted with sugar water. A child's stomach is considered weak as a result of diarrhea; therefore, the child is given diluted milk.
- Sources of protein such as eggs, curds, and meat are avoided because they might adversely affect the baby.
- The mother's diet the first few days is restricted to liquids, rice, gruel, and bread.
- Boiled rice, eggplant, curry, and tamarind juice are added to the diet between 6 months and a year after the birth of the baby.
- Obtain dietary preferences and practices from the family before planning nutritional counseling.

chanting Vedic utterances, the deceased is considered purified and strengthened for the post-mortem journey.

- Hindus prefer to die at home. The eldest son is responsible for the funeral rites. In the absence of the oldest son, determine the spokesperson and decision maker for the family.
- The death rite is called *antyesti,* or last rites. The priest pours water into the mouth of the deceased and blesses the body by tying a thread around the neck or wrist. The priest may anoint with water from the holy Ganges River or put the sacred leaf from the *Tulsi* plant in the mouth. The eldest son completes prayers for ancestral souls, but all male descendants perform the rites; each offers balls of rice on behalf of the deceased.
- The body is usually cremated rather than interred. The ashes are immersed or sprinkled in the holy rivers. Such immersions are of great benefit to the souls of

the dead. Hindus may save their family's ashes to later scatter them in holy rivers when they return to their homeland.

- Women may respond to the death of a loved one with loud wailing, moaning, and beating their chests in front of the corpse, attesting their inability to bear the thought of being left behind to handle situations by themselves. **Offer support and understanding with respect for death and grief behaviors.**

17.1.9   Spirituality

- Hinduism, the largest religion and oldest tradition practiced in India, represents a set of beliefs and a definite social organization. Hinduism denotes belief in the authority of Vedas and other sacred writings of the ancient sages, immortality of the soul and a future life, existence of a Supreme God, the theory of **karma** and rebirth, the theory of the four stages of life, and the theory of four *Purusarthas*, or ends of human endeavor. *Karma* stresses the individual's responsibility for one's actions and is interpreted in terms of past life. One's present condition is seen as a result of one's actions in a past life or lives.
- Orthodox Hindus view society as divinely ordained on the basis of the four castes: (a) *Brahmin*, the highest caste, priests and scholars, emerged from the head of God; (b) *Kshtriya*, warriors, from the arms; (c) *Vaisya*, merchants, from the waist; and (d) *Sudra*, menials, from the feet of God. The four-fold caste system is a theoretical division of society to which tribes, clans, and family groups are affiliated. Although religion does not bestow the caste system with a religious sanction, the great Hindu legal codes are based on the caste system. **Identify Hindu religious beliefs and practices, and incorporate these beliefs into patients' care.**
- **Assess the extent to which religion is a part of the individual's life, how religious beliefs are related to the individual's perception of health and illness, and the individual's daily religious practices. Assessing spiritual life is essential for identifying resources and solutions for therapy.**
- Women often fast one day a week or for a lunar month to fulfill a vow made to a deity in supplication for a particular blessing.
- Shrines may be set up in the living room or the dining room but are most often located in a back room or in a closet. The shrine typically contains representations or symbols of one or more deities.

17.1.10   Health-Care Practices

- Physical examinations are especially traumatic to women who may not have experienced or heard about Pap tests and mammography examinations. **Provide female health-care providers for intimate care.**

- Most believe that illnesses attack an individual through the mind, body, and soul.
- Some believe that excessive consumption of sweets may cause roundworms and that too much sexual activity and worry are associated with tuberculosis. Others believe that diarrhea and cholera are caused by a variety of improper eating habits. **Explain factual information about parasitic and infectious diseases.**
- Suffering of any kind produces hope, which is essential to life. To maintain harmony between self and the supernatural world, the belief that one can do little to restore health by oneself provides a basis for ceremonies and rituals.
- Worshiping goddesses, pilgrimages to holy places, and pouring water at the roots of sacred trees are believed to have medicinal effects in healing the sick person and in appeasing the planets to help prevent illnesses and misfortunes.
- In *Ayurveda*, the traditional system of medicine in India, the primary emphasis is on the prevention of illnesses. Individuals have to be aware of their own health needs. One of the principles of Ayurveda includes the art of living and proper health care, advocating that one's health is a personal responsibility. The key to health is an orderly daily life in which personal hygiene, diet, work, and sleep and rest patterns are regulated. A daily routine has to be established and changed according to the season.
- A common health problem is self-medication. Those migrating to America are accustomed to self-medicating and may bring medications with them or obtain medications through relatives and friends. **Specifically ask patients about folk practices, use of over-the-counter medicines, and medicines brought from India.**
- The traditional healers, *nattuvaidhyars*, use Ayurvedic, Siddha, and Unani medical systems. These systems are all based on the Tridosha theory. The Ayurvedic system uses herbs and roots; the Siddha system, practiced mainly in the southern part of India, uses medicines; and the Unani system, similar to the Siddha, is practiced by Muslims.
- According to the Tridosha theory, the body is made up of modifications of the five elements: air, space, fire, water, and earth. These modifications are formed from food and must be maintained within proper proportions for health. A balance among three elements, or humors—phlegm or mucus, bile or gall, and wind—corresponds to three different types of food required by the body. Table 17.4 shows the three types of foods and the allopathic equivalents of diseases associated with them.
- Because of their religious beliefs of karma, Hindus may attempt to be stoic and may not exhibit symptoms of pain. Pain is attributed to God's will, the wrath of God, or a punishment from God and is to be borne with courage. **Because of stoicism, rely on nonverbal manifestations as well as verbal expressions when assessing pain.**
- Families tend to be protective of ill members. They may not want to disclose the gravity of an illness to the patient or discuss impending disability or death for fear of the patient's vulnerability and loss of hope, resulting in death. **The conflict between medical ethics and patients' values may pose a problem for**

Table 17.4 Bodily manifestations of various foods

• Heat-producing foods: *brinjals*, dried fish, green chilies, raw rice, and eggs. *Pittham* foods include cluster beans, cowgram, groundnuts (peanuts), almonds, millet, oil, and runner beans. Allopathic equivalents of *sudu* diseases include diarrhea, dysentery, abdominal pain, and scabies. Allopathic equivalents of *pittham* diseases include vomiting, jaundice, and anemia.
• Cooling foods: tomatoes, pumpkin, kul, gourds, greens, oranges, sweet limes, carrots, radishes, barley, and buttermilk. Cold, headache, chill, fever, malaria, and typhoid are allopathic equivalents of cool diseases.
• Gas-producing foods: root vegetables such as potato, sweet potato, and elephant yam; plantain; and drumstick. Joint pains, paralysis, stroke, and polio are disorders related to gas-producing foods.
• Heating and cooling effects are produced in the body and hence are not related to the temperature or spiciness of foods. An imbalance leads to disease. If too much heat is in the body from consuming heat-producing foods, then cold foods need to be eaten to restore balance.

health-care providers, who need to be cognizant of the importance of the family members' wishes and values regarding the care of their loved ones.

- The sick role is assumed without any feeling of guilt or ineptness in doing one's tasks. The individual is cared for and relieved of responsibilities for that time.
- Psychological distress may be demonstrated through somatization, which is common, especially in women. The symptoms may be expressed as headaches, a burning sensation in the soles of the feet or in the forehead, and tingling pain in the lower extremities. Because of the stigma attached to seeking professional psychiatric help, many do not access the health-care system for mental health problems. Mental illness is considered to be God's will. **Do not disclose to outsiders that patients are receiving mental health counseling.**
- No Hindu policy exists that prevents receiving blood or blood products. Donating and receiving organs are acceptable.

17.1.11 Health-Care Providers

- Although Hindus in general have a favorable attitude toward American physicians and the quality of medical care received in the US, relatives and friends are usually consulted before health-care providers.
- Physicians are considered omnipotent because God grants cures through physicians. Patients tend to be subservient and may not openly question physicians' behavior or treatment. If they are not pleased with the treatment, they just change physicians. However, they tend to be appreciative of the information that physicians provide about their illness. **Provide factual information about diseases and illnesses. Explain the dangers of "shopping around" for health-care providers.**
- The physician is also viewed like an older person in the family; a protective, authoritative, and responsible relationship; a parent-child relationship; and a

guru-chela (teacher-disciple) relationship. **Patients expect physicians to teach them about the disease and how to get cured in a friend-to-friend relationship.**

- In mental health, traditional healers, such as *Vaids*, practice an empirical system of indigenous medicine; *mantarwadis* cure through astrology and charms; and *patris* act as mediums for spirits and demons. **Specifically ask patients if they are using folk health-care providers and what treatments have been prescribed.**
- Women are especially modest, generally seeking female health-care providers for gynecologic examinations. **Respect modesty by providing adequate privacy and assigning same-gender caregivers whenever possible.**
- **Explain the procedures, provide privacy, and assign a female health-care provider to decrease the stress and discomfort associated with a pelvic examination.**

17.2 Reflective Exercises

1. What are the primary religions practiced by Hindus?
2. What are the two national languages of India? What other languages are common?
3. How are older Hindu family members addressed?
4. Explain why arranged marriages are common and preferred by Hindus.
5. What is the difference between bride price and a dowry?
6. Distinguish between *Indids* and *Melanids.*
7. What health conditions are common to Hindus?
8. Identify at least three common beliefs related to pregnancy and at least three related to postpartum?
9. What are the principles of *Ayurveda?*
10. What are the principles of the Hindu religion?

Bibliography

CIA (2016) World factbook: India. https://www.cia.gov/library/publications/the-world-factbook/geos/in.html

Heart of Hinduism: Family Structure (2014). http://hinduism.iskcon.org/lifestyle/904.htm

Hindu Beliefs Affecting Health Care (n.d.). https://www.health.qld.gov.au/multicultural/support_tools/hbook-hindu-s2.pdf

Hindu Marriages: Purpose and Significance (2015). http://www.hinduwebsite.com/marriage.asp

Jambanathan J (2013) People of Hindu heritage. In: Purnell L (ed) Transcultural healthcare: a culturally competent approach, 4th edn. F.A. Davis Company, Philadelphia

National Center for Complementary and Integrative Health (2015) Ayurvedic medicine in depth. https://nccih.nih.gov/health/ayurveda/introduction.htm

Sen CD (2009) Food culture in India. *Food cultures around the world.* Greenwood Publishiing Group, London

Chapter 18
People of Hmong Heritage

18.1 Overview and Heritage

Hmong (pronounced Mong—the "H" is silent) lived primarily in the mountainous areas shared by China, Burma, Vietnam, Thailand, and Laos. Most Hmong who live in the US are from Laos, which has a population of seven million with only 8% identifying as Hmong. The Hmong are an agrarian society that practiced "slash and burn" agriculture. Mountainous areas were cleared of underbrush, burned, and then used for crops. When the soil became depleted, they moved on, often moving their village as well. They are thought to originate in the Yellow River Valley of China.

Hmong began to immigrate to the US and other countries in 1975 after the Vietnam war. These refugees came from the mountainous regions of Laos where they had fought on the side of the CIA during the war. They were targeted for genocide because they fought against the communist Pathet Lao, and because of this, they fled their county. They escaped to Thailand through the jungles and across the Mekong River. Many Hmong died because of the war, genocide, or while attempting to flee. Many Hmong immigrants in the US bear the scars of war, bullet and shrapnel wounds, and the lasting effects of exposure to biological warfare, something they called "yellow rain."

Although no Hmong initially settled in the Central Valley of California, many now live there with California having the largest numbers of Hmong residents, followed by Minnesota and Wisconsin. Small groups still tend to migrate where they perceive that economic opportunities exist. More recently, some have settled in North Carolina, Oregon, Colorado, and other states. In 2004, the last refugee settlement in Thailand was closed and 15,000 new immigrants were brought to the US.

Hmong in Laos generally had no education; they were primarily illiterate, lived in very primitive and traditional circumstances, and had no access to the modern world or modern medicine. When immigrating to the US, many Hmong experienced shock in a world that was completely foreign to them and they could not understand. Depression, post-traumatic stress disorder, and suicide are common in

© Springer Nature Switzerland AG 2019
L. D. Purnell, E. A. Fenkl, *Handbook for Culturally Competent Care*,
https://doi.org/10.1007/978-3-030-21946-8_18

this population. More recent immigrants have had some education and exposure to the modern world.

Paj ntaub (pan dow) is a form of embroidery that Hmong women do to decorate their clothing and make historical story cloths. Story cloths were the way the family history was passed from generation to generation because literacy was uncommon. Even today, remarkable story cloths are made that show the Hmong fighting the communists in the jungles, Hmong being killed, yellow rain falling on villages, Hmong fleeing through jungles, and Hmong floating across the Mekong river to the refugee camps of Thailand. This aggregate profile may not fit well-acculturated Hmong in the US based on length of time in the US and the variant characteristics of culture (see Chap. 1).

18.1.1 Communications

- Hmong in the US speak either White or Green Hmong, sometimes called Blue. These languages may not be understandable to those who speak the other Hmong language, and the same word can have totally different meanings between the languages.
- **Have a professional interpreter who can speak the patient's language.**
- Hmong did not have a written language until the 1950s when Christian missionaries began to develop a written form of their language. Because of the lack of written Hmong language, most older Hmong are not literate in their own language.
- **Do not provide Hmong language written instructions unless someone in the household can read them. Instructions written in English may be a better choice because school-aged children may be able to read them.**
- In Laos, the Hmong had no calendars or clocks so these concepts were foreign to them. For Hmong born in Laos, their ages are often not known; an age is assigned to them by immigrant officials before they immigrate to the US. Thus, Hmong may appear older or younger than the age on official documents. Appointment times are a difficult concept and Hmong may sometimes arrive early in the morning when they have an afternoon appointment.
- **Be aware that Hmong immigrants born in Laos may not have their true age on their documents. Be flexible with appointments. Scheduling for a morning or afternoon appointment might be helpful.**
- Many Hmong believe that Americans are rude because they look directly in the eyes when speaking and they are too direct with their questions. Proper communication when speaking to a Hmong person is to use fleeting glances without staring. Making light conversation before asking questions about health is proper and important. Hmong also use the word "yes" to indicate that they hear what you are saying. Saying "yes" does not mean that they understand what you are asking or that they will do what you are asking them to do.

- **Avoid direct eye contact for long periods. Do not rush to questions; small talk first if the situation is not of an urgent nature.**
- **To determine if treatments are understood, ask for a demonstration.**
- Hmong in general have a strong desire to be seen positively by people in authority. This social desirability factor may result in them telling the health-care provider what they believe the health-care provider wants to hear, not what is actually happening; this is considered being respectful. **When questioning Hmong about their compliance with treatment recommendations, it is best to ask them to show you what they are doing, for example, how they do the blood sugar testing. For how much medication they are taking, look at the number of pills in the vial and the date the prescription was filled to determine how much was taken.**
- Obtaining informed consent is a legal requirement US, but it is directly oppositional to Hmong traditional decision-making. In the Hmong culture, the male head of the family or clan makes decisions for family members; individuals do not have the right to make decisions for themselves.
- Because the Hmong lack experience with Western medicine and surgical procedures, they often have a great deal of fear of medical situations. They do not understand what is happening and they sometimes distrust medical personnel, especially if the person is a student. Rumors persist among the Hmong that Hmong are used for practice by students and if any treatment is called experimental, this just confirms their beliefs that they are used for experiments; they will most likely refuse treatment. Needing to obtain consent from the head of a family may result in delays for treatment as well. The head of the clan may live in another state.
- **To be successful in obtaining consent, it is always important to respect the wishes of the individual, to wait until family members have arrived, to meet with the family members to explain the situation, and to accept their decision. If the patient's wishes are different from those of the head of the family, most patients will agree with the head of the family's decision.**
- Hmong believe that it is inappropriate to say negative things in front of sick people. In illiterate societies, words have great power and Hmong believe that if you say negative words in front of an ill person, speaking the words can make those things happen.
- **It is very important to have a professional interpreter who can work with the family, imparting information in a culturally congruent manner.**
- Many Hmong have achieved higher education and degrees as registered nurses, physicians, psychologists, and social workers. In Western culture, the achievement of education and position engenders respect and authority. In the Hmong culture, patients may not feel the same deference toward Hmong health-care providers and may treat them according to their position within the Hmong family/community hierarchy. This creates additional stressors for Hmong health-care providers and interpreters who may be expected to defer to the wishes of Hmong patients who have higher status in their community.

- **Be aware that Hmong health-care providers and interpreters may be related to the patient, and they could be placed in an untenable situation because of a clan hierarchy that Western individuals may not understand.**

18.1.2 Family Roles and Organization

- Hmong are organized into 18 clans; each clan has a surname that all men and children have. Although wives and mothers usually retain the clan name from which they were born, they are still considered a member of their husband's clan. Hmong have no single leader; each clan has leaders who are older males. If a husband has died, the oldest son is expected to make decisions for the family. Daughters are expected to marry and to live with their husbands and in-laws.
- Plural marriages were common in Laos and they persist today. Hmong men may marry as many women as they can afford to support. This is becoming less common with Hmong who no longer adhere to traditional beliefs.
- Single Hmong may find mates at the annual Hmong New Year celebrations held between Christmas and New Year. Marriages are thought to be advantageous if they are between first cousins, but not those who retain the same last name. Hmong young women are still victims of "capture bride," a process in which a young girl is taken to her future husband's home overnight where they are then pronounced married. This process is being discouraged because it has resulted in escape with the young women reporting the kidnapping to the police.
- Hmong boys are not considered adults until they marry. Young girls are thought to be marriageable when they become "plump" or enter puberty. This results in young Hmong girls suddenly disappearing from schools because they are now married. Although these girls may be in their early teens, their pregnancies tend to be very healthy with few complications common to teenage pregnancies. The young family lives with their in-laws so multiple generations assist in child rearing. Hmong girls who go to college are sometimes thought to be "too old" to marry and they have reduced choices of Hmong men to marry. Great social pressure exists for very young Hmong girls to marry in their early teens. These marriages are not legal marriages but traditional Hmong ceremonies; therefore, they do not break state laws.
- The Hmong community in the US is well-connected through telephone, e-mail, and frequent family visits. Family "disgraces" are widely known even among Hmong who may live in different parts of the US. "Disgraces" might be birth defects, opium addiction of a family member, or a divorce in the family. Any of these "disgraces" are considered to reflect on every member of a family and, as a result, decrease their chances of making a good marriage. **Do not disclose disgraces to family members.**
- Same-sex relationships are not condoned. **Do not disclose same-sex relationships to family members of others.**

18.1.3 Workforce Issues

- Hmong have the highest unemployment of all recent immigrants. They have the lowest socioeconomic level of all Asians in the US. Employment for first-generation immigrants is difficult because of poor English language skills and few workplace skills. Large family size results in low employment for women.
- For Hmong who gain sufficient English language skills to be employed, they may gain employment in factories or agriculture. Their desire for social acceptance may result in their saying "yes" to questions regarding knowing how to perform something when they actually do not understand. **Requesting demonstrations helps to assess knowledge so that better instructions can be given in the workplace.**
- Young Hmong are rapidly achieving higher education and higher socioeconomic status. This has resulted in their helping other members of their family to live better lives and achieve higher education. Hmong are now in many professional roles and government service. They are hard workers and are loyal to their employers.

18.1.4 Biocultural Ecology

- Hmong men average 5 ft 3 in. tall and women about 5 ft tall. Men may weigh 100–120 pounds and women 85–100 pounds. Since immigration, obesity is very common in children and adults. Skin color is light brown; faces are round with almond-shaped eyes. A few Hmong have blond hair, light skin, and hazel eyes. In Laos, this variation was considered an aberration. Children born in the US are achieving greater heights most likely as a result of improved nutrition.
- Table 18.1 lists commonly occurring health conditions for Hmong in three categories: lifestyle, environment, and genetics. See Chap. 2.
- Health conditions common to Hmong include depression, anxiety, suicide ideation, and post-traumatic stress disorder as a result of experiences during the Vietnamese war, their treacherous flight to safety, and cultural stressors. Many Hmong continue to have nightmares and flashbacks to the terrors they experienced in Laos. Unemployment and reversal of family roles create additional cultural stressors. **Encourage compliance with long-term mental health counseling and adherence to medication regimens.**
- Type 2 diabetes and hypertension are common as a result of the rapidly increasing obesity of both Hmong children and adults. Many Hmong were settled into apartments and because of being on public assistance, there was little to do each day but eat. Their traditional hardworking, agricultural lifestyle was completely changed into a sedentary Western lifestyle. The rice and vegetables they worked so hard to grow were now easily available from grocery stores. They also began to consume American foods such as sugared soft drinks and high-sugar, high-fat pastries.

Table 18.1 Health conditions common to the Hmong

Health conditions	Causes
Cervical cancer	Environment, lifestyle
Depression, anxiety, suicidal ideation, and post-traumatic stress syndrome	Environment, lifestyle
Tuberculosis	Environment, lifestyle
Hepatitis	Environment, lifestyle
Malaria	Environment, lifestyle
Parasitosis	Environment, lifestyle
Thalassemia	Genetic
Liver cancer	Environment, lifestyle
Trichinosis	Environment, lifestyle
Nasopharyngeal cancer	Environment, lifestyle
Lactase deficiency	Genetic
Diabetes	Environment, lifestyle, genetic
Hypertension	Environment, lifestyle, genetic
Asthma	Environment, lifestyle

- **When assessing nutrition, ask what liquids the patient is consuming; this is often not revealed unless asked directly and may show high consumption of sugared juices and drinks.**
- **Do not teach diet using Western measuring amounts. Measuring foods in cups and tablespoons, etc., may not be understood by Hmong patients because baking is uncommon in traditional Hmong households and they may not possess these utensils. A cup may mean a drinking cup to the patient.**
- Hmong of all ages in central California have a very high incidence of asthma. Exacerbations are common when air quality is poor. **Instructing the patients in the proper use of inhalers and other medications is important.**
- Some older Hmong may have breathing difficulties resulting from **paragonimiasis** or opium addiction. Paragonimiasis is a parasite contracted in Southeast Asia. The parasites settle in the lungs causing a diffuse infection. **Consider paragonimiasis with pulmonary infections.**
- Recent immigrants have been diagnosed with tuberculosis at high rates; some were infected with drug-resistant tuberculosis. **Screen all new immigrants for tuberculosis.**
- Hepatitis B occurs at high rates and tends to be present in all members of a family.
- **Assist family to comply with public health monitoring. Screen all new immigrants for hepatitis B. If one family member has hepatitis B, screen all family members.**
- *Helicobacter pylori* also occurs at high rates with peptic ulcers and adenocarcinoma of the gastrointestinal track occurring. Both hepatitis B and *H. pylori* transmission within families may be related to two factors. When eating, Hmong

individuals often serve themselves from communal bowls using eating utensils. When individuals are ill, traditional cupping or coin rubbing is practiced by pricking the skin to release blood; this is thought to release bad spirits. This practice uses sewing needles that are not sterilized and may transmit hepatitis B within the family.

- **Teach the family to sterilize needles and other sharp instruments before practicing coining.**
- The greatest barrier to health care for the Hmong in the US is accurate and culturally appropriate communication. **Ensure the availability of interpreters at all times.**

18.1.5 High-Risk Health Behaviors

- Opium was grown in Laos as a cash crop and was used for pain and by many older Hmong for their aging pains. In the US, opium addiction persists, but it is rare and is considered disgraceful in the Hmong community. Individuals who smoke opium are usually very thin with a cyanotic skin color. They have generalized crackles throughout all lung fields and have frequent problems with lung infections such as bronchitis and pneumonia. **Consider opium addiction with Hmong who are very thin with cyanosis and respiratory problems.**
- Failure to have regular gynecological examinations with Pap smears increases the risk of advanced cervical cancer. **Encourage Pap smears as part of routine examinations as well as other times when women are seen for health care.**
- Many individuals do not know that cigarette smoking can cause cancer. **Explain the adverse effects of cigarette smoking, encourage cessation, and network with schools to teach the hazards of smoking at a young age.**

18.1.6 Nutrition

- Hmong were primarily farmers, so many have developed small farms where they sell fruits and vegetables. Even Hmong living in apartments have small plots for vegetables, and homes may have chicken pens in the backyards.
- Rice is the primary staple of the Hmong diet. Vegetables, fish, chicken, and pork are consumed with the rice. Very hot peppers are made into a condiment that accompanies the meal. Occasionally a special dish called *laub* is made with raw pork and vegetables and spices. This increases risk for trichinosis. American diets have become preferred by many younger Hmong who prefer hamburgers and other fast foods. Hmong eat two to three meals per day. **Discourage the use of raw pork and explain the possibility of hazardous health consequences.**

18.1.7   Pregnancy and Childbearing Practices

- Children are highly valued in Hmong families, and large families are considered an asset. Most Hmong live with extended families, if not in the same household then in the same apartment complex or neighborhood. Older adults often help to take care of their grandchildren. Most women marry early and have pregnancies until menopause. Men tend to marry first when in their early 20s and 30s.
- Women are prohibited from drinking cold beverages and spicy foods during pregnancy. Post-partum, white rice and chicken are the traditional diet for 1 month and the mother may not drink cold drinks. New mothers are expected to rest after delivery. The mother-in-law and husband help the new mother. **New mothers may not want to eat hospital food; allow the family to bring traditional foods from home.**
- Hmong women consider regular menstruation to be a sign of health and may not wish to use birth control that interferes with the regular menstrual cycle.

18.1.8   Death Rituals

- Hmong funerals are distinctive and last many days. The older or more revered the person, the longer the funeral may be. Family members are expected to attend and stay the duration of the funeral. Specific rites are required along with animal sacrifice to honor the deceased. These animals are used to provide food for the people present. A proper burial enables the deceased to enter the spirit world in a positive way, a world leading to reincarnation. Most Hmong believe in multiple souls that reincarnate, one that stays in the area of the body and another that stays in the present world, overlooking and caring for the family. Rituals are conducted so that deceased ancestors are honored.

18.1.9   Spirituality

- Many Hmong who immigrated when young have adopted Christian religious beliefs. Most older and more recent immigrant Hmong hold traditional animist beliefs in which ancestors are revered and spirits are believed to be widely distributed in the world, residing in many inanimate objects or places such as trees, rivers, and houses. Many believe that spirits can cause harm, misfortune, illness, or death, or can be helpful to protect or prevent bad events from occurring. Good spirits are thought to be ancestors who watch over and protect them.
- Christian Hmong have beliefs and practices appropriate to their religion. Some may denounce animist beliefs and traditional Hmong beliefs such as soul loss and soul calling ceremonies performed by shamans.

18.1.10   Health-Care Practices

- Hmong seek Western medical care often; they may also use traditional healers or shamans who perform rituals. They may seek herbalists and take multiple treatments for the same condition. Some practice home remedies such as coining or cupping. These traditional practices cause distinctive bruises, either elongated or round, that are generally over the area where the problem is. For example, a Hmong person with a sore throat will have bruises around the neck and one with chest symptoms may have cupping over the front and back of the chest. Pricking the center of the bruise is done to release bad spirits; unsterilized sewing needles are generally used for this purpose.
- **Do not confuse the regular patterns of coining or cupping for abuse. Encourage sterilization of needles.**
- One cause of illness is thought to be soul loss. Some individuals, such as babies and children, are thought to have souls that have difficulty staying in the body. If the soul(s) leaves for too long, the baby can become ill or die. For this reason, parents may tie a string around the baby's neck or wrist soon after birth. Older children and adults may have strings tied around wrists, waists, or ankles. These must remain on until they fall off naturally, as removing them too soon can result in soul loss and illness or death. Since immigrating to the US, the strings have been replaced with gold necklaces and bracelets.
- **Hmong who practice shaman ceremonies may have amulets in small bags around their neck or waist. These contain objects thought to be protective against evil spirits. These should not be removed without the patient's or family's permission.**
- Some Hmong use herbs prescribed by a Hmong herbalist. Studies have shown that these herbs may have pharmaceutical properties. **Ask patients about their use of herbs upon admission.**
- More traditional Hmong are averse to blood transfusions and organ donation. **Carefully explain the benefits of blood transfusions when required.**
- The traditional herbs found in Laos and Thailand are difficult or impossible to obtain in the US. Traditional herbalists sometimes use plants found in the US only because the plant looks similar to a plant they had used in Laos. Efficacy of these herbs is unknown and may be harmful. **Caution patients about using herbs unknown to them.**

18.1.11   Health-Care Providers

- Shaman ceremonies are used for serious illnesses that have not responded to other treatments. Shamans do not choose to become shamans; the occupation is a calling. Shamans train for many years and learn as apprentices. Shaman ceremonies are conducted within the home with all the family present; they go into the spirit world to find out why the soul was lost or was taken to the spirit world. On a second ritual, the shaman may call the person's spirit home, thereby restoring harmony. A shaman may sacrifice an animal to pay the spirit world to release the soul so it can come home. **Amulets from the ceremony may be placed on the patient to protect them and these should not be removed.**

18.2   Reflective Exercises

1. When and under what conditions did the Hmong begin immigrating to the US?
2. What languages doe the Hmong speak. How might you get an interpreter for a Hmong who does not speak English?
3. Who among the Hmong society makes health-care decision?
4. List at least five diseases and health conditions common among the Hmong?
5. What are the symptoms of opium addition?
6. What concerns might you have for a Hmong patient who practices animism?
7. What conditions do Hmong traditional health treat?
8. Identify post-partum food practices among Hmong women.
9. What symptoms would you see on a patient practicing "cupping" and "coining"?
10. What are some stressors that lead to mental health conditions for Hmong?

Bibliography

BirthAmongHmongCulture(2011).http://sc2218.wikifoundry.com/page/Birth+in+Hmong+culture

Cartaret M (2012) Providing health care to Hmong patients. Dimensions of Culture. http://www.dimensionsofculture.com/2012/01/providing-healthcare-to-hmong-patients-and-families/

CIA (2012) World factbook: Laos. https://www.cia.gov/library/publications/the-world-factbook/geos/la.html

Duffy J, Harmon R, Ranard D, Thao B, Yang K (2004) An introduction to their history and culture. In: The Hmong culture profile no. 18. Center for Applied Linguistics, Washington, DC

Hmong in Minnesota (2016) Common medical issues and cultural concerns of Hmong patients. http://www.culturecareconnection.org/matters/diversity/hmong.html

Hmong Studies Resource Center (2016). www.hmongnet.org

Johnson S (2002) Hmong health beliefs and experiences in the Western health care system. J Transcult Nurs 13(2):127–133

Owens CW (2007) Hmong cultural profile. Ethnomed. https://ethnomed.org/culture/hmong/hmong-cultural-profile

Pinzon-Perez H (2006) Health issues for the Hmong populations in the US: issues for health educators. Int Electron J Health Edu 9:122–133

Purnell L (2013) People of Hmong heritage. In: Purnell L (ed) Transcultural health care: a culturally competent approach. F.A. Davis Co., Philadelphia

Warner ME, Mochel M (1998) The Hmong and health care in Merced, California. Hmong Stud J 2(2):1–30

Chapter 19
People of Iranian Heritage

19.1 Overview and Heritage

The number of Iranians in the US is difficult to determine, with estimates ranging from 400,000 to over one million with half of them living in California. Of the 83 million people living in Iran, over 99% are Muslim. Iranians vary enormously, from highly traditional to highly acculturated, in their reasons for leaving Iran, as well as in the variant characteristics of culture (see Chap. 1) and individuality. The terms *Persian* and *Iranian* are used interchangeably in this chapter because some people call themselves Persian for historical and political reasons. In 1935, the country's name was changed from Persia to Iran to present an image of progress and in an attempt to unify into one nation the enormous diversity of urban dwellers and rural tribes, ethnic groups, and social classes.

Between 1930 and 1970, women were allowed to go unveiled and gained access to university education. After conservative Moslem protests that led to the 1978 violence, martial law was declared and an Islamic government was established. Since then, Iran resembles the more conservative Muslim countries.

Iranians are among the most highly educated immigrant group in the US. Education is greatly valued, and advanced degrees are highly respected. Children are expected to do well in school and to attend college. Many middle-aged physicians, engineers, professors, army generals, and government officials who were unable to find comparable work in the US have gone into business for themselves. Jobs that require manual labor are not respected, and some cannot accept taking such menial jobs often because of their limited skills in English.

© Springer Nature Switzerland AG 2019
L. D. Purnell, E. A. Fenkl, *Handbook for Culturally Competent Care*,
https://doi.org/10.1007/978-3-030-21946-8_19

19.1.1   Communications

- **Farsi** (Persian) is the national language of Iran. Nearly half the population speaks different languages: Turkish, Kurdish, Armenian, Baluchi, or other Iranian dialects. Well-educated and well-traveled immigrants may speak three or more languages, often using French as a cultural language or English in business settings.
- Many foreign invasions and strict control by each ensuing government have influenced Iranians' interaction with outsiders, making them suspicious of foreigners. The disclosure of personal thoughts to strangers is generally perceived to have detrimental consequences. Not verbalizing one's thoughts is considered a customary and useful defense; overt expressions of emotions to strangers may be culturally stigmatized.
- **Establish trust before asking questions of a highly personal nature.**
- Among women, patterns of speech may appear restrained or refined to avoid too much self-disclosure or to prevent loss of face. Most women give considerable detail rather than use blunt and succinct messages.
- Most Iranians are very concerned with respectability, a good appearance of the home, and a good reputation. They are embarrassed by financial troubles and even conceal such problems from relatives. *Zerangi* (cleverness) is valued but only among non-intimates; it means knowing how to manipulate bureaucratic structures.
- The influence of intimate versus public spheres and hierarchical social relationships is seen clearly in the practice of **ta'arof** (ritual expressed courtesy), which is not practiced with intimates. *Ta'arof* expresses the public face, with its respectful forms of speech and behavior, and is used when dealing with individuals whose status is unequal to one's own.
- Communication is structured by social hierarchy and varies according to **baten** and **zaher**. *Baten* (inner self) is the true vulnerable self, a collection of freely expressed personal feelings. In contrast, *zaher* is proper and controlled behavior, a public face to protect and buffer the vulnerable world of the *baten* within. Silence is used to guard confidential matters and to manage impressions.
- Most Iranians refrain from showing anger or other strong emotions to outsiders because self-control is valued; showing anger can produce embarrassment, shame the family, or damage someone's reputation. To avoid embarrassment and to save face, they may agree or say that they understand an outsider, whether they do or not.
- **Because politeness and saving face prevail, do not assume that a positive response means a definite "yes."**
- **At the beginning of any health-care encounter, the provider should take time to "warm up" with social conversation if possible before "getting down to business." Any kind of bad news must be handled carefully by revealing it gently and gradually, in several meetings if possible, or only to the family spokesperson. A person should never be given bad news alone; for example, being informed of a death or serious diagnosis.**

- Among the more traditional, men and women do not hold hands or show affection toward each other in public. However, women often show affection for women and men for men by walking hand in hand or greeting each other with a kiss on each cheek.
- Strangers and health-care providers are greeted with both arms held at the sides or with a handshake. A slight bow or nod while shaking hands shows respect. **Offer something, e.g., a prescription, with both hands, shows respect.**
- Slouching in a chair or stretching one's legs toward another person is considered offensive. It is considered rude to show the sole of one's foot. Beckoning is done by waving the fingers with the palm down. **Do not beckon patients; call them formally by name.**
- Tilting the head up quickly means "no." Tilting the head to the side means "what?" and tilting it down means "yes." Extending the thumb (as in "thumbs up") is considered a vulgar sign. **Refrain from using nonverbal behaviors that may be offensive.**
- Personal distance is generally closer than that of European Americans. Respect for a provider's role and education might be demonstrated by keeping a wide distance.
- **Do not take offense at patients who stand closer to you than you wish.**
- Most individuals maintain intense eye contact between intimates and equals of the same gender, but the traditional tend to avoid eye contact. Younger people and those of lower status do not sustain eye contact with those they perceive as being older or of higher status.
- Temporal relationships are a combination of present and future orientation. **The future orientation enhances the effectiveness of health education.**
- Iranians are not clock-watchers; rather, they are mood- and feeling-oriented. While social time is flexible, Iranians meet the expectations for timeliness in work and appointments. **Stress the importance of timeliness for appointments in the US.**
- In highly traditional families, privacy demands that husbands do not mention their wife's first name to other men.
- Close friends and children may be called by their first name by family members, but others should not do so. Shaking hands with a child shows respect to the parents.
- **A man should wait for a woman to extend her hand first. One is expected to greet every member of the family with a handshake. Heed nonverbal cues when greeting Iranians.**

19.1.2 Family Roles and Organization

- Iranian culture is patriarchal. The father rules the family and expects obedience and submission from family members. In the father's absence, the oldest son has authority.

- Families were traditionally large in Iran, with male children being highly desirable. In more traditional families, older male siblings have the authority to make decisions about younger siblings, even in the father's presence.
- **Never underestimate the power of this hierarchy or of the family in general in decision-making. Traditional individuals usually cannot or will not make decisions.**
- Men see their role as protecting and providing for the family, managing the finances, and dealing with matters outside the home. Women maintain the home; even working women may place their priorities on the family and home.
- In traditional families, both daughters and sons stay at home until married, and in rural areas marriages are often arranged. In the US, young people are free to select their own marriage partners, but families usually want to approve because of the importance of marriage alliances between families.
- Children are expected to be loyal to their families and behave respectfully toward older people. Manners are considered important even outside the home, where children are expected to be clean, well behaved, and refrain from rowdiness.
- Girls are expected to behave and dress more modestly than boys, especially as they approach adolescence. Children and teens are usually included in adult gatherings. Young children are rarely left with babysitters.
- Taboo behaviors for teens include smoking, drugs, alcohol, and sex, in that order.
- Young women are expected to remain virgins until they marry, but sexual activity by men outside marriage is tolerated.
- Family members often live close to each other so they can visit whenever possible. Most maintain strong intergenerational involvement with grandmothers, mothers, and children in the domestic sphere and with male family members. **Take advantage of this living situation when health care becomes a concern for Iranians.**
- Iranians often remove their shoes at the door and wear slippers. Outside the home, they tend to dress conservatively, but more traditional or religious women avoid bright colors, cover their arms and legs, and conceal their heads with head covers or scarves (*hijab*). In Iran, these coverings are mandatory.
- Age is a sign of experience, worldliness, and knowledge. Regardless of kinship or relationship, an older person is treated with respect. When grandparents reach an age when they can no longer care for themselves, they live with and are cared for by the family. Caring for older people is an obligation.
- Older and more traditional Iranians may be uncomfortable with unrelated members of the opposite sex. Although divorce is viewed negatively, its rate has been increasing among Iranians in the US. Cultural mores also advocate ignoring or denying minor marital discord and accepting suffering for the sake of maintaining family stability.
- In Iran, out-of-wedlock teen pregnancy is neither talked about nor prevalent, and it can have a devastating outcome. If it happens in the US, it may be taken care of quietly to preserve the face of the family. **Do not disclose pregnancy outside of marriage to family members.**

- Homosexuality is highly stigmatized and not discussed; gays and lesbians remain "in the closet." Since 1979, when the legal and religious systems became synonymous, homosexuality, which is considered unnatural and sacrilegious, has been a capital offense punishable by death. **Do not expose lesbian and gay relationships to family members or others.**

19.1.3 Workforce Issues

- Iranians may perceive and actually experience a degree of bias at work. More acculturated immigrant professionals respond flexibly in the workplace. Most newcomers may not be familiar with American vernacular or slang. **Do not use idiomatic expressions and slang with Iranian employees.**
- To avoid embarrassment and to save face, they may agree or say that they understand an outsider, whether they do or not. **Because politeness and saving face prevail, do not assume that a positive response means a definite "yes."**
- Personal distance is generally closer than that of Americans or northern Europeans.
    Do not take offense if spatial distancing is closer than what you are accustomed to.
- Most individuals maintain intense eye contact between intimates and equals of the same gender, but the traditional tend to avoid eye contact. Younger people and those of lower status do not sustain eye contact with those they perceive as being older or of higher status.
- **Do not assume that intense eye contact is anger or that avoiding eye contact means that the person is not listening, does not care, or is being less than truthful.**

19.1.4 Biocultural Ecology

- As white Indo-Europeans, coloring ranges from blue or green eyes, light-brown hair, and fair skin to nearly black eyes, black hair, and brown skin. **Because of the variations in skin color, assess jaundice and anemia in Iranians by examining the sclera and oral mucosa rather than by relying solely on skin assessments.**
- Table 19.1 lists commonly occurring health conditions for Iranians in three categories: lifestyle, environment, and genetics (see Chap. 2).
- **Screen newer immigrants for cholera, malaria, and parasitic disease. Screen for glucose-6-phosphate dehydrogenase before administering primaquine. Avoid fava beans, which can cause a hemolytic crisis.**

Table 19.1 Health conditions common to Iranians

Health conditions	Causes
Malaria	Environment, lifestyle
Bacterial meningitis	Environment, lifestyle
Hookworm/parasites	Environment, lifestyle
Cholera	Environment, lifestyle
Ischemic heart disease	Genetic, environment, lifestyle
Birth defects	Genetic, lifestyle
Hemophilia	Genetic
Thalassemia	Genetic
Glucose-6-phosphate dehydrogenase	Genetic
Dubin-Johnson syndrome	Genetic

19.1.5   High-Risk Health Behaviors

- Smoking is more prevalent in Iran than in the US among both men and women. **Encourage decreasing or stopping smoking and network with elementary school teachers to teach smoking hazards at a young age.**
- Some alcohol and street-drug misuse occur in the immigrant population, but the rate is no higher than that of the population at large.
- Alcohol is prohibited by the Qur'an (Holy Book), although many are not religious and drink socially, a few to excess. In some, years of opium use in Iran can create both psychological and physical addiction. Family responses to drug use range from complete support of the family member to disownment.
- Most comply with seat-belt and child-restraint laws, valuing the safety of their children.

19.1.6   Nutrition

- Food is a symbol of hospitality. Iranian food is flavorful and takes hours to prepare. Presentation is important.
- Food is classified into two categories, **garm** (hot) and **sard** (cold), which sometimes corresponds to high-calorie and low-calorie foods. The key is balance. At any given table, there is usually a pleasing mixture of foods of different colors and ingredients, composed of a balance of *garm* and *sard*. Too much of one category can cause symptoms of being "overheated" or "chilled." Symptoms are treated by eating food from the opposite group. Becoming overheated, sweating, itching, and rashes may result from eating too many hot foods, such as walnuts, onions, garlic, spices, honey, or candy. The stomach may become chilled, causing dizziness, weakness, and vomiting, after eating too many cold foods, such as grapes, rhubarb, plums, or yogurt. **Seek assistance from family members in selecting foods while in an inpatient facility.**

- Iranians prefer to use only the freshest foods, although in some cases cost is a factor in their using some dried herbs.
- Hot tea is the most popular drink among Iranians. The most common starchy foods are rice and wheat bread. Bread is usually baked flat like pita. Corn and potatoes are used but less favored. Beans and legumes, for example, lentils, pinto, mung, kidney, lima and green beans, and split and black-eyed peas make up a fairly high proportion of the dietary intake and are commonly used in rice mixtures.
- Dairy products are dietary staples, particularly eggs, milk, yogurt, and feta cheese as well as dairy by-products, such as *doog* (yogurt soda) and *kashk* (milk by-product).
- Favorite meats are beef, chicken, fish, and lamb. Shellfish is sometimes eaten. Fresh fruit is always found in Iranian houses.
- Green leafy vegetables are used in cooking, and herbs such as parsley, cilantro (coriander), dill, fenugreek, tarragon, mint, savory, and green onions are served fresh at a meal.
- Islam has a strict set of dietary prescriptions *(halal)* and proscriptions *(haram)*. Slaughter of poultry, beef, and lamb must be done ritually to make the meat *halal*. **Strict Muslims avoid pork and alcoholic beverages; a few avoid shellfish.**
- Food is eaten with the right hand, and foods or objects are passed with the right hand alone or with both hands. **Make adjustments to accommodate traditional food practices; make provisions for the family to bring food from home if the patient prefers. Incorporate Iranian foods and dietary practices into health teaching to improve compliance with special dietary restrictions.**

19.1.7 Pregnancy and Childbearing Practices

- Menstrual blood is believed to be ritually unclean and physically polluting to the body. Menstruating women are not allowed to touch holy objects or to have intercourse. Menstruation is also considered a time of great fragility when a woman should not exercise or shower excessively, because these activities might cause a hemorrhage. At the end of the menses, the woman must wash and purify herself thoroughly before partaking in any religious rituals.
- Birth control methods used by Iranians include the pill, intrauterine devices, and natural methods. Vasectomies are beginning to gain acceptance.
- A woman's prestige is at its height when she delivers her first child, particularly if the child is a boy. Delivering the first child gives the young wife a more respected and cherished position with her in-laws. **Providing factual information regarding family planning is one area in which health-care providers can improve family care.**
- Food cravings must be satisfied lest a miscarriage occurs from not meeting the fetus's needs.

- Pregnant women avoid fried foods and foods that cause gas; fruits and vegetables are recommended, with special attention to the balance of hot *(garm)* and cold *(sard)* foods.
- In more traditional families, the father is usually not present at the birth. **Do not make the father feel guilty if he chooses not to be at the delivery.**
- To strengthen the postpartum woman, a *ghorse kamar* (a brown, flat disk of dried herbs) is mixed with eggs and placed on her lower back a few hours before bathing.
- New mothers avoid cold water for bathing, ablutions, or cleaning, although they now bathe sooner than the traditional first 30 days.
- Baby boys are considered "hotter" than baby girls, and mothers of sons are considered to have hotter bodies, hotter milk, and hotter temperaments than mothers of girls.
- Mothers of girls are given a mixture of honey and other nutrients, an herbal extract called *taranjebin*, to raise their bodily heat to ensure that the next child is a boy.
- Some families keep an infant home for the first 40 days, at which time the baby is strong enough to fight off environmental pathogens.
- The baby is given a ritual bath between the 10th and 40th days.
- **Ask the childbearing family about prescriptive, restrictive, and taboo cultural practices that they customarily follow, and incorporate nonharmful practices.**

19.1.8   Death Rituals

- Muslims may need only a gentle reminder that death is not a termination of life but rather the beginning of a new and better life. **To discuss termination of life support with a practicing Muslim, begin the conversation by noting God's will for human beings and His power over our destiny, followed by a dialogue about life and death as necessary steps toward immortal life in heaven.**
- Some individuals may oppose stopping life support, viewing it as "playing God," even though they may have no objection to beginning life support, viewing it as the "gift" of medical technology. Some families may demand that strenuous efforts be made to prolong life. **Assess each patient and family individually about life-prolonging medical treatments.**
- Among religious Muslims, the deathbed should be turned to face Mecca so family members can read prayers from the Qur'an to ensure that the dying person hears this at the time of death. **Determine the direction of Mecca so that the dying person can face Mecca.**
- After death, another Muslim should wash the body in a ritual manner, using soap and water and proceeding from head to toe and front to back. All body orifices must be closed and slightly packed with cotton to prevent leakage of bodily fluids (considered unclean). The final rinse is performed with water. The body is

then wrapped in a special white cotton shroud. Prayers and verses from the Qur'an are read during the procedure.

- **A non-Muslim health-care provider should wear gloves when touching the body.**
- Cremation is not practiced in Iran; it is unlikely to practice in the US.
- No specific religious rules exist against autopsy. However, the body of the dead is to be respected. **The reason for the autopsy must be made clear, and some may still refuse. Explain legal requirement of autopsy in the US.**
- Iranians may express their grief over the death of a loved one by crying, wailing loudly, or even striking themselves or an object.

19.1.9 Spirituality

- Specific Muslim practices include prayers, read in the name of one of the 12 imams, to provide peace of mind or to plead for a miracle. Devout Muslims pray five times daily and need privacy and water for ritual washing. **Assist the patient to make arrangements for praying.**
- Most Muslims fast from sunrise to sundown during Ramadan, although pregnant women and the ill are exempt from fasting. **Arrange meals and medications to accommodate Ramadan. See Chap. 8 for Ramadan celebration dates through 2021.**
- Family relationships and friendships are sources of strength and meaning in life for many individuals. **Adjust visiting policies to accommodate family and friends.**
- Death is not a finalization; rather, it is a graduation to a higher level of being.

19.1.10 Health-Care Practices

- Traditional health beliefs and therapeutic processes are a combination of Galenic (humoral), Islamic (sacred), and modern biomedicine. In classic humoral theory, illness arises from an imbalance (excess or deficiency) in the basic qualities such as hot/cold or wet/dry. The purpose of treatment is to restore balance.
- In Galenic-Islamic thought, every individual has a distinctive balance of four humors, resulting in a unique temperament.
- Sacred medicine is from the Qur'an and *hadith*, and holy men are considered able to heal. The sacred tradition includes beliefs in the "evil eye" and *jinns* as disease agents as well as healing by means of manipulating impurity.
- *Narahati* is a general term used to express a wide range of undifferentiated, unpleasant emotional or physical feelings, such as feeling depressed, uneasy, nervous, disappointed, or not fully well. Many individuals somaticize in a sub-conscious effort to communicate *narahati* or distress that cannot be otherwise expressed verbally.

- **By somaticizing, they construct an illness that is culturally sanctioned and socially understood.**
- When *narahati* is caused by fright, the evil eye, or *jinns*, it may be treated with religious cures. If it is considered caused by problems of blood, nerves, or humoral imbalance, it is treated with herbs or biomedicine.
- *Ghalbam gerefteh (narahatiye qalb)* or distress of the heart may be expressed as a feeling that the heart is being squeezed and can range in severity from mild excitation of the heart or palpitations to fainting and heart attack.
- A sudden ailment or symptom of puzzling origin may be attributed to the evil eye, or ***cheshm-i-bad***, which is the belief that the eyes of another person can cause illness. *Cheshm-i-bad* can be unintentional (enthusiastically complimenting someone without saying "In the name of God") or intentional (cast out due to jealousy or enmity).
- Most individuals expect immediate relief or cure from the health-care system and may shop around until they find a provider they like. At the same time, they may seek advice from those who can suggest herbal remedies. **Explain the dangers of "shopping around" and not disclosing all treatments and home therapies used.**
- Most Iranians practice self-medication and use prescription and over-the-counter medications as well as homemade herbal remedies. Antibiotics, codeine-based analgesics, mood-altering drugs in the benzodiazepine family, and intramuscular vitamins are available over-the-counter in Iran, and immigrants often bring these medications with them to the US.
- **Ask patients in a nonjudgmental manner about full disclosure of prescription and nonprescription medicines and herbal treatments.**
- Self-adjusting dosage of prescribed medications is common. **Explain the dangers of self-adjusting medications.**
- When ill, most individuals are inclined to be passive and cared for by health-care providers and family members.
- When an individual is hospitalized, family members are expected to visit frequently or to be present continuously until the patient is discharged from the hospital or is fully functioning again. **Self-care should be implemented by encouraging family members to assist with care of the sick person.**
- Common herbal remedies include dried flowers, seeds, leaves, and berries, steeped in hot or cold water and drunk for a variety of purposes, such as digestive problems, "cleaning the blood or kidneys," coughs, aches and pains, fevers, nerves, or fear. Some common herbal medications include *gol-i-gov zabon* (dried foxglove flowers) for an imbalance in the digestive system or nervous upsets, which is sometimes taken with *nabat*, a concentrated sugar. *Khakshir* (flat, brown rocket seed) is used for stomach problems and "dirty blood"; *razianeh* is used for halitosis; quince seeds are sucked for sore throats; and *sedr* prevents and treats dandruff. *Neshasteh* (wheat starch) is combined with boiling water and drunk for sore throats or coughs and is also used to stop diarrhea. Mint extracts are used to relieve excess stomach gas. *Shatareh* is thought to cure fever.

- Most Iranians are expressive about their pain. Some justify suffering in light of later rewards.
- Mental illness is highly stigmatized and is thought to be genetic. If a family member has a mental illness, it is likely to be called a "neurological disorder" to avoid stigmatizing the family. Psychotherapeutic help may be avoided either because of stigma or because it is perceived as irrelevant.
- Rehabilitation, such as physical therapy and music therapy, is embraced as the way to bring the disabled into the mainstream.
- Blood transfusions, organ donations, and organ transplants are widely practiced. In Iran, donation of organs is often a business transaction—if a kidney is needed, it is purchased.

19.1.11 Health-Care Providers

- Many immigrants think that Iranian doctors make more authoritative and quicker diagnoses, using minimal technology, even if they may be uncertain or wrong. American physicians, who are tentative, ask the patient to describe the problem, or order too many tests may be viewed as incompetent.
- Most Iranians prefer to be cared for by providers of the same sex. Iranian women are modest in front of men. Some very traditional families may consider taking a female patient elsewhere or avoid care until the situation is acute, if only male providers are available. **Male health-care providers should not ask women to undress fully for an examination or procedure.**
- The most respected health-care provider is an experienced, middle-aged to elderly male physician, with several degrees, and preferably with a high position in the hospital or university. He is considered the authority and is expected to act like one, making diagnoses quickly and prescribing remedies that cure the patient.
- Male nurses or women who have grey hair and positions of authority are accorded more respect than young, single, female nurses. Nurses who encourage self-care may be perceived as uncaring or even incompetent. **Provide factual information about the benefits of self-care as a method of rehabilitation.**

19.2 Reflective Exercises

1. What is the national language of Iranians? What other languages spoken by Iranians?
2. What is *zerangi* and how is it displayed?
3. How should traditional Iranian men and women be addressed?
4. Describe Iranian family hierarchy and decision-making practices.

5. How are Iranian gays and lesbians seen in Iran? What ramifications might this have in the US?
6. What are the primary health conditions among Iranians?
7. Explain *garm* and *sard*? What might you do if you are unfamiliar with this category of foods?
8. An Iranian patient is near death. How do you discuss this situation with the Iranian family?
9. How would you vary meals and medication administration during Ramadan?
10. What is *narahati,* what is its significance, and what are its symptoms?

Bibliography

Bahrami H, Sadatsafavi M, Pourshams A, Kamangar F, Nouraei M, Semnani S, Brennan P, Boffetta P, Malekzadeh R (2006) Obesity and hypertension in an Iranian cohort study; Iranian women experience higher rates of obesity and hypertension than American women. BMC Public Health (6):158. https://bmcpublichealth.biomedcentral.com/articles/10.1186/1471-2458-6-158

CIA (2016) World factbook: Iran. https://www.cia.gov/library/publications/the-world-factbook/geos/ir.html

Global Affairs Canada (2014) Cultural information—Iran. https://www.international.gc.ca/cil-cai/country_insights-apercus_pays/ci-ic_ir.aspx?lang=eng

Hafizi H, Hafizi H (2013) People of Iranian heritage. In: Purnell L (ed) Transcultural health care: a culturally competent approach, 4th edn. F.A. Davis Company, Philadelphia

Heisey DR (2011) Iranian perspectives on communication is the age of globalization. http://web.uri.edu/iaics/files/03D.RayHeisey.pdf

Intimate Spaces: Coming Out in Iran (2015) The guardian. https://www.theguardian.com/world/iran-blog/2015/jun/11/iran-gay-coming-out-intimate-spaces

Iran guide: a look at Iranian language, culture, customs, and etiquette (2014) http://www.commisceo-global.com/country-guides/iran-guide

Iranian Americans: Immigration and Assimilation (2014) Public Affairs Alliance of Iranian Americans. http://www.paaia.org/CMS/Data/Sites/1/pdfs/iranian-americans%2D%2D-immigration-and-assimilation.pdf

Mahdi AA (1988) Ethnic identity among second-generation Iranians in the United States. Int Soc Iran Stud 31(1):77–95

Chapter 20
People of Japanese Heritage

20.1 Overview and Heritage

Nihon, or **Nippon**, as Japan is called in the Japanese language, is a 1200-mile chain of islands in the northwestern Pacific Ocean. Japan borders Russia, Korea, and China, and its modern history has, until recently, been shaped by conflict with these countries. Japanese citizens residing in North America tend to locate in large commercial and educational centers.

Education is highly valued; the literacy rate in Japan is nearly 100%. About 40% of all young people go on to higher education. The alumni network primarily provides job placements; the school one attends determines to a great extent where one is employed after graduation. *Issei* (first-generation Japanese immigrants) vary widely in their English-language ability. *Nisei* (second-generation immigrants) and *sansei* (third-generation immigrants) were educated under the American educational system to the extent that they were permitted; for example, educational access was limited or segregated during the World War II internment of American citizens of Japanese ancestry. Acculturation, individuality, and the variant characteristics of culture (see Chap. 1) determine the degree of traditionality of Japanese in the US and elsewhere. Therefore, this profile will not fit all Japanese.

20.1.1 Communications

- Japanese is the language of Japan, with the exception of the indigenous Ainu people.
- Because high school graduates in Japan complete 6 years of English, even newer Japanese immigrants and sojourners can speak, understand, read, and write English to some extent.

© Springer Nature Switzerland AG 2019
L. D. Purnell, E. A. Fenkl, *Handbook for Culturally Competent Care*,
https://doi.org/10.1007/978-3-030-21946-8_20

- **Whereas the language barrier may be an obstacle to verbal instructions or explanations in English-speaking health-care settings, Japanese patients are likely to use written materials effectively.**
- Men tend to speak more coarsely and women with more gentility or refinement.
- Light social banter and gentle joking are a mainstay of group cohesiveness.
- Polite discussion unrelated to business, often over *o-cha* (green tea), precedes business negotiations. Relationship-building and respect for personal privacy are important aspects of working relationships in all sectors.
- Open communication is discouraged, making it difficult to learn what people think. In particular, saying "no" is considered extremely impolite; rather, one should let the matter drop.
- A high value is placed on "face" and "saving face." Asking someone to do something he or she cannot do induces loss of face or shame. For people to be shown wrong is deeply humiliating. People feel shame for themselves and their group, but they are respected when they bear shame in stoic silence. Because of ethnic homogeneity, an ingrained sensitivity to the feelings of others, and close contact with one's family, classmates, and work group, vague and intuitive communication is well understood by fellow group members.
- **Giving a yes answer does not necessarily mean yes as it does in the Western context.**
- Traditional Japanese exhibit considerable control over body language. Anger or dismay may be quite difficult for Westerners to detect.
- Smiling and laughter are common shields for embarrassment or distress.
- Prolonged eye contact is not polite even within families.
- Social touching occurs among group members but not among people who are less closely acquainted. In general, body space is respected.
- Intimate behavior in the presence of others is taboo, although this is changing among the younger generations and in the US.
- When people greet one another, whether for the first time or for the first time on a given day, the traditional bow is performed. The depth of the bow, its duration, and the number of repetitions reflect the relative status of the parties involved and the formality of the occasion.
- An offer to shake hands by a Westerner is reciprocated graciously.
- Overall orientation is toward the future. Punctuality is highly valued.
- Family names are stated first, followed by given names. Seki Noriko would be the name of a woman, Noriko, of the Seki family. The family names of both men and women, married or single, are designated by the suffix *-san*, but one does not use that designation when referring to oneself.
- Women generally assume their husband's family name upon marriage.
- Elders are referred to respectfully. The designation *sensei* (master) is a term of respect used with the names of physicians, teachers, bosses, or others in positions of authority.
- **Greet Japanese patients with a handshake, and address them formally by their last name until invited to do otherwise.**

20.1.2 Family Roles and Organization

- The predominant family structure is nuclear. The role of wife and mother is dominant.
- Children are socialized to study hard, make their best effort, and be good group members. They are taught to take care of each other, and girls are taught to take care of boys. Self-expression is not valued.
- **Be aware of differences in spousal relationships when assessing the quality of family dynamics and communication, sexual health, and sensitivity to risk for sexually transmitted infections.**
- The primary relationship within a family is the mother-child relationship, particularly that of mothers and sons. It is customary for a mother to sleep with the youngest child until that child is 10 years old or older, and when a new baby is born, the older sibling may sleep with the father or a grandparent. **Be aware of Japanese family sleeping practices and refrain from judgmental evaluation.**
- Babies are not allowed to cry; they are picked up instantly. Women constantly hold their babies in carriers on their chests and sleep with them.
- Corporal punishment is acceptable in Japan. **Explain U.S. child abuse laws to patients.**
- Traditional teens and college students generally do not date. They typically join clubs, membership in which is taken seriously; most social activities, such as ski trips, are club activities. Japan has the lowest incidence of teen birth in the world. **Do not assume that dating holds the amount of concern for Japanese young people that it does for American teens. Do not assume that Japanese youth are well informed about sexuality and sexual health risks.**
- Older people are respected and cared for by the family in the home, if at all possible, with the eldest son being the responsible family member. **Be sensitive to Japanese patients' sense of obligation and commitment to older people. Help families network within the Japanese American community for both social support and for resources or for good long-term care facilities.**
- In Japan, a small segment of women have long lived outside the usual constraints for their gender. Women of "the floating world," or the entertainment industry, enjoy a fair amount of autonomy. The most traditional of these, the *geisha*, live in all-female communal arrangements. Geisha are not prostitutes, and they are now recognized as a cultural treasure. **Geisha entertainment industries exist in North America.**
- Tolerance for marriage of a Japanese person to a foreigner is less than in the US.
- A gay and lesbian social network is evident, although they are not generally talked about. **Do not disclose same-sex relationships to family members.**

20.1.3   Workforce Issues

- American practices designed to avoid liability, such as informed consent, are not routinely implemented in Japanese health-care settings. **Explain American practices such as informed consent, the legal and professional requirements of patient autonomy, and the legalities of defensive charting on orientation.**
- **Explain to new employees about the direct communication practices of the American culture.**
- Japanese workers are sensitive to colleagues and superiors. Saying "no" or delivering bad news is extremely difficult; they may avoid issues or indicate that everything is fine rather than state the negative. **Ask less acculturated Japanese to demonstrate procedures instead of asking directly if they understand the procedure.**
- A high value is placed on "face" and "saving face." Asking someone to do something he or she cannot do induces loss of face or shame. For people to be shown wrong is deeply humiliating. **Use tactful communication and approach sensitive and controversial topics carefully to allow the person to "save face."**
- Prolonged eye contact is not polite even within families and among friends. **Do not assume that lack of eye contact means the person is not listening, is not caring, or is being less than truthful.**

20.1.4   Biocultural Ecology

- Racial features include the epicanthal skin folds that create the distinctive appearance of Asian eyes, a broad and flat nose, and "yellow" skin that varies markedly in tone. Hair is straight and naturally black, with differences in shade. **Rely on color changes in the mucous membranes and sclerae to assess oxygenation and liver function.**
- Negative blood types account for less than 1% of the population.
- Table 20.1 lists commonly occurring health conditions for Japanese in three categories: lifestyle, environment, and genetics.
- Asthma, related to dust mites in the *tatami* (straw mats that cover floors in Japanese homes), is one of the few endemic diseases. **These same conditions can affect patients in the US.**
- Drug dosages may need to be adjusted for the physical stature of Japanese adults. Many Asians are poor metabolizers of mephenytoin and related medications, potentially leading to increased intensity and duration of the drugs' effects. **Be aware that most Asians tend to be more sensitive to the effects of some beta blockers, many psychotropic drugs, and alcohol.**
- Most individuals require lower doses of some benzodiazepines and neuroleptics.
- Opiates may be less effective analgesics, but gastrointestinal side effects may be greater than among whites. **Take patients' body mass into consideration in dosing; even with that precaution, patients' responses to drugs need to be monitored carefully.**

Table 20.1 Health conditions common to Japanese Americans

Health conditions	Causes
Heart disease	Genetic, environment, lifestyle
Tuberculosis	Environment, lifestyle
Renal disease	Lifestyle
Asthma	Environment, lifestyle
Vogt-Koyanagi-Harada syndrome	Genetic
Takayasu disease	Genetic
Acatalasemia	Genetic
Cleft lip/palate	Genetic, lifestyle
Oguchi disease	Genetic
Lactase deficiency	Genetic
Stomach cancer	Genetic, environment, lifestyle

20.1.5 High-Risk Health Behaviors

- Smoking rates are high among Japanese and Japanese Americans. **Begin teaching smoking cessation in elementary school.**
- Alcohol (rice wine) is part of many social rituals, such as picnics, to celebrate cherry blossoms, autumn leaves, or moon viewing. Adults commonly drink beer and sake in the home, and children may be seen purchasing alcohol.
- Once alcohol is consumed, people can relax and speak freely; they are forgiven for what they say because of the alcohol. **Be aware of the prevalence of smoking and heavy alcohol consumption, particularly among men. Give specific medical reasons why they must abstain, thus providing a socially acceptable excuse.**
- Mothers' time-honored strategy of rewarding academic diligence with candy and other treats contributes to the issue of the fitness of youth. **Encourage mothers to offer other rewards instead of candy and non-nutritious treats for academic excellence.**
- Public safety consciousness is high; Japanese readily use seat belts and other safety measures such as child safety seats and helmets.

20.1.6 Nutrition

- Dietary staples include rice, beef, poultry, pork, seafood, root vegetables, cabbage, persimmons, apples, and tangerines.
- Rice is the mainstay of the traditional diet and is included in all three meals as well as snacks. A traditional breakfast includes fish; pickles; *nori* (various seaweeds used to flavor or garnish meals); a raw egg stirred into the hot rice; *miso* (soybean-based) soup; and tea. Some people prefer a Western breakfast of toast or cold cereal and coffee.

- Rice has a symbolic meaning related to the **Shinto** religion, analogous to the concept of the "bread of life" among Christians. A staple of schoolchildren's *o-bento* (lunch box) is a white bed of rice garnished with a red plum pickle, reminiscent of the Japanese flag. Meals combine elements of land and sea.
- Schoolchildren eat lunch or their *o-bento*, packed with rice, pickles, and meat or fish. A popular lunch among working people is a steaming bowl of ramen (noodles) in broth or cold noodles on a hot summer day. Instant broth, although high in sodium, is another popular quick lunch.
- A traditional dinner is a pot of boiled potatoes, carrots, and pork seasoned with sweet sake, garlic, and soy sauce or a stir-fried meat and vegetable dish.
- The daily intake of sweets can be high and often includes European-style desserts, pastries and cookies, sweet bean cakes, soft drinks, and heavily sweetened coffee, which may contribute to the high incidence of tooth decay.
- Increasingly, westernized food tastes, resulting in higher fat and carbohydrate intake, have contributed to the rise in obesity and associated increases in diabetes, heart disease, and premature death.
- There is growing public awareness that the sodium content of the traditional soups and sauces contributes to the high rate of cerebrovascular accidents. **Undertake dietary assessment on admission. Incorporate traditional food practices and preparation practices in dietary counseling sessions.**
- Green tea, although high in caffeine, is a good source of vitamin C. Garlic and various herbs are used widely for their medicinal properties.
- Many individuals have difficulty digesting milk products because of lactose intolerance. **Encourage use of reduced-lactose milk, tofu, and unboned fish to meet calcium needs.**
- **Iron deficiency anemia is a concern among young women and can be alleviated with dietary counseling or dietary supplements. Nori is a traditional food source for iron.**

20.1.7 Pregnancy and Childbearing Practices

- Oral contraceptives became legal in Japan in 1999. Condoms remain the most common contraceptive method. Most women have several abortions during their married lives.
- Pregnancy is highly valued within traditional culture as a woman's fulfillment of her destiny. Women may enjoy attention and pampering that they get at no other time. They may prepare themselves for the possibility of pregnancy when they become engaged and eliminate alcohol, caffeine, soft drinks, and tobacco.
- Women often return to their mother's home for the last 2 months of their pregnancy and through the first 2 months postpartum. **Explore a woman's expectations during pregnancy and the possibility that she might return to Japan.**

Finding another Japanese woman who has experienced childbearing in the US and who can share her experiences would support the pregnant patient.

- Loud noises, such as a train or a sewing machine, are thought to be bad for the baby.
- Shinto shrines sell amulets for conception and easy delivery. Husbands do not commonly attend the births of their children. **Do not make the husband feel guilty if he does not want to be in the delivery room.**
- Vaginal deliveries are usually performed without medication. To give in to pain dishonors the husband's family, and mothers are said to appreciate their babies more if they suffer in childbirth. Japan enjoys one of the world's lowest infant mortality rate, at 2.1 per 1000 live births.
- Traditionally, postpartum women do not bathe, shower, or wash their hair for at least the first week.
- Breast-feeding is taken seriously. Maternal rest and relaxation are deemed essential for success. If the mother is asleep, the grandmother feeds the baby formula. Women who give birth in the US may resent the American expectation that they will resume self-care and child-care activities quickly. **Explain the expectations for postpartum care; exercise sensitivity and help plan for assistance upon early discharge.**

20.1.8 Death Rituals

- The taboo against open discussion of serious illness and death is evident. Hospice patients may not want to be told their prognosis in order to allow a peaceful death and to spare both the patient and the family the difficulty of having to discuss the situation.
- **Hospice and palliative care professionals may have to rely on implicit behaviors to admit patients for care.**
- When a person is dying, the family should be notified of impending death so they can be at the dying person's bedside. Traditionally, the eldest son has particular responsibility during this time.
- When death occurs, an altar is constructed in the home. Photographs of the deceased are displayed, and floral arrangements are placed within and outside the home. A bag of money is hung around the neck of the deceased to pay the toll to cross the river to the hereafter. Visitors bring gifts of money and food for the bereaved family.
- The mourning period is 49 days, the end of which is marked by a family prayer service and the serving of special rice dishes. Beliefs are common that the dead need to be remembered and that failure to do so can lead the dead to rob the living of rest; therefore, special prayer services can be conducted for the 1st, 3rd, 7th, and 13th annual anniversaries of the death.

20.1.9   Spirituality

- Shinto, the indigenous religion, is the locus of joyful events such as marriage and birth.
- Many festivals are marked by offerings, parades, and a carnival on the grounds of the shrine.
- Buddhism has permeated artistic and intellectual life. Very few people regularly attend services, but most are registered temple members, if only to ensure a family burial plot.
- Two percent of Japanese people are Catholic or Protestant. Most do not identify themselves solely with one religion or another, and even a baptized Christian might have a Shinto wedding and a Buddhist funeral.
- Many accept the Buddhist belief in reincarnation, and the eternal life of the soul is also recognized in the Shinto faith.
- *Kampo* (healers) often set up shop in the vicinity of the temple or shrine, and a person might be seen scooping incense smoke onto an ailing body part. Prayer boards might bear requests for special healing. Newborns are taken to a shrine for a blessing. Additional blessings take place on November 15, when a child is 3, 5, or 7 years old. Visits to shrines and temples are social, recreational, and spiritual outings. Souvenirs and refreshments are usually available, and the hike into the prayer area provides exercise.

20.1.10   Health-Care Practices

- There is great tolerance for self-indulgence even during minor illnesses. Because Japanese people are less likely to express feelings verbally, this indulgence may be a way for people to affirm caring for one another nonverbally. Termination of pregnancy when the health of the fetus is in doubt is common, and most parents want medically compromised neonates not to be treated aggressively when prognoses are not favorable.
- **Help patients with legal obligations who face value conflicts with termination of pregnancy and not treating medically compromised newborns.**
- The concept of *ki*, the life force or energy and how it flows through the body, is integral to traditional Chinese healing modalities, including acupuncture. Good health requires the unobstructed flow of *ki* throughout the body.
- The concepts of *yin* and *yang* are reflected in modern attitudes, as is the need to balance five energy sources: water, wood, fire, earth, and metal. Strategies that help to restore balance include use of herbal medicines, bed rest, bathing, and having a massage. One traditional form of massage, *shiatsu* (acupressure), involves redirection of energy along the Chinese meridians by application of light pressure to what we might recognize as acupuncture points.
- Whereas Chinese tradition calls for a restoration of balance when one is ill, Shinto calls for purging and purification. Preoccupation with germs and dirt is not likely to interfere with daily life.

Table 20.2 Common Japanese therapies

• *Morita* therapy is an indigenous strategy for addressing *shinkei shitsu*, excess sensitivity to the social and natural environment. Introspection is seen as harmful. *Morita* therapy focuses on constructive physical activity to help patients accept reality as it is
• *Naikan* therapy is one of reflection on how much goodness and love are received from others

- **Health-care providers who visit Japanese homes should note or even ask whether the family removes their shoes upon entering.**
- Many pharmacies stock traditional herbal *kampo* preparations.
- Most individuals make liberal use of both modern medical and traditional providers of health care. Residents in the US have Internet and mail-order access to traditional medications, if they are not available locally. **A complete health assessment includes inquiry about home therapies.** Table 20.2 list common Japanese therapies.
- **Be sensitive to family issues that may underlie illnesses among patients. If psychotherapy is indicated, the therapist must be someone familiar with the Japanese culture. Resources may be obtained through Japanese churches, or other religious organizations where Japanese people gather.**
- Japanese high regard for the status of physicians decreases the likelihood of their asking questions or making suggestions about their care. The idea that patients should be given care options may be alien. **Provide ample opportunity for dialogue, and explain the choices that are offered. Use Japanese and Japanese American health-care providers to bridge gaps in understanding.**
- Some individuals may need assistance in seeking care. Their verbal English skills may be an impediment to making their needs known and to understanding the care they are offered, even though their ability to understand written information may be very good.
- *Itami* (pain) may not be expressed, and bearing pain is considered a virtue and a matter of family honor. Addiction is a strong taboo in Japanese society, making patients reluctant to accept pain medication.
- **Use a schedule of analgesic administration rather than an as-requested or patient-controlled approach to ensure adequate pain management. Explain that physiological status and healing are actually enhanced by pain control.**
- Mental illness carries a significant stigma. Because emotional problems cannot be discussed freely, somatic manifestations are common and acceptable.
- Handicapped people may bring shame or heartache to the family, although they are treated kindly.
- Assumption of the sick role is highly tolerated by families and colleagues, and a long recuperation period is encouraged.
- Donating blood is encouraged in Japanese society.
- Critical care and organ transplantation and donation issues need to be approached sensitively. People rely more heavily on the physician's opinion, and the family may have difficulty negotiating cessation of treatment.
- **The concept of advance directives is unknown to many. Carefully explain advance directives to patients.**

20.1.11   Health-Care Providers

- Physicians, referred to as *sensei*, are highly esteemed.
- Self-care as a philosophy is not evident among most. Being told what to do by the physician or **kampo** health-care provider is expected, and his (or, occasionally, her) authority is not questioned.
- Nurses and nursing are highly regarded, reflecting traditional taboos against illness and impurity as well as the status of women. Currently in Japan, nurses are well respected, even though women in general are not. In the past, nurses were not highly respected because "good women" did not touch people with an illness unless they were immediate family members. If she did touch "sick bodies," the woman would become tainted and less pure.
- Home care and the orchestration of many community-based providers may be resisted by residents who expect long recuperations in the hospital. **Explain how the health-care delivery system works in the US and the functions of the different health-care providers.**

20.2   Reflective Exercises

1. Japanese use the term *sensei* when referring to older people. What does this term mean?
2. Distinguish between male and female communication styles.
3. What does "saving face" mean? How is "saving face" used when communicating with health-care providers?
4. List and explain at least three genetic health conditions common among the Japanese.
5. What is the word for Japanese traditional healers? When might one purchase Japanese herbs?
6. What terms are used to distinguish there generations of Japanese?
7. What is the traditional religion of Japanese? What are some of its principle tenets?
8. Explain the concepts of *yin* and *yang* when referred to in health care.
9. Explain *morita* and *naikan* therapies. For what conditions are they used?
10. What is *itami* and how it is expressed?

Bibliography

CIA (2016) World factbook: Japan. https://www.cia.gov/library/publications/the-world-factbook/geos/ja.html

Duranto PA, Nakayama S (2005) Japanese communication avoidance, anxiety, and uncertainty during initial encounters. https://www2.uni-hamburg.de/oag/noag/noag2005_5.pdf

21.1.2 Family Roles and Organization

- The family is the core of society, and the needs of all family members are respected.
- Jewish school-age children typically attend Hebrew school at least two afternoons a week after public school throughout the school year.
- Children play an active role in most holiday celebrations and services. Respecting and honoring one's parents is one of the Ten Commandments. Children should be forever grateful to their parents for giving them the gift of life.
- In Judaism, the age of majority is 13 years for a boy and 12 for a girl, at which age children are deemed capable of differentiating right from wrong and capable of committing themselves to performing the commandments. Recognition of adulthood occurs during a religious ceremony called a *bar* or *bat mitzvah* (son or daughter of the commandment). This rite of passage is usually accompanied by a family celebration.
- The goal of the Orthodox family is to live their lives as prescribed by *halakhah* (Code of Jewish Law), which emphasizes maintaining health, promoting education, and helping others.
- Ultra-observant women must physically separate themselves from all men during their menstrual periods and for 7 days after. No man may touch a woman or sit where she sat until she has been to the *mikveh*, a ritual bath, after her period is over.
- Older people receive respect, especially for the wisdom they have to share. Honoring one's parents is a lifelong endeavor and includes maintaining their dignity by feeding, clothing, and sheltering them, even if they suffer from senility.
- The Bible, as interpreted by the Orthodox, prohibits homosexual intercourse; it says nothing specifically about sex between lesbians. Some of the objections to gay and lesbian lifestyles include the inability of these unions to fulfill the commandment of procreation and the possibility that acting on the recognition of one's homosexuality could ruin a marriage. The liberal movement within Judaism, however, supports legal and social equality for lesbians and gays. **Do not disclose same-sex relationships to others.**

21.1.3 Workforce Issues

- Jews who observe the Sabbath desire to have off Friday evening and Saturday.
- **They may work on the Sabbath in critical jobs such as health care when needed. They may work on Sundays.**
- Jewish staff should be allowed to request off the major Jewish holidays. **Be sensitive to the needs of Jewish staff and recognize the holiness of the Sabbath, taking into account current U.S. Equal Employment Opportunity Commission laws.**
- Judaism's beliefs are congruent with the values that American society places on the individual and family. For some newer Jewish immigrants English may pose

a challenge. **Speak clearly and ask for clarification of understanding from those for whom English is a second language.**

- Most Jews believe in autonomy and willingly take leadership positions in the workforce when the need arises.

21.1.4   Biocultural Ecology

- Skin coloring for Ashkenazi Jews ranges from fair skin and blonde hair to darker skin and brunette hair. Sephardic Jews have slightly darker skin tones and hair coloring, similar to those from the Mediterranean area. There also are Jewish groups throughout Africa who are black, most notably the *Falasha* from Ethiopia.
- Genetic risk factors vary based on whether the family emigrated from Ashkenazi or Sephardic areas. There is a greater incidence of some genetic disorders among Ashkenazi individuals. Most of these disorders are autosomal-recessive, meaning that both parents carry the affected gene.
- Table 21.1 lists commonly occurring health conditions for Jews in three categories: lifestyle, environment, and genetics.

Table 21.1   Health conditions common to Jews

Health conditions	Causes
Tay-Sachs disease	Genetic
Gaucher's disease	Genetic
Canavan's disease	Genetic
Familial dysautonomia	Genetic
Torsion dystonia	Genetic
Niemann-Pick disease	Genetic
Bloom syndrome	Genetic
Fanconi's anemia	Genetic
Mucolipidosis IV	Genetic
Lactase deficiency	Genetic
Werdnig-Hoffmann disease	Genetic
Kaposi sarcoma	Genetic
Phenylketonuria	Genetic
Ataxia-telangiectasia	Genetic
Metachromatic leukodystrophy	Genetic
Myopia	Genetic
Polycythemia vera	Genetic
Cardiovascular diseases	Genetic, environment, lifestyle
Diabetes mellitus	Genetic, environment, lifestyle
Breast cancer	Genetic, environment, lifestyle
Ovarian cancer	Genetic, environment, lifestyle
Inflammatory bowel disease	Genetic, environment, lifestyle
Colorectal cancer	Genetic, environment, lifestyle

- Ashkenazi Jews have a higher rate of side effects with clozapine; 20% develop agranulocytosis.
- **Test for agranulocytosis when Jewish patients are prescribed clozapine.**

21.1.5 High-Risk Health Behaviors

- Any substance or act that harms the body is not allowed. This includes smoking, suicide, illegal medications, and permanent tattooing.
- Most Jews are health-conscious and practice preventive health care, with routine physical, dental, and vision screening. This is also a well-immunized population. **Encourage these positive health-promotion and disease-prevention practices.**

21.1.6 Nutrition

- For those who follow the dietary laws, much attention is given to the slaughter, preparation, and consumption of food.
- Perhaps the food identified as "Jewish" that receives the most attention is chicken soup, which has frequently been referred to as "Jewish penicillin" and is often served with *knaidle* balls in it (dumplings made of *matzoh* meal). Although it has no intrinsic meaning or religious value, it is a staple in religious homes, especially on Friday evenings to usher in the Sabbath and during times of illness. It is frequently associated with a mother's warmth and love.
- Common foods include *gefilte* fish (ground freshwater fish molded into oblong balls and served cold with horseradish); *challah* (braided white bread); *kugel* (noodle pudding); blintzes (crepes filled with a sweet cottage cheese); chopped liver (served cold); *hamentashen* (a triangular pastry with different types of filling); and lox and bagel sandwiches. Lox is cold smoked salmon, served with cream cheese and salad vegetables, on a bagel.
- Religious laws regarding permissible foods are referred to as **kashrut**. The term **kosher** means "fit to eat." Foods are divided into those that are considered *kosher* (permitted or clean) and those considered **treyf** (forbidden or unclean). A permitted animal may be rendered *treyf* if it is not slaughtered, cooked, or served properly. All blood is drained from the animal before eating it.
- Milk and meat may not be mixed together in cooking, serving, or eating. To avoid mixing foods, utensils used to prepare foods and the plates used to serve them are separated, requiring two sets of dishes, pots, and utensils: one set for milk products and the other for meat. Because glass is not absorbent, it can be used for either meat or milk products, although religious households still usually have two sets. Therefore, cheeseburgers, lasagna made with meat, and grated cheese on meatballs and spaghetti are unacceptable. Milk cannot be used

in coffee if served with a meat meal. Nondairy creamers can be used instead, as long as they do not contain sodium caseinate, which is derived from milk.

- Some foods are *bris* (neutral) and may be used with either dairy or meat dishes. These include fish, eggs, anything grown in the soil (vegetables, fruits, coffee, sugar, and spices), and chemically produced goods.
- Mammals are considered clean if they meet the other requirements for their slaughter and consumption and have split (cloven) hooves and chew their cud. These animals include buffalo, cattle, goats, deer, and sheep. The pig is an example of an animal that does not meet these criteria. Although liberal Jews decide for themselves which dietary laws they will follow, many still avoid pork and pork products out of a sense of tradition and symbolism.
- **Serving pork products to a Jewish patient, unless specifically requested, is insensitive.**
- Poultry is acceptable as well as fish if it has both fins and scales. Nothing that crawls on its belly is allowed, including shellfish, tortoises, and frogs.
- In religious homes, meat is soaked and salted to drain all the blood from the flesh. Broiling is acceptable, especially for liver, because it drains the blood.
- A U with a circle around it is the seal of the Union of Orthodox Jewish Congregations of America and is used on food products to indicate that they are kosher. A circled K and other symbols may also be found on packaging to indicate that a product is kosher. **Do not bring food into the house without knowing whether the patient adheres to kosher standards. If the patient keeps a kosher home, do not use any cooking items, dishes, or silverware without knowing which are used for meat and which are used for dairy products.**
- **Health-care providers must fully understand the dietary laws so they do not offend the patient and can plan medication times accordingly.**
- Care must be taken in serving cheese to ensure that no animal substances are served at the same time.
- Breads and cakes made with lard are *treyf*, and breads made with milk or milk by-products (for example, casein) cannot be served with meat meals.
- Eggs from non-kosher birds, milk from non-kosher animals, and oil from non-kosher fish are not permitted.
- Butter substitutes are used with meat meals. Honey is allowed because it is produced from the nectar of flowers.
- Kosher meals are available in most hospitals. They arrive on paper plates and with sealed plastic utensils. For in-patient facilities that do not have readily available kosher capabilities, frozen meals can be ordered online from a number of places.
- **Do not unwrap the utensils or change the foodstuffs to another serving dish. If health-care providers have difficulty locating a supplier of kosher foods, they should contact a local rabbi. Determining a patient's dietary preferences and practices regarding dietary laws should be done during the admission assessment. Frozen kosher meals can be ordered online if the health-care facility does not have a kosher kitchen.**

- One must always wash one's hands before eating. Religious Jews wash their hands while reciting a prayer. **Make facilities for patients to wash their hands before eating.**
- During the week of Passover, no bread or product with yeast may be eaten. *Matzoh* (unleavened bread) is eaten instead. Any product that is fermented or that can cause fermentation (souring) may not be eaten. Rather than attend synagogue the Jewish house of worship, the family conducts the service, *sedar*, around the dinner table during the first two nights and incorporates dinner into a service that includes all participants and retells the story of Moses and the exodus from Egypt.
- The Jewish calendar has a number of fast days. The most observed is the holiest day of the year, *Yom Kippur* (Day of Atonement). Jews abstain from food and drink as they pray to God for forgiveness for the sins they have committed during the past year. They eat an early dinner on the evening the holiday begins and then fast until after sunset the following day.
- Ill people, the elderly, the young, pregnant and nursing women, and the physically incapacitated are absolved from fasting and may need to be reminded of this exception to Jewish law.
- **If concerns arise about fasting, a consultation with the patient's rabbi may be necessary.**
- **Schedule appointments and treatment to avoid religious days.**

21.1.7   Pregnancy and Childbearing Practices

- Couples who are unable to conceive should try all possible means to have children, including infertility counseling and interventions, and egg and sperm donation. Orthodox opinion is virtually unanimous in prohibiting artificial insemination when the semen donor is not the woman's husband. When all-natural attempts have been made, adoption may be pursued.
- Unless pregnancy jeopardizes the life or health of the mother, contraception is not looked on favorably among the ultra-Orthodox. Condom use is supported, especially when unprotected sexual intercourse poses a medical risk to either spouse.
- To the Orthodox, barrier techniques are not acceptable because they interfere with the full mobility of the sperm in its natural course. The birth control pill does not result in any permanent sterilization, nor does it prevent semen from traveling its normal route. Therefore, use of this method is the least objectionable to most branches of Judaism. Sterilization implies permanence, and Orthodox Jews probably oppose this practice, unless the life of the mother is in danger. **Reform Judaism allows free choice in fertility control.**
- The fetus is not considered a living soul or person until it has been born. Birth is determined when the head or "greater part" is born. If the physical or mental health

of a pregnant woman is endangered by the fetus, all branches of Judaism consider the fetus an aggressor and require an abortion. Random abortion is not permitted by the Orthodox branch because the fetus is part of the mother's body and one must not do harm to one's body. Reform Judaism believes that a woman maintains control over her own body and that it is up to her whether to abort a fetus.

- A Hasidic husband may not touch his wife during labor and may choose not to attend the delivery because he is not permitted to view his wife's genitals. These behaviors should never be interpreted as insensitivity.
- **During the delivery in an ultra-Orthodox family, initiate the following interventions. Give the mother a hospital gown that covers her in the front and back to the greatest extent possible. Provide a surgical cap so that her head remains covered (because the hair is considered a private part of her body). Give the father the opportunity to leave during procedures and during the birth or, if he chooses to stay, drape the mother so that the husband may sit by his wife without viewing her perineum, including by way of mirrors or other means.**
- Pain medication during delivery is acceptable.
- For male infants, circumcision, which is both a medical procedure and a religious rite, is performed on the eighth day of life by a ***mohel***, an individual trained in the circumcision procedure, asepsis, and the religious ceremony. Although a rabbi is not necessary, it is also possible to have the procedure completed by a physician with a rabbi present to say the blessings.
- Attending a *brit milah* is the only mitzvah for which religious Jews must violate the Sabbath so that the brit can be completed at the proper time.
- **Provide a space for a family celebration if the circumcision is done in the hospital.**

21.1.8   Death Rituals

- Traditional Judaism believes in an afterlife where the soul continues to flourish, although many dispute this interpretation. A dying person is considered a living person in all respects.
- Active euthanasia is forbidden for religious Jews. Passive euthanasia may be allowed depending on its interpretation. Nothing may be used or initiated that prevents a person from dying naturally or that prolongs the dying process. **However, anything that artificially prevents death (cardiopulmonary resuscitation, ventilators, and so forth) may possibly be withheld, depending on the wishes of the patient and his or her religious views. The dying person should not be left alone.**
- Taking one's own life is prohibited. To the ultra-religious, suicide removes all possibility of repentance.
- Any Jew may ask God's forgiveness for his or her sins; no confessor is needed. Some Jews feel solace in saying the *Sh'ma* in Hebrew or English. This prayer

confirms one's belief in one God. At the time of death, the nearest relative can gently close the eyes and mouth, and the face is covered with a sheet.

- The body is treated with respect and revered for the function it once filled. **Ask the closest relative of the deceased specifically about the practices to follow after death.**
- For the ultra-Orthodox, after the body is wrapped, it is briefly placed on the floor with the feet pointing toward the door. A candle may be placed near the head. However, this does not occur on the Sabbath or holy days.
- Autopsy is usually not permitted among religious Jews because it results in desecration of the body, and it is important that the body be interred whole. Allowing an autopsy might also delay the burial, something that is not recommended. On the other hand, autopsy is allowed if its results would save the life of another patient who is immediately at hand. Many branches of Judaism currently allow an autopsy if (a) it is required by law; (b) the deceased person has willed it; or (c) it saves the life of another, especially an offspring. **Explain legal requirements for an autopsy.**
- Cremation is prohibited because it unnaturally speeds the disposal of the dead body. Embalming is prohibited because it preserves the dead. However, in circumstances when the funeral must be delayed, some embalming may be approved. Cosmetic restoration for the funeral is discouraged.
- Funerals and burials usually occur within 24–48 h after the death. The funeral service is directed at honoring the departed by only speaking well of him or her. It is not customary to have flowers either at the funeral or at the cemetery. The casket should be made of wood with no ornamentation. The body may be wrapped only in a shroud to ensure that the body and casket decay at the same rate. There is no wake or viewing. The prayer said for the dead, *kaddish*, is usually not said alone.
- After the funeral, mourners are welcomed at the home of the closest relative. Outside the front door is water to wash one's hands before entering, which is symbolic of cleansing the impurities associated with contact with the dead. The water is not passed from person to person, just as it is hoped that the tragedy is not passed. At the home, a meal is served to all the guests. This "meal of condolence" or "meal of consolation" is traditionally provided by the neighbors and friends.
- *Shiva* (Hebrew for "seven") is the 7-day period that begins with the burial. *Shiva* helps the surviving individuals face the actuality of the death of the loved one. During this period when the mourners are "sitting *shiva*," they do not work. When health-care providers are the ones experiencing the loss, it is important for supervisors to understand the mourning customs. In some homes, mirrors are covered to decrease the focus on one's appearance; no activity is permitted to divert attention from thinking about the deceased; and evening and morning services may be conducted in the closest relative's home. Condolence calls and the giving of consolation are appropriate.
- Crying, anger, and talking about the deceased person's life are acceptable.
- A common sign of grief is the tearing of the garment that one is wearing before the funeral service.

- In liberal congregations, a black ribbon with a tear in it is a symbolic representation of mourning. During *shiva*, the mourner sets the tone and initiates the conversation. Because there are such discrete periods of mourning, Judaism tells the mourner that it is wrong to mourn more than 30 days for a relative and for more than 1 year for parents.
- Mourning is not required for a fetus that is miscarried or stillborn. This is also true of any premature infant who dies within 30 days of birth. However, parents are required to mourn for full-term infants who die at birth or shortly thereafter.
- Within Orthodoxy, when a limb is amputated before death, the amputated limb and blood-soaked clothing are buried in the person's future gravesite because the blood and limb were part of the person. No mourning rites are required. **In the case of an amputation, assist with arrangement for burial of the body part as needed.**

21.1.9   Spirituality

- Jews consider only the Old Testament as their Bible. Judaism is a monotheistic faith that believes in one God as the creator of the universe. No physical qualities are attributed to God; making and praying to statues or graven images are forbidden.
- The spiritual leader is the rabbi (teacher). He (or she, in liberal branches) is the interpreter of Jewish law. All Jews pray directly to God. They do not need the rabbi to intercede, to hear confession, or to grant atonement.
- Some of the major principles that guide Judaic bioethics are shown in Table 21.2.
- The practice of Judaism spans a wide spectrum. Although there is only one religion, there are three main branches or denominations of Judaism. The Orthodox are the most traditional. They adhere most strictly to the *halakhah* of traditional Judaism and try to follow as many of the laws as possible while fitting into American society. They observe the Sabbath by attending the synagogue on

Table 21.2   Principles of Judaic bioethics

• Man's purpose on earth is to live according to certain God-given guidelines
• Life possesses enormous intrinsic value, and its preservation is of great moral significance
• All human lives are equal
• Our lives are not our own exclusive private possessions
• The first five books of the Bible, also known as the five books of Moses, are handwritten in Hebrew on parchment scrolls called *Torah*. These scrolls are kept in the "Holy Ark" within each synagogue under an "eternal light." The Torah directs Jews on how they should live their lives; it provides guidance on every aspect of human life
• The 613 commandments within the Torah and the oral law derived from the biblical statutes determine Jewish law. These commandments ask for a commitment in behavior and also address ethical concerns. The commandments reflect the will of God, and religious Jews feel it is their duty to carry them out to fulfill their covenant with God

Friday evening and Saturday morning and by abstaining from work, spending money, and driving on the Sabbath. Orthodox Jews observe the Jewish dietary laws; men wear a *yarmulke* or *kippah* (head coverings) at all times in reverence to God. Women wear long sleeves and modest dress. In many Orthodox synagogues, the services are primarily in Hebrew, and men and women sit separately.

- Ultra-Orthodox men wear a special garment under their shirts year-round.
- A *mezuzah* is a small container with scripture inside. Jewish homes have a mezuzah on the doorpost of the house. Some Jews wear a mezuzah as a necklace. Other religious symbols include the Star of David, a six-pointed star that has been a symbol of the Jewish community, and the *menorah* (candelabrum).
- Whereas Conservative Jews observe most of the *halakhah*, they do make concessions to modern society. Many drive to the synagogue on the Sabbath, and men and women sit together. Many keep a kosher home, but they may or may not follow all of the dietary laws outside the home. Women are ordained as rabbis and are counted in a *minyan*, the minimum number of 10 that is required for prayer. While a *yarmulke* is required in the synagogue, it is optional outside of that environment.
- The liberal or progressive movement is called Reform. Reform Jews claim that post-biblical law was only for the people of that time and that only the moral laws of the **Torah** are binding. They practice fewer rituals, although they frequently have a mezuzah on the doorpost of their homes, celebrate the holidays, and have a strong ethnic identity.
- **Reform Jews may or may not follow the Jewish dietary laws, but they may have specific unacceptable foods (for example, pork), which they abstain from eating.**
- Of the many small groups of ultra-Orthodox fundamentalists, the **Hasidic** (or **Chasidic**) Jews are perhaps the most recognizable. They usually live, work, and study within a segregated area. They are usually easy to identify by their full beards, uncut hair around the ears (*pais*), black hats or fur *streimels*, dark clothing, and no exposed extremities.
- Women, especially those who are married, also keep their extremities covered and may have shaved heads covered by a wig and often a hat as well.
- A relatively new denomination, **Reconstructionism**, is a mosaic of the three main branches. It views Judaism as an evolving religion of the Jewish people and seeks to adapt Jewish beliefs and practices to the needs of the contemporary world.
- The Jewish house of prayer is called a synagogue, temple, or *shul*. Jews may pray alone or may pray as a group anywhere that ten Jews, over the age of 13 who have had their bar/bat mitzvah, are gathered together for prayer. This group is called a minyan. Orthodox Jews pray three times a day: morning, late afternoon, and evening. They wash their hands and say a prayer on awakening in the morning and before meals. Religious patients in hospitals may want their prayer items (*yarmulke* or *kippah, tallit, tzitzit, tefillin*) and may request a minyan. **Disregard hospital policies regarding the number of visitors in the sick person's room in such instances.**

Table 21.3 Jewish holidays: 2019–2023

Jewish holidays begin at sundown the evening before the date recorded on this type of calendar; holidays end at sundown on the date shown					
Holiday	2019	2020	2021	2022	2023
Rosh Hashanah	10/30	9/19	9/7	9/26	9/16
Yom Kippur	10/9	9/28	9/16	10/5	10/25
Sukkot	10/14	10/3	9/21	10/10	9/30
Chanukah	12/23–30	12/11–18	11/29–12/6	12/19–26	12/8–15
Passover	4/20	5/9	3/28	4/16	4/23
Shavuot	6/9	5/29	5/17	6/5	5/27

Retrieved from Jewish Federation of Greater San Gabriel and Pomona Valleys file:///E:/Eight%25 20Year%2520Calendar%2520of%2520Jewish%2520Holy%2520Days%2520%2520-% 25202018-2026.pdf

- The Sabbath begins 18 min before sunset on Friday. During this time, religious Jews do no manner of work, including answering the telephone, operating any electrical appliance, driving, or operating a call bell from a hospital bed. If an Orthodox patient's condition is not life-threatening, medical and surgical procedures should not be performed on the Sabbath or holy days. A gravely ill person and the work of those who need to save him or her are exempted from following the commandments regarding the Sabbath. **Because the orthodox cannot summon an elevator, patient care facilities with a Jewish population should consider a "Sabbath elevator" that automatically stops on every floor.**
- Table 21.3 provides a list of Jewish holidays for the years from 2019 through 2023

21.1.10   Health-Care Practices

- All denominations recognize that religious requirements may be laid aside if a life is at stake or if an individual has a life-threatening illness. **Hospice and palliative care are fully consonant with Jewish beliefs.**
- In ultra-Orthodox denominations of Judaism, taking medication on the Sabbath that is not necessary to preserve life may be viewed as "work" and is unacceptable. This belief may result in some people with conditions such as asthma not recognizing the severity of their condition; they may also be unaware of the laws that allow them to take their necessary medications. **Teach patients about the potential life-threatening sequelae of their condition as well as the exceptions to Jewish law that permits them to take their medications.**
- The verbalization of pain is acceptable and common. Individuals want to know the reason for the pain, which they consider just as important as obtaining relief from pain.
- The sick role for Jews is highly individualized and may vary among individuals according to the severity of symptoms.

- Judaism opposes discrimination against people with physical, mental, and developmental conditions. The maintenance of one's mental health is considered just as important as the maintenance of one's physical health. Mental incapacity has always been recognized as grounds for exemption from all obligations under Jewish law.
- Jewish law considers organ transplants from four perspectives: those of the recipient, the living donor, the cadaver donor, and the dying donor. Because life is sacred, if the recipient's life can be prolonged without considerable risk, then transplant is ordained. For a living donor to be approved, the risk to the life of the donor must be considered. One is not obligated to donate a bodily part unless the risk is small. Examples include kidney and bone marrow donations. Organ donation at the time of death is acceptable if it saves a person's life. **Conservative** and **Reform** Judaism approve using the flat EEG as the determination of death so that organs, such as the heart, can be viable for transplant. Burial may be delayed if organ harvesting is the cause of the delay. **Assist patients to obtain a rabbi when making a decision regarding organ donation or transplant.**
- The use of a cadaver for transplant is usually approved if it is to save a life. No one may derive economic benefit from the corpse. Use of skin for burns is also acceptable, although no agreement has been reached on the use of cadaver corneas.

21.1.11 Health-Care Providers

- Physicians are held in high regard. Although physicians must do everything in their power to prolong life, they are prohibited from initiating measures that prolong the act of dying.
- Only supportive care is required while the patient is dying and includes such care as food and water, good nursing care, and maximal psychosocial support.
- **More traditional Orthodox patients prefer that care be delivered by a same-sex health-care provider.**

21.2 Reflective Exercises

1. What is meant by *halakhah* and what does it emphasize?
2. Name and describe at least three genetic conditions common among Jews.
3. What is meant by *kashrut*? Distinguish between kosher and *treyf.*
4. What are the tenets of the four broad branches of Judaism? What are the primary differences among them?
5. When does Sabbath begin and end?
6. Under what conditions are observant Jews able to work?
7. What are some of the major principles that guide Judaic bioethics?

8. What is meant by *shiva*? Describe its practices.
9. An orthodox Jewish couple is expecting their first child. The expectant father wants to be in the delivery room with his wife. What preparations do you need to make for this to occur?
10. The hospital where you work does not have a kosher kitchen. Where might you purchase kosher meals? Once you have a kosher meal, how do you handle the food?

Bibliography

Ben Ben-Tzion Halevi HY, Lavine JB (2008) The thirteen principles of Jewish medical ethics. http://www.science.co.il/Jewish-studies/Articles/Medical-ethics.php

Death and mourning: Jewish customs, traditions, and practices (2016) http://www.shiva.com/learning-center/death-and-mourning/

Guidelines for health care providers interacting with Jewish patients and their families (2002) https://www.advocatehealth.com/documents/faith/CGJewish.pdf

Jewish perspectives on the birthing experience (2016) https://www.mikvah.org/article/jewish_perspectives_on_the_birthing_experience

National Foundation for Jewish Genetic Diseases (2012) http://www.mazornet.com/genetics/

Pew Research Center (2012) Religious beliefs and practices. http://www.pewforum.org/2013/10/01/chapter-4-religious-beliefs-and-practices/

Selekman J (2013) People of Jewish heritage. In: Purnell L (ed) Transcultural health care: a culturally competent approach, 4th edn. F.A. Davis Company, Philadelphia

U.S. Equal Opportunity Commission (2016) www.eeoc.gov/

Chapter 22
People of Irish Heritage

22.1 Overview and Heritage

The Republic of Ireland, also known as Eire and the Emerald Isle, covers most of the island bearing its name. The remainder of the island, Northern Ireland, is part of Great Britain. During the potato famine between 1846 and 1848, thousands of Irish died from malnutrition, typhus, dysentery, and scurvy, and millions immigrated to America. Over 38 million people of Irish descent live in the US. The Irish in America are a diverse group and vary in their beliefs according to acculturation, individuality, and the variant characteristics of culture as described in Chap. 1.

Most Irish immigrants initially settled in industrial areas in the northeastern US along the Atlantic Coast. The Irish attained success in America because they spoke the same language, had the same physical appearance as other European Americans, and mastered the political system. They have made significant contributions in politics, the labor movement, the Catholic Church, the arts, and service to their country.

22.1.1 Communications

- The major languages spoken in Ireland are English and Irish (Gaelic); the latter is the official language and is spoken primarily in West Ireland. The Irish enjoy puns, riddles, limericks, and storytelling.
- The Irish accent has a nasal quality and is spoken with a strong inflection on the first syllable of a word, resulting in a loss of weak syllables. When one becomes accustomed to hearing the Irish-accented English used by newer immigrants, there is little difficulty in understanding the speaker. The Irish language is low-context, using many words to express a thought.
- Some common Gaelic words and their meanings are *shamrock* for "emblem," *limer* for "folklore character," *colleen* or *lassie* for "girl," *sonsie* or *sonsy* for

© Springer Nature Switzerland AG 2019
L. D. Purnell, E. A. Fenkl, *Handbook for Culturally Competent Care*,
https://doi.org/10.1007/978-3-030-21946-8_22

"handsome," *cess* for "luck," *brogue* for "shoe," *dudeen* for "pipe tobacco," and *paddy* for "Irishman."

- Even though most Irish delight in telling long stories, when discussing personal matters, they are much less expressive, unless they are talking with close friends and family. Even then, many are still reluctant to express their innermost thoughts and feelings.
- Humility and emotional reserve are considered virtues. Displays of emotion and affection in public are avoided and are often difficult in private. To many, caring actions are more important than verbal expressions.
- The Irish use direct eye contact when speaking with each other. Not maintaining eye contact may be interpreted as a sign of disrespect, guilt, or evidence that the other person cannot be trusted.
- Personal space is important to the Irish, who may require greater distance in spatial relationships than other ethnocultural groups.
- Although the Irish may be less physically expressive with hand and body gesturing, facial expressions are readily displayed, with frequent smiling even during times of adversity.
- The Irish in America, with their strong sense of tradition, are typically past-oriented. They have an allegiance to the past, their ancestors, and their history. While respecting the past, they balance "being" with "doing," and they plan for the future by investing in education and saving money. Many Irish see time as being elastic and flexible. **Explain the importance of punctuality in health appointments.**
- *Mac* before a family name means "son of," whereas the letter *O* in front of a name means "descended from."

22.1.2 Family Roles and Organization

- The traditional Irish family is nuclear, with parents and children living in the same household.
- Irish families emphasize independence and self-reliance in children. Boys are allowed and expected to be more aggressive than girls, who are raised to be respectable, responsible, and resilient. Children are expected to have self-restraint and self-discipline and to be respectful and obedient to their parents, elders, and church and community figures.
- Adolescent years are a time for experiencing emotional autonomy, independence, and attachment outside the family while remaining loyal to the family and maintaining the traditional Irish belief in the importance of family. Peer-group pressures at school may have a significant influence and are often incongruent with the belief systems of Irish Americans. **Because it may be difficult for some Irish Americans to express their feelings, encourage openness between parents and teenagers.**
- Provisions are made in Irish homes for care of older family members, a task that becomes increasingly difficult when both parents work outside the home. Irish respect the experience of older people and seek their counsel for decision making.

- **Include older people into decision making regarding health-care activities.**
- The Irish value physical strength, endurance, work, the ability to perform work, children, and the ability to provide their children with the needed education to attain respectable socioeconomic status and professional accomplishments. Same-sex relationships continue to carry a stigma for some; however, they are becoming more acceptable in the US and in Ireland. **Do not disclose same-sex relationships to family members or others.**

22.1.3 Workforce Issues

- Because cultural differences between Ireland and the US are minimal, Irish assimilate into the workforce easily.
- When change is necessary to improve the status quo, the Irish readily relinquish traditional beliefs and adjust to the workforce. Even though the Irish are typical of past-oriented groups in other ways, they tend to question the status quo of the American workforce. The Irish in America have taken many leadership positions and helped with unionization and other activities to express their collective autonomy.
- The low contextual use of language, in which most of the message is in an explicit mode rather than an implicit mode, enhances pragmatic communications in the workforce.
- Personal space is important and may require greater distance in spatial relationships than other ethnocultural groups. **Do not take offense if the Irish stand farther from you than expected.**

22.1.4 Biocultural Ecology

- Most Irish have dark or red hair, ruddy cheeks, and fair skin; however, other variations exist in hair and skin color. The fair complexion of the Irish places them at risk for skin cancer. **Explain the harmful effects of the sun, and encourage the use of sun block and other types of protection from the sun.**
- Because mining is an important economic activity in Ireland, miners are at increased risk for respiratory diseases. In addition, the cool maritime climate of Ireland increases susceptibility to respiratory diseases. **Assess newer immigrants who worked in mining industries for respiratory illnesses.**
- Table 22.1 lists commonly occurring health conditions for Irish Americans in three categories: lifestyle, environment, and genetics.
- **Provide education and counseling regarding lifestyle and dietary changes to reduce risks associated with cardiovascular diseases.**
- Screen for osteoporosis in women in their late 60s or even younger if assessment data indicate it.

Table 22.1 Health
conditions common to Irish
Americans

Health condition	Cause
Coronary heart disease	Genetic, environment, lifestyle
Phenylketonuria	Genetic
Osteoporosis	Genetic, environment, lifestyle
Alcoholism	Genetic, environment, lifestyle
Skin cancer	Genetic, environment, lifestyle

- The major cause of infant mortality in Ireland is congenital abnormalities. Other conditions with a high incidence among Irish newborns are phenylketonuria (PKU), neural tube defects, and fetal alcohol syndrome.
- **Encourage women who give birth at home to seek PKU screening for their infants.**
- **Explains the hazards of alcohol intake during pregnancy.**

22.1.5 High-Risk Health Behaviors

- Smoking has been identified as a major risk factor causing premature mortality from cancer in Ireland and among the Irish in America. **Encourage patients to stop smoking, and assist them in finding smoking cessation programs with group and one-to-one counseling. Partner with teachers beginning in elementary school and teach smoking cessation.**
- The use of alcohol and IV drugs are major health problems among Irish Americans. Alcohol problems in Ireland are among the highest internationally. Alcoholism researchers generally agree that Irish ancestry puts individuals at risk for developing drinking problems. Irish pubs are popular establishments that have become synonymous with alcohol intake, lively music, and a vivacious time.
- **This image of the Irish pub perpetuates the stereotype of Irish as heavy drinkers; do not ascribe this label to all Irish Americans. Because drinking may be a way of coping with problems, assist Irish patients in exploring more effective coping strategies, and caution them against the dangers of mixing alcohol with medications.**

22.1.6 Nutrition

- Irish food is unpretentious and wholesome if eaten in recommended proportions. Food is an important part of health maintenance and celebrations. Vitamins are commonly used as a dietary supplement.
- Meat, potatoes, and vegetables are dietary staples. Lamb, mutton, pork, and poultry are common meats. Seafood includes salmon, mussels, mackerel, oysters, and scallops. Popular Irish dishes include Irish stew made with lamb, potatoes, and onions.

- Potatoes are used in a variety of ways. *Colcannon* is made with hot potatoes, mashed with cabbage, butter, and milk, and seasoned with nutmeg. This dish may be served at Halloween. *Champ* is a popular dish made with mashed potatoes and scallions. The scallions are cut in small pieces, including the green tops, boiled in milk until tender, and then added to the mashed potatoes and served with butter. Potato cakes, made with mashed potatoes, flour, salt, and butter, are shaped into patties and fried in bacon grease. Potato cakes are served hot or cold with butter and sometimes molasses or maple syrup. Another popular dish is Dublin coddle, made with bacon, pork sausage, potatoes, and onions. Oatmeal is popular in Ireland. Soda bread, another popular food, is made with flour, baking soda, salt, sugar, cream of tartar, and sour milk.
- Mealtimes are important occasions for Irish families to socialize and discuss family concerns. Meals are eaten three times a day, with a large breakfast in rural areas, lunch around noon, and a late dinner. Some Irish Americans continue the afternoon tradition of "tea," a light sandwich or biscuit with hot tea.
- **Because the Irish diet has the potential for being high in fats and cholesterol, assist patients with balanced food selections and preparation practices that reduce their risk of cardiovascular disease.**

22.1.7 Pregnancy and Childbearing Practices

- The Irish believe that pregnant women who do not eat a well-balanced diet or do not eat the right kinds of food may cause the baby to be deformed.
- The Irish share the belief, common to many other ethnic groups that the mother should not reach over her head during pregnancy because the baby's cord may wrap around its neck. A taboo behavior in the past, which some women still respect, is that if the pregnant woman sees or experiences a tragedy, a congenital anomaly may occur. **Provide factual information and dispel myths regarding pregnancy.**
- Eating a well-balanced diet after delivery continues to be a prescriptive practice for the Irish to ensure a healthy baby and maintain the mother's health. Plenty of rest, fresh air, and sunshine are also important for maintaining the mother's health. Going to bed with wet hair or wet feet causes illness in the mother.

22.1.8 Death Rituals

- The Irish are fatalists and acknowledge the inevitability of death. The American emphasis on technology and dying in the hospital may be incongruent with the Irish belief that family members should stay with the dying person. **Make arrangements for family to stay with the patient in the hospital and in long-term care facilities.**

- Whereas men are expected to be more stoic in their bereavement, women are more expressive. **Expect and accept a wide variety of grief expression among the Irish.**
- A wake continues as an important phenomenon in contemporary Irish families and is a time of melancholy, rejoicing, pain, and hopefulness. The occasion is a celebration of the person's life.
- Cremation is an individual choice, and there are no proscriptions against autopsy if required.

22.1.9   Spirituality

- The predominant religion of most Irish is Catholicism, and the church is a source of strength and solace. Other religions common among Irish in America include various Protestant denominations, such as the Church of Ireland, Presbyterian, Quaker, and Episcopalian.
- In times of illness, Irish Catholics receive the Sacrament of the Sick, which includes anointing, communion, and a blessing by the priest. The Eucharist, a small wafer made from flour and water, is given to the sick as the food of healing and health. Family members can participate if they wish.
- **Inquire whether sick individuals want to see a member of the clergy, even if they have not been active in church.**
- Attending mass daily is a common practice among many traditional and devout Irish Catholic families. Prayer is an individual and private matter. In times of illness, the clergy may offer prayers with the sick as well as with the family. **Give patients privacy for prayer whether or not a clergy member is present.**
- For those who practice their religion regularly, holy day worship begins at 4 p.m. the evening preceding the holy day; all Sundays are considered holy days. The obligation to fast and abstain from meat on specified days is relinquished during times of illness.
- Some wear religious medals to maintain health. **Religious emblems provide solace and should not be removed.**

22.1.10   Health-Care Practices

- Many Irish use denial as a way of coping with physical and psychological problems. The Irish view of life is illustrated in the belief that life is black with long suffering and the less said about it, the better.
- Many Irish ignore symptoms and delay seeking medical attention until symptoms interfere with the ability to carry out activities of daily living. Irish Americans limit and understate problems and handle problems by using denial. Because Irish may not be very descriptive about their symptoms, treatment may be more difficult.

- **Encourage early intervention and seeking health professionals for illnesses. Factually explain the importance of early intervention.**
- Illness or injury may be linked to guilt and is considered to be the result of having done something morally wrong. Restraint is a *modus operandi* in the Irish culture; temptation is ever-present and must be guarded against.
- Most Irish believe one is obligated to use ordinary means to preserve life. Extraordinary means may be withheld to allow the person to die a natural death. The sick person and family define extraordinary means; finances, quality of life, and effects on the family usually influence the decision. **Arrange for a family conference to address issues and concerns in hospice and palliative care.**
- In most Irish families, nuclear family members are consulted first about health problems. Mothers and older women are usually sought for their knowledge of folk practices to alleviate common problems such as colds. The Irish believe that having a strong religious faith, keeping one's feet warm and dry, dressing warmly, eating a balanced diet, getting enough sleep, and exercising are important for staying healthy. **Optimize compliance for reducing health risks by emphasizing these important cultural values.**
- Irish practices include wearing religious medals to prevent illness, using cough syrup made from honey and whiskey, taking honey and lemon for a sore throat, drinking hot tea with whiskey for a cold, drinking hot tea for nausea, drinking tea and eating toast for a cold, and putting a damp cloth on the forehead for a headache. Some folk practices may be harmful, such as the use of senna to cleanse the bowels every 8 days, eating a lot of oily foods, and avoiding seeing a physician. **Encourage patients to reveal all traditional and home remedies being used to treat symptoms. Provide factual information about potentially harmful folk and traditional practices.**
- The behavioral response of the Irish to pain is stoicism, usually ignoring or minimizing it. **Encourage pain medication, explaining that it will help the healing process.**
- One explanation for high rates of mental illness may be associated with the Irish having difficulty describing emotions and expressing feelings. **Encourage the expression of emotions and feelings before symptoms become a problem.**
- Blood transfusions are acceptable to most Irish Americans. Many participate in organ donation and indicate their willingness to do so on their driver's licenses.
- **Obtain organ-donor status on an individual basis, be sensitive to patients and family concerns, explain procedures involved with organ donation and procurement, answer questions factually, and explain the risks involved.**

22.1.11   Health-Care Providers

- The Irish respect all health-care professionals.
- Although the Irish are not noted for being overly modest, some may prefer to receive intimate care from someone of the same gender. In general, men and

women may care for each other in health-care settings as long as privacy and sensitivity are maintained.

22.2 Reflective Exercises

1. Identify four common health problems among Irish Americans. Provide a recommendation to address each problem.
2. What are some childrearing practices common among the Irish?
3. How are emotions displayed among the traditional Irish?
   - What does *Mac* before a family name mean? What does the letter *O* in front of a name mean?
4. Identify two high risk behaviors common among the Irish. What health education would you provide to decrease these behaviors?
5. The family of a deceased patient invites you to a wake. What might you expect from a traditional wake?
6. What is the dominant religion among the Irish? How might you partner with the church to improve the health of the Irish?
7. Most Irish speak low-contexted English. What does this mean and how might it be displayed on an intake assessment?
8. Identify at least three common traditional health-care practices.
9. List at least three common traditional beliefs practiced by pregnant women.
10. Identify at least common health conditions among newborns. What health teaching would you do to decrease these conditions?

Bibliography

CIA (2016) World factbook: Ireland. (2007). https://www.cia.gov/library/publications/the-world-factbook/geos/ei.html

Fitzpatrick M, Newton J (2005) Profiling mental health needs: what about your Irish patients? Br J Gen Pract 55(519):739–740

Ireland guide: a look at Irish language, culture, customs, and etiquette (2016) Commisceo Global. http://www.commisceo-global.com/country-guides/ireland-guide

Irish potato famine: gone to America (2010) The History Place. http://www.historyplace.com/worldhistory/famine/america.htm

Myers Shim S (2013) People of Irish heritage. In: Purnell L (ed) Transcultural health care: a culturally competent approach, 4th edn. F. A. Davis, Philadelphia

Parents' perspective on parenting styles and disciplining children (2010) National Children's Strategy Research Series. https://www.tcd.ie/childrensresearchcentre/assets/pdf/Publications/Parents'_Perspectives_on_parenting_styles.pdf

Purnell L, Foster J (2003a) Cultural aspect of alcohol use: part I. Drug Alcohol Prof 3(3):1723

Purnell L, Foster J (2003b) Cultural aspect of alcohol use: part II. Drug Alcohol Prof 2(3):3–8

Rapple BA (n.d.) Irish Americans. Countries and their culture. http://www.everyculture.com/multi/Ha-La/Irish-Americans.html

Chapter 23
People of Korean Heritage

23.1 Overview and Heritage

This chapter focuses on the commonalities among Koreans from the Republic of South Korea. The Korean peninsula is separated at the 38th parallel. Korea is one of the two oldest continuous civilizations in the world, second only to China.

The first major immigration from Korea occurred between 1903 and 1905, when more than 7000 men arrived in Hawaii. Today, Koreans, almost exclusively from South Korea, immigrate to America to increase socioeconomic opportunities and improve educational opportunities. Much diversity exists among Koreans who live in the US according to acculturation, individuality, and the variant characteristic of culture as outlined in Chap. 1. They place a high value on education. Their reputation for hard work, independence, and self-motivation has earned them the label of "model minority."

23.1.1 Communications

- The dominant Korean language, *han'gul*, was the first phonetic alphabet in East Asia.
- Most Koreans in America can speak, read, write, and understand English to some extent. However, some Americans may have difficulty understanding their English, especially those who learned English from Koreans who spoke with their native intonations and pronunciations.
- A high value is placed on harmony and the maintenance of a peaceful environment. Most are comfortable with silence. Small talk may appear senseless and insincere. Most stand close when conversing.
- Touch in the realm of health care is readily accepted. Touching among friends and social equals is common and does not carry a sexual connotation, as it might

© Springer Nature Switzerland AG 2019

L. D. Purnell, E. A. Fenkl, *Handbook for Culturally Competent Care*,
https://doi.org/10.1007/978-3-030-21946-8_23

in Western societies. Hugging and kissing are uncommon among parents and children as well as among children and older aunts or uncles.

- Age, gender, and social status determine the use of eye contact. Respect for those in senior positions is shown by not looking them directly in the eye.
- Feelings are infrequently communicated in facial expressions.
- Koreans prefer indirect communication because they perceive direct communication as an indication of intention or opinions as rude. Koreans may agree with the health-care provider in order to avoid conflict. **A response of "yes" may not have the same connotation as it does in the European American culture.**
- More traditional Koreans are past-oriented. Much attention is paid to the ancestry of a family. Yearly, during the Harvest Moon in Korea, *chusok* (respect) is paid to ancestors by bringing fresh fruits from the autumn harvest, dry fish, and rice wine to gravesites.
- The younger and more educated generation is more futuristic and achievement-oriented.
- **Punctuality is the norm for keeping important appointments, making transportation connections, and reporting to work.**
- The number of surnames in Korea is limited, with the most common ones being Kim, Lee, Park, Rhee or Yi, Choi or Choe, and Chung or Jung. Korean names contain two Chinese characters, one of which describes the generation and the other the person's given name.
- The surname comes first; however, because this may be confusing to many Americans, some Koreans in the US follow the Western tradition of using the given name first, followed by the surname. **Address Koreans by their surname, with the title Mr., Mrs., Miss, Ms., Dr., or Minister, unless otherwise indicated. Determine Korean clients' language ability, comfort level with silence, use of eye contact, and spatial distancing practices when completing health assessments.**

23.1.2 Family Roles and Organization

- Men are the primary financial providers. Women are expected to stay home and care for the children and domestic affairs unless they are professionals or it is necessary to work for economic purposes.
- Women have long been degraded in Korean society and seen as appendages of males. In earlier times, a woman's identity was determined by her role as someone's daughter, wife, or mother. Although many still practice these gender relationships, more educated women and men no longer adhere to these Confucian values.
- Parenting in Korea is authoritative, although class differences play a more influential role in determining parenting styles and family roles.
- Children are expected to be well behaved because the whole family is disgraced if a child behaves in an embarrassing manner.

- Discussing domestic violence violates Korean cultural norms. **When family violence is suspected, approach the topic in an indirect manner.**
- Dating is uncommon among high school students.
- Once young adults have entered a university, they receive their freedom and are permitted to make their own decisions about personal and study time. Group outings are common for meeting the opposite sex.
- With rapid acculturation, children often take on the values of the dominant society or culture, challenging parents who support traditional values and ideals. **Assist parents in exploring alternative parenting practices.**
- Parents expect their children to care for them in old age. *Hyo* (filial piety) is the obligation to respect and obey parents, care for them in old age, give them a good funeral, and worship them after death. The obligation to care for one's parents is written into civil code in Korea.
- Older people are frequently consulted on important family matters as a sign of respect for their life experiences.
- Old age begins when one reaches the age of 60 years.
- Women who divorce may suffer social stigma. Living together before marriage is not customary in Korea. If pregnancy occurs outside marriage, it may be taken care of quietly and without family and friends being aware of the situation.
- In the past, status was customarily assigned, but presently achieved status is more common.
- Lesbian and gay relationships are frowned upon. Personal disclosure to friends and family jeopardizes the family name and may lead to ostracism.
- **Do not disclose same-sex relationships to family members.**

23.1.3 Workforce Issues

- The skills and work experiences Koreans bring from their home country are often not accepted in the American workforce, forcing them to take jobs in which they may be over skilled.
- **Orientation programs need to address language skills, practice differences, and communication and interpersonal relationships to help Koreans adjust to the American workforce.**
- A supervisor is treated with much respect in work and in social settings. Informalities and small talk may be difficult for Korean immigrants.
- For an employee to refuse a request of an employer is unacceptable, even if the employee does not want or feel qualified to complete the request. **Asking Korean employees to demonstrate procedures is better than asking them if they know how to perform them.**
- American slang and colloquial language is difficult for Koreans to understand. **Be clear in communication style and refrain from using slang and idiomatic expressions.**

- Most stand close when conversing. **Do not take offense if a Korean stands closer than how you are accustomed.**
- Age, gender, and social status determine the use of eye contact. Respect for those in senior positions is shown by not looking them directly in the eye. **Do not assume that lack of eye contact means the person is not listening, does not care, or is being less than truthful.**

23.1.4   Biocultural Ecology

- Common physical characteristics include dark hair and dark eyes, with variations in skin color and hair darkness. Skin color ranges from fair to light brown, with those residing in the southern part of South Korea being darker.
- Epicanthal skin folds create the distinctive appearance of Asian eyes.
- Table 23.1 presents commonly occurring health conditions for Koreans in three categories: lifestyle, environment, and genetics.
- Korea continues to manufacture and use asbestos-containing products.
- **Assess for stomach and liver cancer, tuberculosis, hepatitis, renal impairment, hypertension, parasites, and asbestos-related health problems.**
- **Be aware that many Koreans require lower dosages of psychotropic drugs.**

23.1.5   High-Risk Health Behaviors

- Smoking by women in public is taboo, but some women smoke at home.
- Men have a high incidence of alcohol consumption.

Table 23.1   Health conditions common to Koreans

Health conditions	Causes
Schistosomiasis	Environment, lifestyle
Renal failure	Genetic, environment, lifestyle
Asbestosis	Environment, lifestyle
Hypertension	Genetic, environment, lifestyle
Tuberculosis	Environment, lifestyle
Hepatitis	Environment, lifestyle
Stomach cancer	Environment, lifestyle
Liver cancer	Environment, lifestyle
Lactase deficiency	Genetic
Osteoporosis	Genetic, lifestyle
Peptic ulcer disease	Environment, lifestyle
Insulin autoimmune deficiency disease	Genetic

- Seat belts are worn infrequently in South Korea.
- **Encourage moderate alcohol intake and smoking cessation. Explain the legal mandates of seat-belt and child restraint laws in the US.**

23.1.6   Nutrition

- The traditional Korean diet includes steamed rice; hot soup; *kimchee*; and side dishes of fish, meat, or vegetables served in some variation for breakfast, lunch, and dinner. Breakfast is traditionally considered the most important meal.
- Rice is served with 5–20 small side dishes of mostly vegetables and some fish and meats.
- Food is flavorful and spicy. Cooking includes a variety of seasonings such as red and black pepper, garlic, green onion, ginger, soy sauce, and sesame seed oil.
- Most Korean Americans are at high risk for calcium deficiencies due to lactose intolerance. **Determine individual cultural food choices that are high in calcium.**
- A cultural treatment for the common cold is soup made from bean sprouts, anchovies, garlic, and other hot spices.
- Common Korean American dishes are listed in Table 23.2.

23.1.7   Pregnancy and Childbearing Practices

- Pregnancy is a highly protected time for women. Both pregnancy and the post-partum period are ritualized. See Table 23.3 for some possible rituals.
- Once a woman is pregnant, she starts practicing **Tae-Kyo**, which literally means "fetus education." The objective of *Tae-Kyo* is to promote the health and well-being of the fetus and mother by having the mother focus on art and beautiful objects.

Table 23.2 Common Korean American dishes

• *Kimchee*: a spicy fermented cabbage made from a variety of vegetables but made primarily from a Chinese, or Napa, cabbage. Spices and herbs are added to the previously salted cabbage, which is allowed to ferment over time, and it is served with every meal in a variety of forms
• *Beebimbap*: rice, finely chopped mixed vegetables, and a fried egg served in a hot pottery bowl. Hot pepper paste is usually added
• *Bulgolgi*: thinly sliced pieces of beef marinated in soy sauce, sesame oil, green onions, garlic, and sugar and served after being barbecued
• *Chopchae*: clear noodles mixed with lightly stir-fried vegetables and meats
• Adapt health teaching to Korean American food choices and practices

Table 23.3 Pregnancy beliefs

• Pregnant women who handle unclean objects or kill a living creature may experience a difficult birth
• Women wear tight abdominal binders beginning at 20 weeks' gestation or work physically hard toward the end of the pregnancy to increase the chance of having a small baby
• Expectant mothers should avoid chicken, fish with scales, squid, or crab because eating these foods may affect the child's appearance. For example, eating duck may cause the baby to be born with webbed feet
• Women attribute a variety of complaints to *naeng* (chill), a cold imbalance of the womb that brings on a heavy vaginal discharge and can cause women who experience it to become sterile
• Eating blemished fruit causes a skin disease on the infant or an unpleasant face
• Women commonly labor and deliver in the supine position. After the delivery, women are traditionally served seaweed soup, a rich source of iron, which is believed to facilitate lactation and to promote healing of the mother
• Bed rest is encouraged after delivery for 7–90 days
• Women are encouraged to keep warm by avoiding showers, baths, and cold fluids or foods in order to prevent chronic illnesses such as arthritis
• The baby should be wrapped in warm blankets to prevent harm from cold winds
• Postpartum women should care for their bodies by augmenting heat and avoiding cold, resting without working, eating well, protecting the body from harmful strains, and keeping clean. To improve postpartum care, develop bilingual pamphlets that include medical terms used in the U.S. health-care system

23.1.8   Death Rituals

- Death and dying are fairly well accepted in the Korean culture.
- Prolonging life may not be highly regarded in the face of modern technology. Families are expected to stay with family members and assist in feeding and personal care around the clock. **Include family in caring for their hospitalized family member.**
- Many believe that patients should not be told they have a terminal illness. **Take cues from the patient and, if necessary, disclose terminal illnesses to the oldest son or designated family spokesperson, who will inform the patient when the family deems necessary.**
- Crying and open displays of grief are common and signify respect for the dead.
- Relatives and friends pay respect by viewing photographs of the deceased instead of viewing the body.
- An ancestral burial ceremony follows death, with the body being placed in the ground facing south or north. Rice wine is sprinkled around the gravesite.
- The eldest son or male family member sits by the deceased, sometimes holds a cane, and makes a moaning noise to display his grief. The cane is a symbol of needing support.
- **Provide a private room so family may grieve in culturally congruent ways.**

23.1.9 Spirituality

- Organized religions include Christianity, Buddhism, and **Chondokyo**. The church is a powerful social support group for Korean immigrants.
- Christians believe the spirit goes to heaven; Buddhists believe the spirit starts a new life as a person or an animal. **Individually assess each patient's spiritual practices.**
- Family and education are central themes that give meaning to life.

23.1.10 Health-Care Practices

- Many prescription drugs in the US, such as antibiotics, anti-inflammatory and cardiac medications, and certain pain-control medications, can be purchased over-the-counter in Korea.
- Herbal medicine may be used in conjunction with Western biomedicine. Herbal remedies include ginseng, seaweed soup, and *haigefen* (clamshell powder), which has high levels of lead, causing abdominal colic, muscle pain, and fatigue.
- **Ask about the use of traditional medicine, and include nonharmful practices in prescriptions.**
- *Hanbang* is the traditional Korean medical care system and works on the principle of restored energy. Complementary health practices include acupuncture, acumassage, acupressure, and moxibustion therapy.
- Some Korean Americans are stoic and are slow to express emotional distress from pain. Others are expressive and discuss their smallest discomforts.
- **Monitor nonverbal cues and facial expressions for pain.**
- Mental illness may be stigmatized. *Hwa-Byung*, a traditional Korean illness, occurs from the suppression of anger or other emotions. These emotions are expressed as physical complaints, ranging from headaches and poor appetite to insomnia and lack of energy.
- Disability may carry a stigma, especially for older, more traditional Koreans.
- **Assist families to access resources to care for parents and children with disabilities.**
- Whereas blood donation and transfusions are acceptable, organ donation and organ transplantation are rare, reflecting traditional attitudes toward integrity and purity.

23.1.11 Health-Care Practitioners

- More traditional individuals frequently prefer health-care providers who speak Korean and are older.

- Because of modesty, women prefer women for performing Pap smears, mammography, and breast examinations.
- **Provide a same-gender health-care provider for intimate care whenever possible.**

23.2   Reflective Exercises

1. For Koreans a response of "yes" may not have the same connotation as it does in the European American culture. What does it mean in the Korean culture?
2. Explain the name format for Koreans and how you would address them.
3. When family violence is suspected, how might you approach the topic?
4. What is *hyo* and how is it displayed?
5. What needs to be included in an orientation program for Koreans to be successful in the workforce?
6. Name at least three common health conditions among Koreans.
7. What is *Tae-Kyo* and what are its purpose?
8. What is the term for traditional Korean medicine and what are its principles?
9. What is *Hwa-Byung* and what is its causes and symptoms?
10. How do more traditional Koreans see disability?

Bibliography

An ancient medical tradition (2012) http://www.korea.net/NewsFocus/Society/view?articleId=99249

CIA (2016) World factbook: Korea. https://www.cia.gov/library/publications/the-world-factbook/geos/ks.html

Connor P (2014) 6 facts about South Korea's Growing Christian population. Pew Reseaerch Center. http://www.pewresearch.org/fact-tank/2014/08/12/6-facts-about-christianity-in-south-korea

Health problems of Korean Americans: access to health care (2016) Stanford School of Medicine. https://geriatrics.stanford.edu/ethnomed/korean/patterns/access.html

Im E-O (2013) People of Korean heritage. In: Purnell L (ed) Transcultural health care: a culturally competent approach, 4th edn. F.A. Davis Company, Philadelphia

Kim SS (2017) A culturally adapted smoking cessation intervention for Korean American: preliminary findings. J Transcult Nurs 28(1):24–32

Kim-Godwin YS (2003) Postpartum beliefs and practices among non-western cultures. MCN Am J Matern Child Nurs 28(2):74–78

Merkin RS (2009) Cross-cultural communication patterns—Korean and American Communication. http://www.immi.se/intercultural/nr20/merkin.htm

South Korea guide: a look at South Korean language, culture, customs, and etiquette (2016) http://www.commisceo-global.com/country-guides/south-korea-guide

Chapter 24
People of Mexican Heritage

24.1 Overview and Heritage

Great variations exist among people of Mexican heritage, depending on accultura-tion, individuality, and the variant characteristics of culture as described in Chap. 1. However, a core of beliefs and practices is shared by those who self-identify as Mexican, Mexican American, **Latino**, Chicano, or **Hispanic**. Hispanic and Latino are used interchangeably; however, Latino includes Brazilians who are not Hispanic. In addition, some who identify as Mexican have no Hispanic heritage because they are from indigenous Indian groups who have never mixed with/married people from Spanish backgrounds; however, they are still considered Mexican.

Mexican Americans include newer immigrants and people who have been in the US for six or seven generations. Some originate from high mountains and plateaus, some from low-lying swamp areas, and others from sea- and ocean-side communities. The topography of the region of origin in Mexico may be the key to accurate health assess-ments for dietary preferences, illnesses, and disease. Many Mexicans come to the US because of the high poverty rate of over 18% in Mexico. As of 2012, more Mexicans are leaving the US and returning to Mexico than Mexicans who are migrating to the US.

24.1.1 Communications

- Great diversity exists in the Spanish language; many dialects are spoken in Mexico. Although most Mexicans use Spanish as their primary language, some speak only English, some speak both English and Spanish, and some speak only an indigenous Indian language.
- Significant importance is placed on verbal communication. Conversants may stand very close, even with people who are not well known to them. Loud voice volume in formal settings may connote anger.

© Springer Nature Switzerland AG 2019
L. D. Purnell, E. A. Fenkl, *Handbook for Culturally Competent Care*,
https://doi.org/10.1007/978-3-030-21946-8_24

- **Whereas more traditional and older individuals do not maintain eye contact, acculturated and more educated people usually do maintain eye contact.**
- Many Mexicans tend to be fatalistic and present-oriented. Time is relative, not categorically imperative; thus, many arrive late for appointments and delay seeking health care until the condition is more serious. **Stress the importance of punctuality for health-care appointments in the US.**
- Touch is common between people of the same sex. However, men and women rarely touch in public. **Explain the necessity and ask permission before touching during a physical examination.**
- Formal names may be extensive and include a middle name and the mother and father's surnames. A married woman may take her husband's surname, thereby having three surnames. **Greet adults formally with Señor, Señora, or Señorita unless told to do otherwise. Ask which name is preferred as well as which name is used for legal purposes.**

24.1.2 Family Roles and Organization

- In traditional families, men are expected to provide financial support for the family.
- The stereotype of "**machismo**," in which men are the primary decision-makers in the household, is not accurate for all people of Mexican heritage. **Do not stereotype machismo.**
- Women care for children and maintain the home. In most homes where the man and woman both work outside the home, they tend to share household chores and child-rearing responsibilities.
- Some households are patriarchal, some are matriarchal, and some are egalitarian. Regardless of who makes the decision, the male traditionally is expected to be the spokesperson for the family. **Ask specifically who makes which decisions for the family.**
- Priorities for children are getting an education, having good manners, and respecting older people and people in high-status positions.
- Traditionally, adults expect children to live with their parents until marriage.
- Children born out of wedlock are loved regardless of the parents' marital status. Men usually remain involved with their children who are born out of wedlock.
- When possible, extended family members prefer living close to each other. The elderly frequently live with their children when self-care becomes a concern; in addition, the elderly are always respected for their wisdom.
- **Extended family members are a good resource for home care.**
- Social status is gained through formal academic education, having a respected position, and having children and family members with good manners. The extensive name format represents status.
- Although some states in Mexico have sanctioned same-sex marriages and Mexico's Supreme Court has ruled that all 31 states must recognize same-sex

marriages performed in the capital; it does not force those states to marry gays and lesbians in their territories. Many continue to be stigmatized. **Do not disclose sexual orientation status to family members.**

24.1.3 Workforce Issues

- People educated in Mexico are likely to have been exposed to pedagogical approaches that include rote memorization and an emphasis on theory with little practical application. American educational systems usually emphasize an analytical approach, practical applications, and a narrow, in-depth specialization.
- **Teach practical application of knowledge during orientation and ongoing.**
- Because family is a first priority for most Mexicans, activities that involve family members usually take priority over work issues. **Be flexible with work hours when feasible.**
- Most Mexican Americans tend to shun confrontation for fear of losing face. Many are very sensitive to differences of opinion, which are perceived as disrupting harmony in the workplace.
- Truth is tempered by diplomacy and tact. When a service is promised for tomorrow, even when they know the service will not be completed tomorrow, it is promised to please, not to deceive. Truth is seen as a relative concept, not an absolute value, and people are expected to give direct yes and no answers.
- **Explain the value of "absolute" truth in the American workforce.**
- For most Mexicans, titles and positions may be more important than money.
- Many Mexicans believe that time is relative and elastic, with flexible deadlines, rather than stressing punctuality and timeliness. **Explain the expectation of punctuality in the American workforce and repercussions for chronic tardiness.**
- Many are used to having traditional autocratic managers who assign tasks but not authority. A Mexican worker who is not accustomed to responsibility may have difficulty assuming accountability for decisions.
- More traditional and some older individuals do not maintain eye contact, whereas acculturated and more educated people usually do maintain eye contact. **Do not assume that the lack of eye contact means the person is not listening, does not care, or is being less than truthful.**

24.1.4 Biocultural Ecology

- Most Mexicans have dark hair and dark eyes, but some may have blond hair and blue eyes. Many have a diverse gene pool that includes indigenous Indian groups.
- **Assess for cyanosis and jaundice in those with dark skin by observing the sclera and conjunctiva, palms of the hands, soles of the feet, and buccal mucosa and tongue rather than relying on skin tone.**

Health conditions	Causes
Lactase deficiency	Genetic
Diabetes mellitus	Genetic, environment, lifestyle
Parasitic diseases	Environment, lifestyle
Malaria	Environment, lifestyle
Dengue fever	Environment, lifestyle
Cleft lip/palate	Lifestyle
Dental caries	Lifestyle
Cardiovascular disease	Genetic, environment, lifestyle
Hypertension	Genetic, environment, lifestyle
Tuberculosis	Environment, lifestyle
Pesticide/herbicide poisoning	Environment, lifestyle

Table 24.1 Health conditions common to Mexicans

- Table 24.1 lists commonly occurring health conditions for Mexicans in three categories: lifestyle, environment, and genetics.
- **Assessments need to include exposure to pesticides, parasitic diseases, and illnesses that are common in their home country.**
- Mexicans may require lower doses of antidepressants and experience more intense side effects than non-Hispanic white populations. Many are also poor metabolizers of debrisoquine. **Observe for side effects of antidepressants.**

24.1.5   High-Risk Health Behaviors

- Smoking is more common among the more acculturated, but they smoke fewer cigarettes than European Americans when they do smoke. **Encourage smoking cessation beginning in elementary schools.**
- Some Mexicans use alcohol to make them emotionally and socially extroverted and are more likely to engage in binge drinking than other groups. **Assess alcohol consumption by determining the amount of alcohol used both daily and during celebrations.**
- Many newer immigrants may be reluctant to use seat belts and helmets because they are not used in their home country. **Explain the legal requirements of using seat belts and helmets.**
- Men may object to condom use because it decreases sensitivity. Women may object to the use of condoms because of the suggestion that the woman is "dirty."
- **Explain that condoms help prevent pregnancy and decrease the incidence of HIV/AIDS (*SIDA*) and other sexually transmitted infections.**

24.1.6   Nutrition

- Food choices vary widely, depending on the region of Mexico from which the person comes. Rice, beans, and tortillas are food staples.

- **Determine preparation practices because some people use coconut oil and animal fat for flavoring. Determine flavorings and ingredients as well as the major food choices because these can add significant calories to the diet. Encourage use of corn tortillas instead of flour tortillas, leafy green vegetables, and soups and stews that use bones in them to increase calcium content in the diet and avoid side effects with lactose intolerance.**
- Being overweight according to the U.S. height/weight tables is seen as positive. **Patients should be co-participants in deciding an acceptable weight.**
- Sweetened fruit drinks are popular, and adding sugar to fresh fruit juice is common.
- **Discouraging the use of sweetened fruit drinks and encouraging the use of natural juices without added sugar is one means of reducing calorie consumption.**
- Mealtimes may not coincide with the dominant American schedule. For many, the noontime meal is the largest meal of the day, and the evening meal may be served very late in the evening. **Determine the individual's and family's mealtimes, and adjust medication schedules accordingly.**
- Many individuals balance food choices according to the "**hot**" and "**cold**" theory. Variations occur in the assignment of different foods for hot and cold properties. Herbal teas are commonly used to maintain health and treat illnesses.
- **Determine the patient's use of the hot and cold theory of food choices, which herbs and teas are used, how frequently they are used, and the amount used each time. Nonharmful practices should be incorporated into the plan of care.**

24.1.7   Pregnancy and Childbearing Practices

- Fertility practices are primarily connected with religious beliefs usually associated with the Catholic Church. The church discourages the use of condoms, believing they promote promiscuity.
- The rhythm method or Norplant is usually deemed acceptable because they are more natural means of helping to prevent an unwanted pregnancy. Foams, creams, and intrauterine devices may not be acceptable because women are not supposed to touch their genitals except for purposes of bathing.
- Abortion is considered morally wrong. A few individuals may see sterilization and other methods of birth control as acceptable.
- **Determine on an individual basis acceptable methods of birth control.**
- Multiple births are more common among Mexicans. Pregnancy is considered a natural condition, with most advice for health care coming from family members; this may deter prenatal care. **Stress that medical evaluation during pregnancy is necessary to help ensure the health of both the mother and the baby. Encouraging female relatives to accompany the pregnant woman for prenatal checkups may be helpful.**

Table 24.2 Common cultural beliefs and customs

• Walking in the moonlight while pregnant may cause birth defects
• Safety pins, metal keys, or other metal amulets are worn to prevent birth deformities
• Raising one's arms over the head while pregnant will cause the cord to wrap around the infant's neck
• Postpartum women are discouraged from taking baths, sitting in a bathtub, taking sitz baths, or washing their hair for 6 weeks
• Bathing with a cloth, including the hair, is acceptable. Encourage use of warm compresses instead of sitz baths
• Cutting the baby's hair or nails in the first 3 months is believed to cause blindness and deafness in the baby. Ask permission before cutting an infant's nails or hair
• Placing a key, coin, or other metal object on the infant's umbilicus is believed to promote healing. Teach how to clean the object to prevent infection
• Wearing an abdominal binder can prevent air from entering the uterus in postpartum women
• Covering their ears, head, shoulders, and feet can prevent blindness, mastitis, frigidity, or sterility in women
• Recognize which practices are harmless, and respect the woman's beliefs as appropriate
• Using videos and literature in Spanish and with pictures of Hispanics may help compliance with health interventions

- For many, the delivery room is not a place for men, and many believe that allowing the father to see the woman or baby during the delivery may harm the mother or baby.
- Various cultural customs and beliefs are included in Table 24.2.

24.1.8 Death Rituals

- Death is an extended family affair, and members are obligated to visit in the hospital or long-term care facility. Expect many visitors in the inpatient setting. Among more traditional Mexicans, a family member must remain with the dying person.
- **Find a place where family members can be together, preferably in the unit with the dying person. In alternative settings, find another quiet place where the family can gather. Elicit a family spokesperson to assist with crowd control.**
- Some families may want to have lighted candles in the room of the dying patient. Although inpatient facilities will not permit open candle flames because of fire safety, electric candles are acceptable to most.
- Death is considered part of life, and some, especially men, may approach death stoically.
- **Just because someone does not openly display emotions about death does not mean the person does not care about the deceased.**
- Some, especially women, may have an *ataque de nervios* on hearing about the death of a loved one. In this culture-bound syndrome, the person exhibits

hyperkinetic and seizure-like activity that releases strong emotions. **This is a normal reaction, and treatment is usually not necessary except to remain with the person and provide emotional support. Use family members for assistance when possible.**

- When a person dies, family and friends gather for a *velorio*, a festive occasion to celebrate the person's life.
- Burial practices vary, but the more traditional do not practice cremation. Autopsy is not welcomed. **Explaining the legalities of autopsy is conducive to having the family accept an autopsy.**

24.1.9 Spirituality

- Although many other religions are practiced, the predominant religion is Roman Catholicism. However, do not assume that the practice of the Catholic religion is the same for all people. Each person's religious practices must be assessed on an individual basis. Many families have altars in their homes.
- The family is foremost. The individual's source of strength comes from being with family members. Priests and other religious leaders are major sources of emotional strength and support for most Mexicans. **On admission to the health-care facility, inform the patient and family of the availability of religious leaders.**

24.1.10 Health-Care Practices

- The family is considered the most credible source of health-care information among Mexicans. This practice can impede health-seeking behaviors. For many, good health means being free of pain and is largely due to the "will of God."
- Many migrant workers are not aware of the necessity of protecting themselves from pesticides and herbicide poisoning because they are not common in Mexico.
- **Health teaching on herbicides and pesticides should be a family affair.**
- Almost all Mexicans use herbal medicines and teas. The specific herbs and teas vary among families. Most of these teas and herbs are beneficial or at least not harmful. A few can be harmful by themselves, whereas others may be harmful when included with prescription and over-the-counter medications. **Ascertain if patients are using over-the-counter medicines and explain the hazards of excessive use of over-the-counter medications as well as using medicines that were originally intended for use by another family member.**
- Assess for use of herbs and teas. **Two herbs commonly used by Mexican Americans are *azarcon* and *greta*, which are used for colic and stomach conditions in children. Both of these herbs contain lead and can be toxic, especially in children.**

- Many Mexicans practice the hot and cold theory, according to which many diseases and illnesses are caused by a disruption in the hot and cold balance of the body. Thus, if too many cold forces in the body cause an illness, treatment is aimed at balancing the condition by introducing hot treatments and foods that are considered hot. *Hot and cold do not always coincide with temperature*, but perceived metabolic properties.
- Hot conditions include infection, diarrhea, and sore throats and, therefore, are treated with such cold foods as fruits, vegetables, and dairy products. Cold conditions, such as cancer, malaria, and earaches, are treated with such hot foods as liquor, beef, pork, and spicy foods. Tremendous variations exist between and among hot and cold conditions, depending on the family.
- **Include the entire family in health promotion and health teaching to increase compliance with health prescriptions and interaction.**
- Frequently seen cultural illnesses or conditions are shown in Table 24.3.
- Perform an individual pain assessment, determine usual treatment modalities used for pain, ask what the patient thinks caused the pain, and determine what the patient usually does to relieve pain. Have pain scales in Spanish as well as visual scales for those who do not read Spanish or English. **Explaining that pain medication will promote healing may encourage the patient to accept pain medication.**
- The sick role is easy to enter without personal feelings of inadequacy or blame. Family members readily take on the sick person's responsibilities. Family members usually care for the ill family member at home, if at all possible. **Query the family to determine if they have the resources to care for a member at home. Not all Mexicans in the US have the extended family to care for relatives at home.**

Table 24.3 Common cultural illnesses among Mexicans

• **Empacho** (blocked intestines) may result from an incorrect balance of hot and cold foods causing a lump of food to stick in the gastrointestinal tract. Treatment includes massaging the stomach and back to dislodge the food bolus
• **Mal de ojo** (evil eye) occurs when an older person looks at a younger person in an admiring fashion. Such eye contact can be voluntary or involuntary. Symptoms are numerous, including fever, anorexia and vomiting, or irritability. The spell can be broken if the person doing the admiring touches the person while admiring him or her. Allopathic health-care providers are usually unable to cure *mal de ojo*. Making a referral to a folk health-care provider is advisable
• **Caida de mollera** (fallen fontanel) has numerous causes, which may include removing a nursing infant too harshly from the nipple or handling an infant too roughly. Symptoms vary from failure to thrive to irritability. The usual treatment is to hold the infant upside down by the feet. Assess for dehydration. A referral to a folk health-care provider is also recommended
• **Susto** (magical fright or soul loss) is associated with epilepsy, tuberculosis, and other infectious diseases as well as "overwhelming feelings of loss." Symptoms may be physical or psychological in nature and may also be consistent with depression. Gear treatment to the underlying causes, and treat physical and psychological symptoms

Table 24.4 Major Mexican folk health-care providers

• **Curanderos** receive their gift from God or serve an apprenticeship. Some even prescribe over-the-counter medications. They usually treat traditional illnesses not caused by witchcraft
• **Espiritistas** (spiritualists) treat conditions caused by witchcraft. Amulets and prayer are a large part of the treatment. Seeing an *espiritista* may carry a stigma among some Hispanics
• **Yerberos** or **Jerberos** use herbs, teas, and roots to prevent or treat illnesses. Patients usually purchase the herbs from a **botanica**, a specialist herb shop that also sells religious figurines
• *Sobadores* treat muscle and joint problems using massage and manipulation; usually they do not have formal training
• Ask patients if they are using folk health-care providers and the reasons why they are using them, and have them fully disclose all treatments prescribed

- **Because long-term care facilities are either nonexistent or are of very poor quality in Mexico, family members may be very reluctant to place family members in such a facility. Suggesting a visit to one may be an option.**
- Extraordinary means being used to preserve life are frequently frowned upon and are often determined by finances, education, and availability of services.
- Blood transfusions are acceptable, but some may be reluctant to accept transfusion or blood products for fear of HIV/AIDS *(SIDA)*. The belief that the body must be buried whole deters organ donation. Organ transplantation may not be acceptable because of the belief that *mal aire*, bad air, will enter the body and increase one's risk for developing cancer. **Dispel myths related to organ donation and organ transplantation. Eliciting the assistance of a priest may be helpful.**

24.1.11 Health-Care Providers

- Many Mexicans use a number of folk health-care providers who are usually well known to the patient. See Table 24.4.
- Health-care providers with the right qualities are well respected by Mexicans.
- **Important qualities include addressing the patient formally unless told to do otherwise, showing respect by asking questions, accepting the patient's ideas, taking an interest in the entire family, and being well groomed. Always ask permission, and explain the reason for the necessity of touching the patient during a physical examination.**

24.2 Reflective Exercises

1. Describe the extensive name format for Mexican women. How would you address her? What is her legal name?
2. Describe eye contact protocol among traditional Mexicans.

3. How do you assess for jaundice and cyanosis among dark-skinned Mexicans?
4. List three common health conditions of Mexican that have a lifestyle component.
5. Describe Mexican fold healers and describe their treatment modalities.
6. What are the deciding factors for treating Mexicans with extraordinary illnesses?
7. Describe the "hot" and "cold" theory of illnesses among Mexicans.
8. Describe the "hot" and "cold" theory of food among Mexicans. If you are unfamiliar with foods in each category, what are your resources for finding out more?
9. Two herbs commonly used by Mexican Americans are *azarcon* and *greta*, which are used for colic and stomach conditions in children. What are the hazards of giving them to children?
10. What is the folk illness *ataque de nervios*. What are its symptoms? How would you treat it?

Bibliography

Carbaugh D (2007) Cultural discourse analysis: communication practices and intercultural encounters. J Intercult Commun Res 36(3):167–182

Centers for Disease Control and Prevention (2015) Hispanic Health. Accessed from https://www.cdc.gov/vitalsigns/hispanic-health/

CIA (2016) World factbook: Mexico. Accessed from https://www.cia.gov/library/publications/the-world-factbook/geos/mx.html

Clark Callister L, Khalaf I (2010) Spirituality in childbearing women. J Perinatal Educ 19(2):16–24

Cultural Information – Mexico (2014) Global affairs Canada. Accessed from https://www.international.gc.ca/cil-cai/country_insights-apercus_pays/ci-ic_mx.aspx?lang=eng

Eggengerger SK, Grassley J, Restrepo E (2006) Culturally competent nursing care for families: listening to the voices of Mexican-American women. Online J Issues Nurs 11. Accessed from http://www.nursingworld.org/MainMenuCategories/ANAMarketplace/ANAPeriodicals/OJIN/TableofContents/Volume112006/No3Sept06/ArticlePreviousTopics/CulturallyCompetentNursingCare.html

National Center for Health Statistics, Centers for Disease Control and Prevention (2016) Health of Mexican American population. Accessed from https://www.cdc.gov/nchs/fastats/mexican-health.htm

Smith AB (2003) Mexican cultural profile. EthnoMED. Accessed from https://ethnomed.org/culture/hispanic-latino/mexican-cultural-profile

Varela ER, Vernberg EM (2004) Parenting style of Mexican, Mexican American, and Caucasian–Non-Hispanic families: social context and cultural influences. J Fam Psychol 18(4):651–657

Zoucha R, Zamarripa C (2013) People of Mexican heritage. In: Purnell L (ed) Transcultural health care: a culturally competent approach, 4th edn. F.A. Davis, Philadelphia

Chapter 25
People of Polish Heritage

25.1 Overview and Heritage

Poland was conceived near the middle of the tenth century and reached its golden age in the sixteenth century. In a series of agreements between 1772 and 1795, Russia, Prussia, and Austria partitioned Poland among themselves. Poland regained its independence in 1918 only to be overrun by Germany and the Soviet Union in World War II. It became a Soviet satellite state following the war, but its government was comparatively tolerant and progressive. Labor turmoil in 1980 led to the formation of the first independent free trade union in Eastern Europe, *Solidarnosc* (Solidarity) that over time became a political force and by 1990 had swept parliamentary elections and the presidency.

In 1988, formal negotiations between the Polish Communist Party leaders and the unofficial opposition, called the "Round Table talks," resulted in partially free Parliamentary elections. Solidarity won a landslide victory in the 1989 elections, and the newly elected Parliament changed the country's name and constitution, establishing the Third Republic of Poland.

Polish and their descendants who have immigrated to America for many generations have maintained their ethnic heritage by promoting their culture, attending Catholic churches, attending parades and festivals, maintaining ethnic food traditions, speaking the Polish language, and promoting interest in their home country through media events as well as economic and political channels. For newer immigrant Poles, maintaining ethnic heritage means learning English and obtaining a good job. Newer immigrants are less concerned with raising consciousness over Polish American issues than they are with financially helping families who remain in Poland and raising concerns over the political and economic climate in their homeland.

Polonia is the name given to Polish communities in the US. Poles are a heterogeneous group. Much of the diversity within this ethnic group is due to the degree of acculturation, individuality, and the variant cultural characteristics as described in Chap. 1.

© Springer Nature Switzerland AG 2019
L. D. Purnell, E. A. Fenkl, *Handbook for Culturally Competent Care*,
https://doi.org/10.1007/978-3-030-21946-8_25

25.1.1   Communications

- The dominant language of people living in Poland is Polish, although there are some regional dialects and differences.
- The Polish language was influenced by the countries surrounding Poland and by the Latin of eleventh- and twelfth-century kings. Depending on the regional and cultural background of the speaker, Polish may sound German, Russian, or French.
- Poles are an animated group, and facial expressions generally convey the tone of the conversation.
- Poles use touch as a form of personal expression of caring. Touch is common among family members and friends, but Poles may be quite formal with strangers and health-care providers.
- Handshaking is considered polite. In fact, failing to shake hands with everyone present may be considered rude. **Greet patients formally and with a handshake.**
- Most Poles feel comfortable with close personal space, but distances increase when interacting with strangers.
- First-generation Poles commonly kiss "Polish style," that is, once on each cheek and then once again.
- Two women may walk together arm in arm, or two men may greet each other with an embrace, a hug, and a kiss on both cheeks.
- Many Polish Americans consider the use of spoken second-person familiarity rude. Polish people speak in the third person. For example, they might ask, "Would Martin like some coffee?" rather than "Would you like some coffee?" Although the first expression might sound awkward, the latter expression may be considered impolite and too informal, especially if the person being asked is older.
- Polish Americans also use direct eye contact when interacting with others. Many Americans may feel uncomfortable with this sustained eye contact and feel it is quite close to staring, but to Poles, it is considered ordinary. **Sustained eye contact does not necessarily mean anger.**
- Poles, as a group, tend to share thoughts and ideas freely.
- Punctuality is important to Polish Americans. To be late is a sign of bad manners. Even in social situations, people are expected to arrive on time and stay late.
- Polish Americans are both past and future oriented. The past is very much a part of Polish culture, with the families passing on their memories of World War II, which still haunt them in some way. A strong work ethic encourages Poles to plan for the future.
- Many Polish names are difficult to pronounce. Even though a name may be mispronounced, a high value is placed on the attempt to pronounce it correctly.
- Some examples of common Polish names include *Kowal*, meaning "blacksmith." Numerous suffixes, such as "icz," "czyk," "iak," and "czak," which mean "son of,"

can be added. Two of the most common suffixes are "ski" for male and "ska" for female, which originally were added to many names because they were associated with nobility. The suffix "cki" became the phonetic version of "ski." Surnames ending in "y," "ow," "owo," and "owa" are usually derived from names of places. The "ak" suffix is typical of western Poland, whereas "uk" is found in the east.

25.1.2 Family Roles and Organization

- Polish culture centers on family. Each member has a certain position, role, and related responsibilities. All members are expected to work, make contributions, and strive to enhance the entire family's reputation and social and economic position.
- In most families, the father is perceived as the head of the household. Depending on the degree of assimilation, the father may rule with absolute authority in first-, second-, and even third-generation Polish American families. However, among some third- and fourth-generation Polish Americans and second- and third-wave immigrants, more egalitarian gender roles are becoming the norm.
- **Determine the spokesperson and decision maker for health-care decisions.**
- Historically, large families were expected and commonplace among Poles, with women often experiencing between 5 and 10 pregnancies.
- The most valued behavior for children is obedience. Taboo child behaviors include anything that undermines parental authority. Parents are quite demonstrative with young children, but they resist showing much affection toward them once they are older than toddler age. This is the parents' way of teaching children to be strong and resilient.
- Many parents praise children for self-control and completing chores. Little sympathy is wasted on failure, but doing well is openly praised. Children are taught to resist feelings of helplessness, fragility, or dependence.
- Traditional family values and loyalty are strong in most households. The elderly are highly respected. They attend church regularly and carry on Polish traditions.
- The elderly play an active role in helping grandchildren learn Polish customs and in assisting adult children in their daily routine with families.
- For some families, one of the worst disgraces, as seen through the eyes of the Polish community, is to put an aged family member in a nursing home. Third- and fourth-generation Polish Americans may consider an extended-care or assisted-living facility because of work schedules and demands of care, but first-generation immigrants rarely perceive this as an option.
- Extended family, consisting of aunts, uncles, and godparents, is very important to Poles. Longtime friends become aunts or uncles to Polish children.
- Alternative lifestyles are seen as part of assimilation into the blended American culture. Same-sex couples are frowned upon and may even be ostracized,

depending on the level of assimilation. **Do not disclose same-sex relationships to others.**

- When divorce occurs, single heads of households are accepted in the Polish American community.

25.1.3 Workforce Issues

- Some Polish immigrants are underemployed and may have difficulty working with authority figures who are less educated. Poles quietly comment that they are disrespected for their educational background and that they must endure decreased status to stay in America.
- The Polish characteristic of praising people for their work makes Poles strong managers.
- Because Poles learn deference to authority at home, in the church, and in parochial schools, some may be less well suited for the rigors of a highly individualistic, competitive market.
- The strong Polish work ethic, exhibited as volunteering for overtime, being punctual, and rarely taking sick days, is valued by employers.
- Native-born Polish Americans have little, if any, difficulty with the English language. Foreign-born Poles frequently have some difficulty understanding the subtle nuances of humor. Recent Polish immigrants, who had experience working under a Communist bureaucratic hierarchy, may have some difficulty with the structure, subtleties, and culture of the American workplace. **Carefully explain workforce rules and expectations in the US.**

25.1.4 Biocultural Ecology

- Most Poles are of medium height with a medium-to-large bone structure. As a result of foreign invasions over the centuries, Polish people may be dark and Mongol-looking or fair with delicate features, blue eyes, and blonde hair. Those with fair complexions are predisposed to skin cancer. **Teach patients about protecting themselves from sun exposure and cancer.**
- In the Polish culture, a common belief is that enduring pain without complaining or asking for relief demonstrates virility in men and self-control in women. **Encourage acceptance of pain medication, explaining that it will promote healing.**
- Risk factors for newer Polish immigrants are connected with their employment in industries in their homeland. Heavy industry in Poland produced prolonged, significant air pollution and environmental neglect. Living in polluted environments led to an increase in premature deliveries, low-birth-weight children, diseases of the pulmonary and circulatory systems, and various forms of cancer.

Table 25.1 Health conditions common to Polish people

Health conditions	Causes
Cardiovascular disease	Environment, lifestyle
Diabetes mellitus	Genetic, lifestyle
Alcohol misuse	Environment, lifestyle
Cancer	Environment, lifestyle
Pulmonary disorders	Environment, lifestyle

- In Poland, air pollution remains a serious problem because of sulfur dioxide emissions from coal-fired power plants. In 1986, the Chernobyl incident in Russia contaminated the land and water systems of eastern Poland.
- **Assess new immigrant Poles for respiratory conditions and cancer resulting from radiation exposure.** Table 25.1 lists commonly occurring health conditions for Poles in three categories: lifestyle, environment, and genetics.

25.1.5 High-Risk Behaviors

- Poles have had a long history of excessive smoking. **Encourage anti-smoking beginning in elementary school.**
- In Poland, a high rate of alcoholic psychosis, cirrhosis of the liver, and acute alcohol poisoning exists. **Encourage socially responsible alcohol intake and refer patients to support groups for those with alcohol misuse.**
- **Children of immigrants should especially be targeted for counseling regarding the health effects of smoking and alcohol consumption.**
- **Screen Polish immigrants for diseases common in their home country: hypertension, CVDs, respiratory conditions, cancer (particularly leukemia), and thyroid disorders.**

25.1.6 Nutrition

- Meats and vegetables are cooked for a very long time, resulting in the destruction of B and other vitamins.
- The strong Catholic influence is evidenced by attending many food-laden celebrations, festivals, and rituals, each of which has its own traditional high-calorie foods.
- **Complete a dietary assessment on admission. Help patients structure a diet that is culturally acceptable, promotes healthy food choices, and is sustainable.**
- Staples of the Polish diet are millet, barley, potatoes, onions, radishes, turnips, beets, beans, cabbage, carrots, cucumbers, tomatoes, apples, and wild mushrooms.

Table 25.2 Polish foods

Common name	Description	Ingredients
Babka	Coffee cake	Yeast bread
Barszcz	Beet soup	Served plain or with sour cream
Bigos	Hunter's stew	Stew with game, sausage, sauerkraut
Chrusciki	Polish bowties	Fried egg dough
Golabki	Cabbage rolls	Cooked cabbage stuffed with chopped meat and rice in tomato sauce
Kielbasa	Sausage	Ground pork and spices
Ogorki smietanie	Sour cream and cucumbers	Sour cream, cucumbers
Pierogi	Boiled dumplings	Dumplings filled with potatoes, cheese, or sauerkraut
Sledzie	Herring	Pickled fish

- Common meats are chicken, beef, and pork. Traditional high-fat entrees include pigs' knuckles and organ meats such as liver, tripe, and tongue. *Kapusta* (sauerkraut), *golabki* (stuffed cabbage), *babka* (coffee cake), *pierogi* (dumplings), and *chrusciki* (deep-fried bowtie pastries) are common ethnic foods. Table 25.2 lists a variety of traditional Polish foods.

25.1.7   Pregnancy and Childbearing Practices

- Because family is very important, most Poles want children. In Poland, the Catholic Church strongly opposes abortion, which is the prevailing attitude of many Poles in the US and elsewhere.
- Fertility practices are balanced between the needs of the family and the laws of the Church.
- Pregnant Polish Americans are expected to seek preventive health care, eat well, and get adequate rest to ensure a healthy pregnancy and baby. Pregnant women usually follow the physician's orders carefully.
- The emphasis on food and "eating for two" is a common philosophy. **Counsel pregnant Polish-American women on weight gain and the long-term hazards.**
- Many consider it bad luck to have a baby shower, and even now, many Polish grandmothers may be reluctant to give gifts until after the baby is born.
- The birthing process is considered the domain of women. Newer Polish immigrants may feel uncomfortable with men in the birthing area or with family-centered care. **Do not make men feel guilty if they do not wish to be in the delivery room.**
- Women are expected to rest for the first few weeks after delivery. For many, breast-feeding is important. **Provide lactation education about care during breast-feeding and help the woman understand the balance between diet, rest, and exercise after delivery.**

25.1.8 Death Rituals

- Most Poles have a stoic acceptance of death as part of the life process.
- Family and friends stay with the dying person to negate any feelings of abandonment.
- Make arrangements for family to remain with the dying patient.
- Home hospice care is acceptable to most Poles.
- Health-care providers may encounter difficulty in convincing the family that the dying member may choose to refuse food as a result of the illness. **Provide factual information about end-of-life care.**
- Polish American family members follow a funeral custom of having a wake for 1–3 days, followed by a Mass and religious burial.

25.1.9 Spirituality

- The Catholic Church, with its required attendance at Mass on Sundays and holy days, is an integral part of the lives of most Polish people.
- Polish Americans have had a renewed interest in their ethnic roots. For example, their attendance at language classes, festivals, and Polish Catholic churches has become very widespread.
- Primary spiritual sources are God and Jesus Christ, with many Polish immigrants praying to the Virgin Mary, saints, and angels to ward off evil and danger. **Contact a chaplain if the patient wishes.**
- Honor and special attention are paid to the Black Madonna or Our Lady of Czestochowa. During times of illness and serious family concerns, one might hear a Pole evoking *Matka Boska*, which literally translated means "Mother of God."
- Many older Polish people believe in the special properties of prayer books, rosary beads, medals, and consecrated objects. **Do not remove religious items and amulets from patients.**
- Polish Americans commonly exhibit devotions to God, such as crucifixes and pictures of the Virgin Mary, the Black Madonna, and Pope John Paul II, in their homes.

25.1.10 Health-Care Practices

- Most Poles put a high value on stoicism and doing what needs to be done. Many go to health-care providers only when symptoms interfere with function; then they may carefully consider the advice provided before complying.
- Many Poles are reluctant to discuss their treatment options and concerns with health-care providers and routinely accept the proposed care plan.

- If Poles believe they are unable to pay the medical bill, they may refuse treatment unless the condition is life-threatening. Many have a strong fear of becoming dependent and resist relying on charity.
- Because many Poles consider Medicare, Medicaid, and managed care as forms of social charity, they are reluctant to apply for them. **Encourage patients and families and help them access social services when needed.**
- Poles usually look for a physical cause of disease before considering a mental disorder. If mental health problems exist, home visits are preferred. Few Poles turn to psychiatrists or mental health providers for help. Those who seek help from mental health providers do so as a last resort. Many individuals choose their priest or seek assistance from a Polish volunteer-run agency before going to a health provider for psychiatric help.
- Talk-oriented interventions and therapies without pharmaceutical or suitable psychosocial strategies are dismissed unless interventions are action oriented. In addition, Poles consult other family members and the community to assess the appropriateness of treatments.
- Polish Americans often seek self-help groups such as Alcoholics Anonymous before seeing a health-care provider.
- Assimilated Poles respect the health-care system and tend to seek specialized care when necessary.
- Given the continuation of limited access to care and the strong work ethic of this cultural group, health promotion practices are often undervalued by Polish Americans.
- Explain the benefits of health promotion activities and wellness checkups.
- Breast self-examination and Pap smear tests are poorly understood by many newer female immigrants. **Explain the benefits of breast self-examination and the necessity of Pap smears.**
- The Polish ethic of stoicism discourages the use of over-the-counter medications unless a symptom persists.
- Polish Americans may use certain remedies to cure an illness, such as tea with honey and spirits to "sweat out" a cold. Herbs and rubbing compounds may also be used for problems associated with aches, pains, and inflammation from overworked joints and muscles. Some additional common cultural practices included treating the symptoms of colds with herbs or poultices made from goose grease or fat. **Specifically ask about the use of home remedies and over-the-counter medications.**
- When a Pole is asked to undress for a physical examination, pay special attention to any medals pinned to the patient's undergarments. Most of these medals have special religious significance to the wearer. **Do not remove medals from clothing without asking permission from the patient or family.**
- Being unable to speak and understand English and the cost of health care and its complexity are the greatest barriers to health care for newer Polish immigrants. **Obtain interpreters when needed.**
- Poles are polite to authority figures and avoid offending a health-care worker by disagreeing with them. Thus, they may be reluctant to ask for clarifications on questionable issues.

- Owing to their strong sense of stoicism and fear of being dependent upon others, many Polish Americans use inadequate pain medication and choose distraction as a means of coping with pain and discomfort. Many Poles either deny or minimize their pain or level of discomfort. Poles with chronic illnesses may have similar attitudes; thus, persevering with pain is common. **Explain that pain medicine will promote healing. Use a visual analog scale to assess pain and assist patients with distraction techniques.**
- The ethic of being useful, independent, and a good Catholic influences one to refrain from using extraordinary means to keep people alive. The individual or family determines what means are considered extraordinary. Receiving blood transfusions or undergoing organ transplantation is acceptable.

25.1.11 Health-Care Providers

- Immigrant Poles often assess health-care providers by their demeanor, warmth, and show of respect.
- Health advice may be sought from chiropractors and local pharmacists as well as neighbors and extended family.
- **When caring for Polish patients, particularly older adults, make every attempt to address individuals by their surname. Attempting to pronounce the name demonstrates respect for the patient.**
- Polish women are modest and self-conscious and may refuse health care when asked to disrobe in front of a male health-care provider. **Provide a same-sex health-care provider when possible.**
- Physicians are held in high regard in Polish communities. Poles may change physicians if they believe their recovery is too slow or if a second opinion is needed.
- Immigrant Poles may be unfamiliar with the advanced roles of American nurses, who are expected to know about, plan, and be directly involved in the patients' care. Thus, many Poles may still want only the physician to explain all aspects of their care.

25.2 Reflective Exercises

1. Describe Polish communication practices as they relate to the practice of using third person.
2. What do Poles expect in terms of timeliness?
3. Describe taboo behaviors for Polish children.
4. Name and describe at least three diseases common among Poles.
5. What are some of the foods a Pole might eat?
6. Describe some prescriptive and restrictive Polish prenatal/postpartum practices.

7. What might you do to encourage a stoic Pole to accept pain medication?
8. How is mental illness perceived by Poles?
9. Identify at least three common folk remedies commonly used by Polish people.
10. Describe a traditional Polish funeral.

Bibliography

CIA (2016) World factbook: Poland. Accessed from https://www.cia.gov/library/publications/the-world-factbook/geos/pl.html

Mendoza SL, Halualani R, Dzewiecka JA (2002) Moving the discourse on identities in intercultural communication: structure, culture, and resignifications. Commun Q 50(3–4):312–327

Pietkiewics I (2015) Burial rituals and cultural changes in the Polish community: a qualitative study. Accessed from http://www.academia.edu/2277040/Burial_rituals_and_cultural_changes_in_the_polish_community_a_qualitative_study

Polish Culture Profile (2006). Accessed from http://www.diversicare.com.au/wp-content/uploads/2015/10/Polish.pdf

Public Opinion Research Center (2010) Poliah values: what is important, what is permitted, what must not be done. Accessed from http://www.cbos.pl/PL/publikacje/public_opinion/2010/07_2010.pdf

Purnell L (2013) People of Polish heritage. In: Purnell L (ed) Transcultural health care: a culturally competent approach. F.A. Davis, Philadelphia

Rempusheski VF (1998) Caring for self and others: second generation Polish American elders in an ethnic club. J Cross Cult Gerontol 3:223–271

Ryndyk O, Johannesses L (2015) Review of research literature on parenting styles and childrearing practices among poles. Historical and contemporary perspectives. Accessed from http://migrationnavigator.org/ftp/Sik-report%202015-3.pdf.pdf

Chapter 26
People of Puerto Rican Heritage

26.1 Overview and Heritage

The Island of Puerto Rico has a population of over 3.5 million. Approximately three million Puerto Ricans live in the mainland US; more than half live in the northeastern US. They self-identify as *Puertorri-queños* or **Boricua** (the Taíno Indian word for Puerto Rican) or **Niuyoricans** for those born in New York. Diversity among Puerto Ricans includes assimilation, individuality, and the variant characteristics of culture as presented in Chap. 1.

Puerto Rico is a commonwealth of the US. Citizenship status has created a controversial *va y ven* (go and come) circular migration in which individuals and families are often caught in a reverse cycle of immigration, alternately living a few months or years in the US and then returning to Puerto Rico. Puerto Ricans fear that the dominant American culture is a potential threat to the Puerto Rican culture, language, and future.

Although the educational system in Puerto Rico is similar to that in the US, when children emigrate from Puerto Rico to the mainland US, many educational organizations place them one grade below their previous academic year as a result of language barriers.

26.1.1 Communications

- English and Spanish are both official languages.
- Most Puerto Ricans speak with a melodic, high-pitched, fast rhythm, which is maintained when speaking English. **Avoid making comments about accent, use caution when interpreting voice pitch, and seek clarification when in doubt about the content and nature of a conversation that may seem confrontational.**

© Springer Nature Switzerland AG 2019
L. D. Purnell, E. A. Fenkl, *Handbook for Culturally Competent Care*,
https://doi.org/10.1007/978-3-030-21946-8_26

- Many individuals prefer to read or share sensitive information, options, and decisions with close family members. Some obtain assistance from extended family or community members who are knowledgeable in health matters. **Seek clarification of the information provided, ask for language preference in verbal and written information, and allow time for the exchange of information with questions and answers when critical decisions need to be made.**
- Great value is placed on interpersonal interactions such as *simpatía*, when an individual is perceived as likeable, attractive, and fun-loving.
- Most individuals enjoy sharing information about their families, heritage, thoughts, and feelings and expect the health-care provider to exchange personal information when beginning a professional relationship.
- If *confianza* (trust) is established, health-care providers can establish open communication channels with individuals and families.
- Most Puerto Ricans readily express their physical ailments and discomforts, with the exception of sexuality, which is taboo. **Set boundaries with discretion, emphasizing personal, rather than impersonal and bureaucratic, relationships.**
- Among younger generations and those born in the US, eye contact is maintained and is often encouraged. **With the more traditional, limited eye contact is a sign of respect, especially with older people.**
- Hand, leg, head, and body gestures are commonly used to augment messages expressed by words. Feelings and emotions are also expressed through touch.
- Gesturing includes an affirmative nod with an "Aha!" response, which does not necessarily mean agreement or understanding related to the conversation.
- Women greet each other with a strong, familiar hug and if among family or close friends, a kiss is included. Young women may take offense to verbal and nonverbal communications that portray women as nonassertive and passive. Men may greet other men with a strong right handshake and a left hand stroking the greeter's shoulder.
- **Greet Puerto Ricans formally and with a friendly handshake.**
- Most individuals are present-oriented, having a relativistic and serene view and way of life. **Respect this view, and assist in identifying options, choices, and opportunities to empower individuals to change health-risk behaviors.**
- **Carefully explain appointment times, and explain time limits at the beginning of an interview.**
- Respect is reflected in the way children talk, look, and refer to adults and older people. Rather than *Señora* (Mrs.) and *Señor* (Mr.), children and adults are expected to use the term *Doña* (Mrs.) and *Don* (Mr.) for most adults. Aunts and uncles have their name preceded by *tití* or *tío* (auntie/uncle) and *madrina* or *padrino* (godmother or godfather).
- **Address Puerto Ricans as *Sr.*, *Sra.*, *Don*, and *Doña*.**
- A single woman may use her name as follows: Sonia López Mendoza, with López being her father's surname and Mendoza her mother's. When she is married, the husband's last name, Pérez, is added with the word *de* to reflect that

she is married. This woman's married name would be Sonia López de Pérez; the mother's surname is eliminated. In business and health-care organizations, Señora López de Pérez is the correct formal title to use when promoting conversation or building a relationship. **Ask patients their complete name as well as their legal name.**

26.1.2   Family Roles and Organization

- Traditional families may view women as lenient, submissive, and always wanting to please men. Nevertheless, women play a central role in the family and the community, and the family is moving toward more egalitarian relationships. **Accept family decision-making styles without judgment.**
- Traditional cultural norms discourage an overt sexual-being image for women, but with family assimilation many of these traditional values disappear. When topics such as sex, sexually transmitted infection, or other infections, establish an environment built on *confianza* and *personalismo.*
- Most families expect their children to stay home until they get married or pursue a college education. The mother is expected to assume an active role disciplining, guiding, and advising children. Most fathers expect to be consulted, but they mainly see themselves as financial providers.
- Teen pregnancy, substance abuse, delinquent behaviors, and depression have been associated with the conflict between traditional values and those of mainstream American culture. **Address family conflict within the context of the family to resolve adolescents' mental health issues rather than using individual approaches.**
- Many traditional families socialize male children to be powerful and strong. This macho behavior encourages dominance over women.
- Female children are socialized with a focus on home economics, family dynamics, and motherhood, which places women in a powerful social status.
- Many Puerto Rican mothers use threats of punishment, guilt, and discipline, which can create stress and difficulties for adolescents as they struggle with the more permissive cultural patterns of the US, such as dating. **Provide counseling, and explain U.S. child abuse laws.**
- The family structure may be nuclear or extended. Family members include grandparents, great-grandparents, married children, aunts, uncles, cousins, and even divorced families with their children.
- **Use and encourage older Puerto Ricans to introduce health promotion and disease prevention education within their families.**
- Dependent older people are expected to live with their children and be cared for emotionally and financially. Placements in nursing homes and extended-care facilities may be considered inconsiderate and cause family members to feel guilty and to experience depression and distress.

- **Explore alternatives for long-term care, and provide information to all family members involved in this decision-making process. Address discharge planning and hospice and palliative care in a "conference-style" approach to develop strategies for providing emotional support and assistance to family members.**
- A family member is expected to be at the bedside of the sick person. **Ask the name of the family spokesperson, and document it in the patient's chart.**
- Less educated families may have difficulty educating young women about sexuality and reproductive issues. Thus, adolescents depend on educational organizations to learn about menstruation and the reproductive system. **Educating the family about sexuality issues gains respect and entrance into the trusted family environment.**
- Homosexuality continues to be a taboo topic that carries a great stigma among Puerto Ricans. Homosexual behavior is often undisclosed to avoid family rejection and preserve family links and support. Some men may perceive that sexual intercourse with men is a sign of virility and sexual power rather than a homosexual behavior.
- **When caring for gays and lesbians, inquire about their "disclosed" or "undisclosed" status. Do not disclose same-sex relationships to family.**

26.1.3 Workforce Issues

- The educational system in Puerto Rico emphasizes theoretical and practical content as well as neatness.
- Some have a relativistic view of time and may not value regular attendance and punctuality in the workforce. **Explain the requirement of punctuality in the dominant American workforce and any repercussions for chronic tardiness.**
- Most Puerto Ricans value personal relationships at work. Work is perceived as a place for social and cultural interactions, which may include listening to background music while performing job activities. This practice can lead to loud, cheerful, and noisy conversations that may require the employer's attention. **Explain the workforce culture of the dominant American workforce and the allowable noise level.**
- For many women, family responsibilities, pregnancy, and the health of their children and other family members is a priority over work. **Negotiate flexible work hours when possible.**
- Although most Puerto Ricans are bilingual, some may speak broken English, street English, or Puerto Rican *Spanglish.*
- Among younger generations and those born on the mainland US, eye contact is maintained and is often encouraged. With the more traditional, limited eye contact is a sign of respect, especially with older people. **Do not assume that the**

Table 26.1 Health conditions common to Puerto Ricans

Health conditions	Causes
Heart disease	Genetic, environment, lifestyle
Lung cancer	Environment, lifestyle
Ovarian and cervical cancer	Genetic, environment, lifestyle
Asthma	Environment, lifestyle
Prostate cancer	Environment, lifestyle
Stomach and esophageal cancer	Environment, lifestyle
Pancreatic cancer	Environment, lifestyle
Diabetes mellitus	Genetic, environment, lifestyle
Dengue fever	Environment, lifestyle

lack of eye contact means the person is not listening, does not care, or is being less than truthful.

26.1.4 Biocultural Ecology

- Given the mixed heritage of Native Indian, African, and Spanish, some Puerto Ricans have dark skin; thick, kinky hair; and a wide, flat nose; others are white-skinned with straight auburn hair and hazel or black eyes.
- Table 26.1 presents commonly occurring health conditions and diseases for Puerto Ricans in three categories: lifestyle, environment, and genetics. See Chap. 2.
- **Advise patients and families traveling to Puerto Rico to avoid exposure to endemic areas and to use mosquito repellent and protective clothing at all times. Become familiar with the signs, symptoms, and current treatment recommendations for dengue fever.**

26.1.5 High-Risk Health Behaviors

- Alcoholism is the precursor of increased unintentional injuries, family disruption, spousal abuse, and mental illness among families. **Help families seek resources for counseling.**
- Many women smoke tobacco. **Encourage smoking cessation and network with schools starting in elementary schools.**
- The longer one lives in the mainland US, and the more acculturated one becomes, the greater the use of marijuana, tobacco, and cocaine.
- Lack of condom use may be a significant risk behavior. Some men fear that if they use condoms, they portray a less macho image, have decreased sexual satisfaction, or portray that they have a sexually transmitted infection or HIV. Some women may refuse a condom, believing that the man thinks she is "dirty."

26.1.6 Nutrition

- Puerto Ricans celebrate, mourn, and socialize around food. Food is used (a) to honor and recognize visitors, friends, family members, and health-care providers; (b) as an escape from everyday pressures, problems, and challenges; and (c) to prevent and treat illnesses. Patients may bring home-made goods to health-care providers as an expression of appreciation, respect, and gratitude for services rendered. **Refusing food offerings may be interpreted as a personal rejection.**
- Being overweight is a sign of health and wealth. Some individuals eat to excess believing that if they eat more, their health will be better. **Negotiate an acceptable weight with patients.**
- Diets include a variety of roots called *viandas*, vegetables rich in vitamins and starch. The most common are celery roots, sweet potatoes, dasheens, yams, breadfruit, breadnut, green and ripe plantains, green bananas, tanniers, cassava, and chayote squash or christophines. **Become familiar with traditional foods and their nutritional content to assist families with dietary practices that integrate their traditional or preferred food selections.** Table 26.2 lists a variety of common Puerto Rican foods and their ingredients.
- Many families believe that a healthy child is one who is *gordita* or *llenito* (diminutive for fat or overweight) and has red cheeks. Mothers are often encouraged to add cereal, eggs, and *viandas* to their infant's milk bottles. **Educate mothers about healthy nutrition and the health risks for children who are overweight.**
- A traditional breakfast includes hot cereal such as oatmeal, cornmeal, or rice and wheat cereal cooked with vanilla, cinnamon, sugar, salt, and milk. Although less

Table 26.2 Common Puerto Rican meals and fritters

Puerto Rican meal	English translation
Alcapurrias	Green plantain fritters filled with meat or crab
Arepas de maíz y queso	Cornmeal and cheese fritters
Arroz con pollo	Rice with chicken
Arroz con gandules	Rice with pigeon peas
Arroz blanco (con aceite)	Plain rice (with oil)
Arroz guisado básico	Plain stewed rice
Asopao de pollo	Soupy rice with chicken
Bacalaitos	Codfish fritters
Bocadillo	Grilled sandwich
Mondongo	Tripe stew
Paella de mariscos	Seafood paella
Pastelillos de carne, queso o pasta de guayaba	Turnovers filled with meat, cheese, or guava paste
Pollo en fricase con papas	Stewed chicken with potatoes
Relleno de papa	Potato ball filled with meat
Sancocho	Viandas and meats stew
Sofrito	Condiment (1 tbsp)
Surullo de queso	Cornmeal fritters filled with cheese

common, many people also eat corn pancakes or fritters for breakfast. Lunch is served by noon, followed by dinner at around 5 or 6 p.m.

- Rice and stew *habichuelas* (beans) are the main dishes. Rice may be served plain or cooked and served with as many as 12 side dishes. Rice cooked with vegetables or meat is considered a complete meal. *Arroz guisado* (rice stew) is seasoned with *sofrito*, a blend of spices such as cilantro, *recao* (a type of cilantro), onions, green peppers, and other nonspicy ingredients. Rice is cooked with chicken, pork, sausages, codfish, calamari, or shrimp. It is also cooked with corn, several types of beans, and *gandules* (green pigeon peas), a Puerto Rican bean that is rich in iron and protein.
- A great variety of pastas, breads, crackers, vegetables, and fruits are eaten. Fritters are also common foods. Fried green or ripe plantains are a favorite side dish.
- Many ascribe to the hot-cold classifications of foods for nutritional balance and dietary practices during menstruation, pregnancy, the postpartum period, infant feeding, lactation, and aging. Table 26.3 identifies many of the foods that are

Table 26.3 Puerto Rican hot-cold classification of health-illness status, medications, herbs, and selected foods

Hot-cold classification	Health/illness status	Western medications	Traditional herbs	Foods
Hot	GI illnesses (constipation, diarrhea, Crohn's colitis, ulcer bleeding) Gynecologic issues (pregnancy, menopause) Skin disorders (rashes, acne) Neurological disorders (headache) Heart disease Urologic illnesses	Syrups Dark-colored pills Aspirin Anti-inflammatory agents Prednisone Antihypertensives Castor oil Cinnamon Vitamins (iron) Antibiotics	Teas: Cinnamon Dark-leaf teas	Cocoa products Alcoholic beverages Caffeine products Hot cereals (wheat, corn) Salt Spices and condiments Beans Nuts and seeds
Cold	Osteomuscular illnesses (arthritis, rheumatoid arthritis, multiple sclerosis) Menstruation Respiratory illnesses	Diuretics Bicarbonate of soda Antacids Milk of magnesia	Teas: Orange-lemon, chamomile Linden Anise	Rice Rice and barley water Milk Sugar and sugar products Mint Root vegetables Avocado Fruits Vegetables White meat Honey Onions

considered either "hot" or "cold." **Become familiar with these food practices when planning culturally congruent dietary alternatives.**

- Understanding that iron is considered a "hot" food that is not usually taken during pregnancy can help health-care providers in negotiating approval. **Educate women about the importance of maintaining adherence to daily iron recommendations, even during pregnancy and lactation.**
- A summary of food habits, reasons for practices, and recommendations for health-care providers during such developmental stages is included in Table 26.4.
- Black cohosh, evening primrose, St. John's wort, gingko, ginseng, valerian root, sarsaparilla, chamomile, red clover, and passionflower are the most common herbs. **Discuss the safety and efficacy of the most frequently used alternatives. Include use of complementary and alternative therapies in routine health assessments.**

Table 26.4 Puerto Rican cultural nutrition and health beliefs and practices during health stages

Behavioral period	Dietary and health practices	Cultural justification	Recommendation for health-care providers
Menstruation	Food taboos: Avoid spices, cold beverages, acid-citric fruits and substances, chocolate, and coffee	May induce cramps, hemorrhage clots, and physical imbalance. May produce acne during menstruation	Assess individual beliefs and acknowledge them. Incorporate traditional beliefs with treatments as required in the use of nonsteroidal anti-inflammatories for dysmenorrhea
	Foods encouraged: Plenty of hot fluids such as cinnamon tea, milk with cinnamon and sugar. Teas such as chamomile, anise seed, linden tea, mint leaves	Fluids encourage body cleaning of impurities. Hot beverages encourage circulation and reduce abdominal colic, cramps, and pain. Teas are soothing to all body systems	Encourage passive exercise. Provide information about the role of exercise in the reduction of menstrual pain. Support other practices
	Health practices: Avoid exercise and practice good hygiene. Do not walk barefoot. Avoid wind and rain. Stay as warm as possible	Exercise may increase pain and bleeding. Good hygiene is important for health. Walking barefoot during menstruation may cause rheumatoid arthritis and other inflammatory diseases. Warm temperatures promote circulation and the health of the reproductive system as well as prevent cramps	

Table 26.4 (continued)

Behavioral period	Dietary and health practices	Cultural justification	Recommendation for health-care providers
Pregnancy	Food taboos: Hot foods, sauces, condiments, chocolate products, coffee, beans, pork, fritters, oily foods, and citric products	May cause excess flatus, acid indigestion, bulging, and constipation. Chocolate and coffee may cause darker skin in fetus. Some believe citric products may be abortive	Encourage healthy food habits. Provide information about chocolate and coffee myths. Encourage fruits Discourage the use of raw eggs in beverages because of possibility of salmonella poisoning
	Food encouraged: Milk, beef, chicken, vegetables, fruits, ponches	Considered healthy and nutritious. Increase hemoglobin, strengthen and promote good labor	Encourage use of food recommended for pregnancy. Provide information about sexual activity
	Health practices: Rest and get plenty of sleep. Eat plenty of food. Follow diet cautiously. Many avoid sexual intercourse early in pregnancy. Practice good hygiene and take warm showers	Enhances health and prevents problems during birth. Sex may cause problems with baby or preterm labor	Encourage a balanced plan of exercise with emphasis on weight control and the health of the baby
Lactation	Food taboos: Avoid beans, cabbages, lettuce, seeds, nuts, pork, chocolate, coffee, and hot food items at all times	These foods cause stomach illnesses for the infant and mother, including baby colic, diarrhea, and flatus	Include a dietary plan that is balanced with substitute food items. Clarify any myths about infant diarrhea, colic, and flatus
	Food encouraged: Milk, water, ponches, chicken soup, chicken, beef, pastas, hot cereals	Improve health and increase hemoglobin and essential vitamins. Protect mother and infant from illnesses. Fluids and ponches increase milk supply. Red meats reduce cravings	As above with raw eggs. Provide information about reasons for stroke and facial paralysis. Provide time to ask questions and reduce anxiety during winter season deliveries
	Health practices: Avoid cold temperatures and wind. A few may avoid showering for several days after birth. Great attention is paid to health of the mother	Cold temperatures and winds are believed to cause stroke and facial paralysis in a new mother. Showering may cause respiratory diseases. Mother is believed to be at risk and fragile	

(continued)

Table 26.4 (continued)

Behavioral period	Dietary and health practices	Cultural justification	Recommendation for health-care providers
Infant feeding	Food taboos: Beans, too much rice, and uncooked vegetables	Believed to cause stomach colic, flatus, and distended abdomen. Too much rice causes constipation	Provide information about appropriate dietary patterns for infant
	Foods encouraged: Hot cereals, ponches, chicken broth or *caldos*. Fresh fruits, cooked vegetables, *viandas* (raw eggs, cereals, baby foods in milk bottle). Fresh fruit juices. Mint, chamomile, and anise tea. Sugar and honey used for hiccups	Believed to be nutritious, healthy, and to decrease hunger. *Caldos* are fortifying and prevent illness. Cooked vegetables are healthy and prevent constipation. Bottle food fills the baby. Fresh juices and fruits refresh the stomach. Teas help baby sleep and cure flatus. Sugar and honey have curing properties	Instruct about infant diet and timely introduction of food items to diet. Explain consequences of excessive weight in infants. Discourage food in bottle to prevent choking. Discourage raw eggs because of the risk of salmonella and egg allergies and the use of honey because of the risk of botulism. Teas are harmless and provide additional fluid when used in moderation without sugar
	Health practices: Keep baby warm while feeding	Warm babies eat, chew, and digest food better, and choking is decreased	Provide information about choking and babies

- If the individual is believed to have low blood pressure and is weak or tired, a small portion of brandy may be added to black coffee to enhance the work of "an old heart."
- **Do not criticize folk practices; it deters patients from seeking follow-up care and decreases trust and confidence. Inquire about nutritional practices, and incorporate harmless or non-conflicting practices into the diet.**
- During illness, chicken soups and *caldos* (broth) are used as a hot meal to provide essential nutrients.
- A mixture of equal amounts of honey, lemon, and rum is used as an expectorant and antitussive. A malt drink, *malta* (grape juice), or milk is often added to an egg yolk mixed with plenty of sugar to increase the hemoglobin level. Ulcers, acid indigestion, and stomach illnesses are treated with warm milk, with or without sugar.
- Herbal teas are used to treat illnesses and to promote health. Most herbal teas do not interfere with medical prescriptions. **Incorporating herbal tea with traditional Western medicine may enhance compliance.**

26.1.7 Pregnancy and Childbearing Practices

- Because the Catholic Church condones only the natural family planning methods and sexual abstinence, women do not commonly use birth control methods such as foams, creams, and diaphragms, which are perceived as immoral. Traditionally, abortion has never been accepted except in cases in which the life of the mother was in danger. Nevertheless, this view is changing, and many women now ascribe to this practice. Many women do not begin prenatal care until later in their pregnancies. **Encourage early prenatal visits and explain the importance of them in having a healthy pregnancy and newborn.**
- Men are socialized to be tolerant, understanding, and patient regarding pregnant women and their preferences. Women are encouraged to rest, consume large quantities of food, and carefully watch what they eat. Some individuals expect women to "get fat" and place little emphasis on weight control. Strenuous physical activity and exercise are discouraged, and lifting heavy objects is prohibited.
- Many women refrain from *tener relaciones* (having sexual intercourse) after the first trimester to avoid hurting the fetus or causing preterm labor.
- Loud and verbally expressive behaviors are a culturally accepted and an encouraged method of coping with pain and discomfort. Pain medications are welcomed. Most women oppose having a caesarean section because it portrays a "weak woman."
  Discuss the possibility of a caesarean section early in the pregnancy. Explain reasons for the necessity of invasive interventions during labor.
- The first postpartum meal is homemade chicken soup to provide energy and strength.
- Women are encouraged to avoid exposure to wind and cold temperatures, not to lift heavy objects, and not to do housework for 40 days after delivery. Some traditional women do not wash their hair during this time.
- Mothers who breast-feed are encouraged to drink lots of fluids such as milk and chicken soup. If feeling weak or tired, they drink beverages consisting of milk or fresh juices mixed with a raw egg yolk and sugar. Hot foods such as chocolate, beans, lentils, and coffee are discouraged because they are believed to cause stomach irritability, flatus, and colic for the mother and infant.
- Some women believe that breast-feeding increases their weight, disfigures the breast, and makes them less sexually attractive and undervalue the benefits of breast-feeding.
- **Provide factual information about breast-feeding, and educate women about myths and misconceptions. Because maternal grandmothers have a great influence on practices related to breast-feeding, maternal grandmothers should be included along with significant others in programs that encourage breast-feeding.**

26.1.8   Death Rituals

- Death is perceived as a time of crisis. The body is considered sacred and is guarded with great respect.
- Give news about the deceased to the head of the family first, usually the oldest daughter or son. **Use a private room to communicate such news, and have clergy or a minister present when the news is disclosed. Allow time for the family to view, touch, and stay with the body before it is removed.**
- Burial rituals may be delayed until all close family members can be present. Cremation is rarely practiced.
- Among Catholics, religious ceremonies such as praying, *velorio* (wake), novenas, and 9 days of saying the rosary follow the death.
- Families freely express themselves through loud crying and verbal expressions of grief. Some may talk in a thunderous way to God. Others may express their grief through a sensitive but continuous crying or sobbing. Some believe that not expressing their feelings could mean a lack of love and respect for the deceased. Some develop psychosomatic symptoms, and others may experience nausea, vomiting, or fainting spells as a result of an *ataque de nervios* (nervous attack). **Be nonjudgmental with patients' expressions of grief. Provide a private environment to help to minimize interruptions.**

26.1.9   Spirituality

- Most Puerto Ricans are Catholic; however, many have joined Evangelical churches because they offer a more personal spiritual approach.
- A few individuals practice *espiritismo*, a blend of Native Indian, African, and Catholic beliefs that deal with rituals related to spiritual communications with spirits and evil forces. *Espiritistas* (individuals who communicate with spirits) may be consulted to promote spiritual wellness and treat mental illnesses. *Espiritistas* are often consulted to determine folk remedies compatible with Western medical treatments.
- Because older persons may consider illness a result of sins, a cure should be sought through prayer or by the "laying on of hands."
- **Inquire about the family's wishes regarding the Sacrament of the Sick. Clergy are a resource for spiritual wisdom and help with a host of spiritual needs.**
- Although amulets have lost their popularity, some still use them. An *azabache* (small black fist) or a rabbit's foot might be used for good luck, to drive away bad spirits, and to protect a child's health. Rosary beads and patron saint figures may be placed at the head or side of the bed or on the patient to protect him or her from outside evil sources. **Ask permission before removing, cleaning, or moving amulets; a benediction may be requested before removing amulets or religious objects, giving the Sacrament of the Sick, or providing spiritual support.**

26.1.10 Health-Care Practices

- Many Puerto Ricans tend to underuse health promotion and preventive services such as regular dental or physical examinations and Pap smears. The variant characteristics of culture (see Chap. 1) influence health-seeking beliefs and behaviors.
- **Integrate individual, family, and community resources to encourage a focus on health promotion and enhance early health screening and disease prevention. Offering weekend, evening, and late-night health-care services increases the use of preventive services.**
- Most Puerto Ricans believe in "family care" rather than self-care. Women are considered the main caregivers and promoters of family health and are the source of spiritual and physical strength. **Incorporate the participation of the family in the care of the ill.**
- Natural herbs, teas, and over-the-counter medications are often used as initial interventions for symptoms of illness. Many consult family and friends before consulting a health-care provider.
- Although Puerto Rico is subject to U.S. drug administration regulations and practices, many are able to obtain controlled prescriptions from their pharmacists in Puerto Rico.
- Patients who use folk practices visit *botánicas* (folk religious stores), and use natural herbs, aromatic incenses, special bathing herbs, prayer books, prayers, and figurines for treating illness and promoting good health. **Refrain from making prejudicial comments that may inhibit collaboration with folk healers.**
- Most families prefer to keep chronically or terminally ill family members at home.
- Some place a damp cloth on the forehead to refresh the "hot" inside the body and relieve nausea. Some put the head between the legs to stop vomiting. Mint, orange, or lemon-tree leaves are boiled and used as tea to relieve nausea and vomiting.
- Barriers to using health-care services include poor English-language skills, low socioeconomic status, lack of insurance, and lack of transportation and child care. **Provide interpreters as needed, and offer transportation services, if available.**
- Most Puerto Ricans tend to be loud and outspoken in expressing pain. Some prefer oral or intravenous medications for pain relief rather than intramuscular injections or rectal medications. Herbal teas, heat, and prayer are often used to manage pain.
- **Do not censure expression of pain, or judge it as an exaggeration. This expressive behavior is a socially learned mechanism to cope with pain. *¡Ay bendito!* is a common verbal moaning expression for *dolor* (pain).**
- **Rural elderly individuals might have difficulty interpreting and quantifying pain, the use of numerical pain-identifying scales may be inappropriate.**
- Because mental illness carries a stigma, obtaining information or talking about mental illness may be difficult. Some families might not disclose the presence or history of mental illnesses, even in a trusting environment.

- Symptoms of mental illness are often perceived as the result of *nervios* (nerves), having done something wrong, or failing God's commandments. When someone is anxious or overcome with emotions or problems, he or she is just *nervioso*. Similarly, someone who is experiencing despair, anorexia, bulimia, melancholy, anxiety, or lack of sleep may be *nervioso(a)* or suffering from an *ataque de nervios* (attack of nerves) rather than being clinically depressed, manic-depressive, or mentally ill.
- **Acknowledge the confidentiality when obtaining a history. If trust is developed, health-care providers may get a more accurate response to their questions.**
- **Community-based settings such as churches, schools, and child-care centers are excellent environments for promoting physical and mental health.**
- **Organ donation is considered an act of goodwill and a gift of life. Autopsy may be considered a violation of the body. When discussions regarding autopsies and organ donations are necessary, proceed with patience, and provide precise and simple information. A priest or minister is often helpful and may be expected to be present at the time of death.**
- Many individuals are reluctant to receive or donate blood for fear of contracting HIV.
  Explore beliefs about blood donation and transfusions to dispel myths.

26.1.11 *Health-Care Providers*

- Most hold health-care providers in high regard. Distrust may develop if the health-care provider (a) lacks respect for issues related to traditional health practices, (b) ignores *personalism* in the relationship, (c) does not use advanced technological assessment tools, and (d) has a physical or personal image that differs from the traditional "well-groomed white attire" image.
- Many Puerto Ricans use traditional and folk healers such as *espiritistas* and *santeros* in conjunction with Western health-care providers. Some *espiritismo* practices are used to deal with the power of good and evil spirits in the physical and emotional development of the individual. *Santeros*, individuals prepared to practice *santería*, are consulted in matters related to the belief of object intrusion, diseases caused by evil spirits, the loss of the soul, the insertion of a spirit, or the anger of God.
- Men prefer male physicians and may feel embarrassed and uncomfortable with a female physician.
- A few individuals discount the academic and intellectual competencies of female physicians and may distrust their judgment and treatment.
- Some women feel uncomfortable with a male physician; a few prefer a male doctor.

- Older people may prefer older health-care providers because they are considered wise and mature in matters related to health, life experiences, and the use of folk practices.
- **Younger and female health-care providers must demonstrate an overall concern for the patient and develop respect and understanding by acknowledging and incorporating traditional healing practices into treatment regimens.**

26.2 Reflective Exercises

1. Identify three common health conditions among Puerto Ricans. What educational recommendation would you make to decrease them?
2. Explain the "hot" and "cold" theory of foods and illnesses. If you are unfamiliar with the foods, what might you do to learn more about them?
3. Describe Puerto Rican communication practices, including temporality and punctuality.
4. What are the roles for men and women in the traditional Puerto Rican culture?
5. What are acceptable fertility control practices for traditional Puerto Ricans? How might they view the use of condoms?
6. Explain Santería. For what conditions would a Puerto Rican cult a practitioner of Santería?
7. What are some traditional practices for pregnancy?
8. How to most Puerto Ricans see allopathic providers? Is the sex of the provider a concern?
9. What might you do to get allopathic and traditional practitioners to work together?
10. Identify death and funeral practices of traditional Puerto Ricans?

Bibliography

Centers for Disease Control and Prevention (2015) Hispanic Health. Accessed from https://www.cdc.gov/vitalsigns/hispanic-health/

Centers for Disease Control and Prevention (2011) Maternity practices in infant nutrition and care in Puerto Rico. Accessed from https://www.cdc.gov/breastfeeding/pdf/mpinc/states/mpinc2011puertorico.pdf

CIA (2016) World factbook: Puerto Rico. Accessed from https://www.cia.gov/library/publications/the-world-factbook/geos/rq.html

Diaz L (2012) A structuralist analysis of Puerto Rican Santería. Accessed from http://future-non-stop.org/c/b684d4ce0746140c5c08714153843d32

Munet-Vilaro F (1998) Grieving and death rituals of Latinos. Oncol Nurs Forum 25(10):1761–1763

Patcher L (1995) Puerto Rican health beliefs and practices: exploring the boundaries between ethnomedicine and biomedicine. Accessed from https://practicalbioethics.org/files/members/documents/Pachter_11_2.pdf

Purnell L (2013) People of Puerto Rican heritage. In: Purnell L (ed) Transcultural health care: a culturally competent approach, 4th edn. F.A. Davis, Philadelphia

Puerto Rico (2016) Countries and their culture. Accessed from http://www.everyculture.com/No-Sa/Puerto-Rico.html

Puerto Rico: family culture (n.d.-a) Accessed from http://knowpuertorico.weebly.com/family-culture.html

Puerto Rico: non-verbal communication (n.d.-b) Accessed from http://knowpuertorico.weebly.com/-non-verbal-communication.html

Chapter 27
People of Russian Heritage

27.1 Overview and Heritage

The Russian Federation (Russia), the largest country in the world with a population of over 142 million, is composed of 21 republics and covers parts of two continents, Asia and Europe. The diversity of Russians reflects the degree of acculturation, individuality, and the variant characteristics of culture (see Chap. 1) as well as the physical size of the country. Under Communism all media were controlled, disseminating only information that the government wanted people to know. This chapter focuses on Russians who are immigrants to the US, most of whom are well educated. Given the complicated history of Russian immigration, most of the information in this chapter pertains to immigrants who emigrated in the latter part of the twentieth century.

Under Communism, everyone could attend higher educational institutions, resulting in a well-educated population. Many scientists, physicians, and other professionals who have immigrated to the US have difficulty in practicing their profession, necessitating employment in occupations that lower self-esteem.

27.1.1 Communications

- The official language of Russia is Russian. However, most educated Russians in the US speak English to some extent because professional literature in Russia was in English.
- Many do not understand medical jargon and have difficulty communicating abstract concepts. **Speak slowly and clearly without exaggerated mouthing or using a loud voice volume, which changes the tone of words. Even though**

© Springer Nature Switzerland AG 2019
L. D. Purnell, E. A. Fenkl, *Handbook for Culturally Competent Care*,
https://doi.org/10.1007/978-3-030-21946-8_27

the patient may appear to understand the fundamentals of the English language, provide an interpreter if in doubt.

- Many older Russian Jewish immigrants speak **Yiddish**. Younger Jewish immigrants usually do not speak Yiddish because it was strongly discouraged in Russia.
- Punctuality is the norm, and many arrive early. Temporality is toward present and future orientation because in Russia the concern for many was to have food and other necessities not just for that day but also for the following days and weeks ahead. Thus, some may take medicine until the symptoms disappear and then save the remainder for future use. **Explain the necessity of taking all medicine as prescribed and that some prescriptions can be refilled.**
- Direct eye-to-eye contact is the norm among family, friends, and others, without distinction between genders. Tone of voice may be loud, extending to those nearby who are not part of the conversation.
- **Some may avoid eye contact when speaking with government officials, a practice common in Russia where making eye contact with government officials could lead to questioning. Do not interpret direct eye contact as aggression or a loud voice volume as anger.**
- Most individuals accept touch regardless of age and gender.
- Russians do not appreciate gestures such as standing with hands inserted into pockets or arms crossed over the chest; neither do they appreciate slouching postures when being interviewed. Until trust is established, many Russians stand at a distance and are aloof when speaking with health-care providers.
- **Greet patients with a handshake, and call them by their surname and title such as Mr., Mrs., Miss, Ms., or Dr.**
- Many educated women keep their maiden name when they marry.

27.1.2   Family Roles and Organization

- Family, children, and older adults are highly valued. Russians, accustomed to extended family living in their home country, continue the practice in the US.
- Decision-making among current immigrants is usually egalitarian, with decisions being made by the parents or by the oldest child. **Ask patients whom they want included in making medical decisions.**
- While parents work, grandparents care for grandchildren. **Include grandparents and older family members in health education.**
- Older people live with their children when self-care is a concern. Nursing homes are of poor quality in Russia; thus, children may fear placing parents in long-term care facilities. **Help families find long-term care facilities if needed, and encourage visiting the facility before a decision is made to place a family member there.**
- Children of all ages are expected to do well in school, go on for higher education, help care for older family members, and tend to household chores.

- Teens are not expected to engage in sexual activity. Sex and contraceptive education are not traditionally provided.
- Singleness and divorce are accepted without stigma. The Russian penal code was revised in 1997 and homosexuality is no longer a crime. However, gay and lesbian relationships are not always recognized or discussed and are still stigmatized by a large part of the population. **Do not disclose same-sex relationships to family members.**

27.1.3 Workforce Issues

- Russian nurses have as their practice motto the relief of suffering. The concept of teamwork is new to Russian nurses as are critical thinking and sensitive care-giving.
  Teach critical thinking skills during orientation.
- When communicating in the workplace, Russians promote the value of positive politeness, a technique that employs rules of positive social communication. The employee using positive politeness will say nice things that show that the person is accepted, while simultaneously providing support and empathy, and avoiding negative discourse with coworkers.
- When negotiating, Russians express emotion and invest considerable time and effort in supporting decisions.
- With colleagues and friends, Russians communicate directly, which is considered a sign of sincerity.
- Some may avoid eye contact when speaking with government officials, a practice common in Russia where making eye contact with government officials could lead to questioning. **Do not assume that the lack of eye contact means the person is not listening, does not care, or is being less than truthful.**
- Tone of voice may be loud, extending to those nearby who are not part of the conversation. **Do not interpret a loud voice volume as a sign of anger.**

27.1.4 Biocultural Ecology

- Russians who immigrate to the US are predominately white and have a physical structure similar to other white Americans, making them prone to skin cancer.
  Encourage the use of sunblock, which is unknown in Russia, and instruct all family members about the dangers of skin cancer.
- Table 27.1 lists commonly occurring health conditions for Russians in three categories: lifestyle, environment, and genetics. See Chap. 2.

Table 27.1 Health
conditions common to
Russians

Health conditions	Causes
Alcoholism	Genetic, environment, lifestyle
Depression	Environment, lifestyle
Gastrointestinal disorders	Environment, lifestyle
Respiratory diseases	Environment, lifestyle
Cardiovascular diseases	Genetic, environment, lifestyle
Cancer due to radiation	Environment
Dental disease	Environment, lifestyle
Tuberculosis	Environment, lifestyle
Diabetes mellitus	Genetic, environment, lifestyle
Hyperlipidemia	Genetic, environment, lifestyle

27.1.5 High-Risk Health Behaviors

- Both men and women have high smoking rates. **Encourage ceasing smoking. Assist with finding smoking cessation programs. Network with schools to stress anti-smoking beginning in elementary.**
- Some who have a heritage from Mongolian invaders cannot metabolize alcohol, making them more susceptible to alcoholism.
- Domestic violence is common and is related mostly to high rates of alcohol consumption. Domestic violence support services are not available in Russia; thus, patients are reluctant to report or seek help for domestic violence in the US.
  Explain support services in the US for domestic abuse. Help patients access them.
- **Screen all newer immigrants for conditions common in their home country. Provide scientific, factual information and assist in finding appropriate services and resources.**
- Many who come from Eastern Europe were exposed to the radiation effects of the Chernobyl disaster, resulting in a high incidence of cancer among this immigrant group.

27.1.6 Nutrition

- Common foods include cucumbers in sour cream, pickles, hard-boiled eggs as well as eggs served in a variety of other ways, marinated or pickled vegetables, soup made from beets (borscht), cabbage, buckwheat, potatoes, yogurt, soups, stews, and hot milk with honey. Cold drinks are not favored.
- Meat choices include pickled herring, smoked fish, anchovies, sardines, cold tongue, chicken, ham, sausage, and salami. Bread is a staple with every meal. The diet overall is high in fat and salt.
- **Determine preferred foods and preparation practices before providing dietary counseling. Incorporate traditional foods into prescriptions.**

27.1.7 Pregnancy and Childbearing Practices

- Contraception for Russian women is allowed without sanctions or taboos. Many new immigrants may not be aware of different methods of fertility control. Abortion is very common in Russia, and some may choose this option in the US. Russian condoms are made of thick rubber, discouraging their use by men. Explain that condoms in the US are made from latex (the cheapest), lambskin, polyurethane, and polyisoprene.
  Inform patients of the variety of birth control methods common in the US.
- Pregnant women have regular prenatal checkups, which are mandatory in Russia. During pregnancy, women are discouraged from heavy lifting and from engaging in strenuous physical activities; they are also protected from bad news that can be harmful to the fetus. They are encouraged to eat foods that are high in iron, calcium, and vitamins. Strawberries, citrus fruits, peanuts, and chocolate are avoided to prevent allergies in the newborn.
- As labor approaches, women take laxatives and enemas to facilitate delivery. Traditionally in Russia, husbands and relatives could not participate in the delivery or visit the hospital postpartum. There are no cultural restrictions for fathers or female relatives not to participate in delivery. **Ask patients who they wish to be involved in the delivery.**
- The delivery room should not have bright lights because many believe that bright lights will harm the newborn's eyes.
- Many women breast-feed until the infant reaches the toddler stage. Many women believe the breasts must be kept warm during feeding lest the mother get breast cancer later in life.
- Peri-care with warm water is important, and a binder is worn to help the mother's figure return to normal.
- In Russia, women were accustomed to 8 weeks of maternity leave before delivery and up to 3 years of leave following delivery. **Explain family leave practices in the US.**

27.1.8 Death Rituals

- Families want to be notified about impending death first, before the patient is told.
- Most families prefer to have the dying family member cared for at home.
  Facilitate transferring the dying person to the home if the family wishes and has the means to provide terminal care. Assist with obtaining hospice home care.
- Do-not-resuscitate orders are appropriate; many families want their loved one to die in comfort. However, consents for withholding or withdrawing treatment are usually declined by Russian patients.

- Russian Orthodox spiritual leaders institute a prayer, called *panikhida*, over the deceased, a vigil that includes chants, prayers, hymns, and gospel readings. **Ask permission from the family before contacting clergy.**
- Few families believe in cremation; most prefer interment.
- Both men and women may wear black as a sign of mourning. Black wreaths are hung on the door of the deceased's home. Expression of grief varies greatly. **Recognize a wide variety of grieving and bereavement behaviors.**

27.1.9   Spirituality

- Most who practice a religion are Eastern Orthodox or Jewish, with smaller numbers of Molokans, Tartar Muslims, Seventh Day Adventists, Pentecostals, and Baptists. Sixty percent of Russian people are nonreligious.
- The state-controlled Russian Orthodox Church was the only accepted religion in Russia (other religions were prohibited) until *perestroika* and *glasnost*. Russian Americans pray in their own way, which may be different from that of the dominant religion with which they identify. **Discuss individual religious practices with patients and family on admission.**

27.1.10   Health-Care Practices

- Most Russians define health as the absence of disease.
- Because health care is free at the point of entry in Russia, newer immigrants might not be aware of the need for insurance in the US. **Explain health reimbursement procedures in the US, and elicit assistance from social workers as needed.**
- Hospital stays in Russia average 3 weeks. Some patients may expect this in the US. **Explain shorter hospital stays in the US. Elicit the help of home health nurses, and teach extended family how to care for members at home.**
- Unmarried women are not accustomed to Pap tests; in Russia only married women get them. Mammography is uncommon in Russia. **Explain the importance of Pap tests and mammography, regardless of the woman's marital status.**
- Many individuals are preoccupied with remaining warm to prevent colds and other illnesses. Most do not want breezes from fans or drafts from an open window to blow directly on them. They may also be reluctant to apply ice at the recommendation of a health-care provider. **Explain the necessity of treatments with ice in factual terms. Do not assume that patients want ice in their drinks.**
- Some individuals may be reluctant to wash their hair for fear of catching a cold if the room is not warm or has a draft.
- Most Russians are stoical with pain and may not ask for pain medicine.

Offer and encourage pain medicine, and that it will help the healing process.

- Because of high radiation in parts of Russia, many fear having an x-ray. **Explain the necessity of x-rays and the lower radiation x-rays used in the US.**
- Patients are not accustomed to being told about cancer, terminal illnesses, or grave diagnoses because it is believed to make the condition worse. **Inform patients about grave diagnoses gradually, on their terms, and preferably in several meetings. Consult with family members, and listen for verbal and nonverbal cues for readiness to receive news about a grave diagnosis.**
- A primary treatment for a variety of respiratory illnesses is cupping. Physicians, nurses, and family members use cupping in Russia and the US. **Do not mistake round ecchymotic areas on children or adults as abuse. Ask if they have been practicing cupping.** A small glass cup, a *bonzuk* or *bonki*, has alcohol-saturated cotton or other materials in it. The material is lighted, and then the lighted cup is turned upside down on the patient's back. The skin is drawn into the cup. The cup is then pulled off the skin, leaving round ecchymotic areas 1–1½ inches wide.
- Common cultural practices include taking vodka with sugar for a cough; soaking one's feet in warm water for a sore throat; aromatherapy for a variety of respiratory illnesses; mud and mineral baths to promote healing; and herbs and teas for fever, and colds. **Incorporate nonharmful folk practices into allopathic prescriptions.**
- Most Russians are accustomed to seeing more than one health-care provider without the other's knowledge. Each provider may give prescriptions. Patients rarely inform the provider about seeing other providers. **Specifically ask patients if they are seeing other health-care providers and taking any prescription medicines, over-the-counter medicines, or traditional herbs and teas.**
- People are accustomed to not telling health-care providers about depression or any other emotional or mental health concerns because mental illness carries a significant stigma and mental health facilities are very poor in Russia. **Be alert for post-refugee stress and anxieties. Establish trust before eliciting information about mental health concerns.**
- Based on inadequate screening of blood in Russia, many have a great fear of contracting HIV from blood transfusions. Most Russians do not believe in organ donation. **Explain U.S. safety procedures with blood donations and transfusion.**

27.1.11   Health-Care Providers

- Health-care providers are respected. Because nurses function in higher roles in the US and some other countries than in Russia, they may be mistaken for physicians. **Explain the roles and education of nurses in the US.**

- Men and women are accustomed to living together in very small physical quarters; thus, most do not have a problem with privacy. Gender is not generally a concern in delivering care.
- **Always ask about gender concerns before providing intimate care to patients.**

27.2   Reflective Exercises

1. What are traditional Russians verbal and nonverbal communications practices
2. What is the temporality of most traditional Russians, including punctuality?
3. What are the primary decision making practices of current immigrant Russians?
4. List one health care conditions of Russians in each of the three categories: environment, lifestyle, and genetics. What might you do to decrease the incidence of these conditions?
5. What are common high-risk health behaviors of Russian immigrants? Include some tactics for decreasing them.
6. What are fertility control practices for Russians? What education would you provide for those who are interested?
7. Many Russians are afraid of x-rays. Why? What might you do to ameliorate their fear?
8. Identify a traditional treatment that Russians use for respiratory ailments. How would you know that this treatment is being used?
9. Identify some practices women use for a health pregnancy?
10. Identify some home traditional cultural practices used by Russians. Which of them might have deleterious effects?

Bibliography

Aorian K, Khatutsky G, Dashevskaya (2013) People of Russian heritage. In: Purnell L (ed) Transcultural health care: a culturally competent approach, 4th edn. F.A. Davis, Philadelphia

Callister CC, Getmanenko N, Farvish G, Eugeneyna MO, Vladimiorova ZN, Lassetter J, Turkina N (2007) Giving birth: the voices of Russian women. MCN: Am J Matern Child Nurs 32(1):18–24

CIA (2016) World factbook: Russia. Accessed from https://www.cia.gov/library/publications/the-world-factbook/geos/rs.html

End-of life care: the Russian culture (2002) Accessed from https://depts.washington.edu/pfes/PDFs/End%20of%20Life%20Care-Russian.pdf

Russia behind the headlines: why Russians prefer alternative medicine (2013) Accessed from http://rbth.com/arts/2013/08/22/why_russians_prefer_alternative_medicine_29141.html

Russia guide: a look at Russian language, culture, customs, and etiquette (2016) Accessed from http://www.commisceo-global.com/country-guides/russia-guide

Russians in Minnesota (2012) Accessed from http://www.culturecareconnection.org/matters/diversity/russian.html

Shpilko I (2006) Russian-American health care: Bridging the communication gap between physicians and patients. Patient Educ 64(1):331–341

Traditional medicine in Russia (2011) Accessed from http://blogs.dickinson.edu/russenviro/2011/12/15/278/

Chapter 28
People of Somali Heritage

28.1 Overview and Heritage

Somalia, with a population of over 10.8 million occupies the "horn" of Africa, a large land mass jutting into the Indian Ocean just south of the Arabian Peninsula, wrapping around Ethiopia and Kenya to the west. With a long coastline on the Indian Ocean, and also on the Gulf of Aden, which leads into the Red Sea and Suez Canal, Somalia has a long history of trade and interaction with other cultures. Most of the country is arid and sparsely settled by nomadic herders. The southern part has a tropical climate supporting agriculture and settled communities, including the large capital city, Mogadishu.

Somali people are united by a common language, Somali; religion, Sunni Muslim; and culture, which is influenced by their Arabic neighbors. However, as in many countries, tribal allegiance is primary.

After the collapse of the Somali government in 1991, at least one million Somali people fled to neighboring counties, such as Ethiopia, Djibouti, Yemen, and especially Kenya. Many of these refugees have now resettled in the US, Canada, the United Kingdom, Holland, Sweden, Denmark, Germany, Australia, and New Zealand.

Immigrants to Western countries were initially scattered, but many have now joined family and clan members in urban centers throughout the US, and especially Minneapolis–St. Paul, Minnesota. Many have built prosperous lives, opened businesses, and taken advantage of educational opportunities. In some Minnesota high schools and community colleges, more than half the students are Somali. Residual conflict persists in their new homes between members of different clans that are still at war in Somalia. Diversity among Somali varies according to acculturation, and the variant characteristics of culture (see Chap. 2).

© Springer Nature Switzerland AG 2019   307
L. D. Purnell, E. A. Fenkl, *Handbook for Culturally Competent Care*,
https://doi.org/10.1007/978-3-030-21946-8_28

28.1.1 Communications

- Somali is the universal language understood by all Somali people despite minor differences in dialect. Many people speak Arabic. A few older and educated Somalis speak Italian or English, depending on experiences with former colonial powers. In 1972, during a literacy campaign, the Somali language began to be written using phonetic spelling and European alphabet and taught in primary schools. Older Somalis are generally not able to read this language. **Determine language preference, Somali, Arabic, or English, upon admission and obtain an interpreter as necessary.**
- Somalis are polite and appreciate full introductions of everyone in the room, including interpreters and family members. **Formally introduce yourself by title and purpose.**
- Women and men are not shy about the process of obtaining medical care, and often will ask detailed questions until they are satisfied. This may include challenging questions such as "why is this medicine not working?" However, they are almost always respectful and usually conclude a medical or nursing interaction by expressing thanks.
- **Explain the purpose for every intervention.**
- Somalis dress up for medical appointments. Women and female children as young as 7 or 8 generally wear the Muslim headscarf and dress discreetly, often with a strong sense of color and fashion. Men dress in slacks and shirts with good quality leather belts and shoes. One rarely encounters sloppy or worn clothing. **Sloppy or worn clothing may be a symptom of other problems besides the reason for the appointment.**
- Traditional Somalis follow Muslim traditions regarding social touching. A man or woman may not touch a person of the opposite sex except for spouses or close relatives. However, Somalis are practical people; in the exam room it is understood that touching is permitted because there is a specific reason. **Because shaking hands when greeting a patient of the opposite sex is inappropriate, saying hello while putting your hands in your pocket or clasping them together puts a patient at ease.**
- Many Somali patients are uncomfortable having a genital exam by a provider of the opposite sex. **Provide a same-sex provider for intimate exams.**
- Eye contact and voice volume are not noticeably different from the European American cultural norms. If a patient is exceptionally shy and soft-spoken, the explanation may be depression or post-traumatic stress disorder rather than culture.
- **Smiling, informality, expressing friendship, and engaging in humor by the health-care provider are acceptable.**
- Naming is by the usual Muslim patronymic system. Both girls and boys are given a specific name at birth. This is then followed by the name of their father (for both sexes) and then the name of their grandfather. Jibrill Ibrahim Mohamud would be a boy named Jibrill whose father was Ibrahim and grandfather was Mohamud.

- Punctuality is not expected in a social context, but Somali people are generally quick to adapt to new cultures and learn that they are expected to be on time for appointments. **Explain the necessity of being on time for appointments.**

28.1.2 Family Roles and Organization

- A Somali nuclear family is considered to include father, mother, children, and grandchildren. The father is the head of the family and the final decision maker in family affairs. Family relations are based on hierarchal respect, as the **Qur'an** expects children to be unconditionally obedient to their parents. Children are also expected to care for their parents when they are in need, especially in times of sickness, aging, or financial hardship. There was no social security or pension back home in Somalia. **Accept family dynamics and hierarchy without judgment.**
- The father is the main producer of income and manager of family finances. Other family members including his wife and their grown children and sometimes grandchildren work and help out as they are able. The mother is a homemaker in charge of cooking, cleaning, shopping, and caring for children. She is revered and respected in this role, even more so than the father. **Accept gender roles without judgment.**
- When children reach the age of marriage, the father has the final say regarding to whom and when his daughters marry. When a son marries, his father participates in decision-making and the marriage ceremony but does not have the same level of authority as the father of the bride.
- Marriage in Somalia was an entirely religious ceremony conducted by a religious official versed in *Sharia* law, with two male witnesses and male members of the bride's family. Civil authorities sometimes issued certificates of marriage, but these were not essential.
- Divorce under *Sharia* can simply be declared by the husband in the presence of two male witnesses. If a wife wishes to divorce, she must apply to a court for judgment. In rural areas, she may be represented by her male relatives instead. Grounds for divorce include lack of financial support or sexual impotence. These traditions do not coincide with Western civil law, confusion and misunderstanding can occur, especially in newly arrived immigrants. **Carefully explain the legalities of divorce in the US.**
- Extended families are very important. Most Somali in the US help support their family members in Somalia or in refugee camps in neighboring countries.
- **Include extended family in health teaching.**
- Clan relationships are important. Wealthy and productive members of a clan are expected to help others, especially in the event of major illness or legal trouble. Administered by the clan leader and persists in Western countries.
- The family is becoming less cohesive in the US. Adolescents are less obedient to their parents. Fathers may believe they are losing their traditional positions as providers and heads of the family when two incomes are needed to survive.

Mothers may feel long work hours interfere with their role as homemakers. Large families add to financial stress. These factors and the American culture of individualism are loosening the tight relationships and traditional roles that have been characteristic of the collectivistic Somali family for generations. **Assist the family to understand the dominant American culture and refer for parenting skills when appropriate.**

- Same-sex relationships are hidden and not discussed. Marrying and producing children is the expectation. **Do not disclose same-sex relationships to family or outsiders.**

28.1.3 Workforce Issues

- In general, Somali culture values practicality and assimilation.
- Sometimes Somali employees will join each other and speak Somali, but they are generally polite about switching to English for the comfort of English-speaking co-workers. **Explain organizational policies as to when English only is required.**
- Somali culture also values hard work, self-reliance, and entrepreneurship. When employers fail to permit prayer at times prescribed by the Qur'an (imagine an assembly line), conflict may ensue. **Employers may not be aware that prayer at these times is mandatory and not something that can be done when convenient.**
- Conflict between Somali and African American teenagers occurs in some high schools and may carry over into the workplace.
- In the US, Somali immigrants are generally quick to learn American culture and the English language. Keeping apart from mainstream culture and language, as seen in some immigrant communities, is distinctly unusual.

28.1.4 Biocultural Ecology

- Somali people tend to have thin frames and light skin. Men are quite lean, women less so. Table 28.1 lists commonly occurring Somali health conditions in three categories: lifestyle, environment, and genetics.
- Tuberculosis is by far the most important infectious disease in Somali immigrants. Providers should not assume that testing for this has been accomplished during the immigration process. Extrapulmonary tuberculosis, especially in lymph nodes, is more common than pulmonary tuberculosis and often presents with fever alone or with vague other symptoms. **A tuberculin skin test should be performed unless the patient gives a history of a previously positive test. Screen every Somali patient for tuberculosis.**

Table 28.1 Health conditions common to Somalis

Health condition	Causes
Depression, post-traumatic stress disorder	Environment, lifestyle
Tuberculosis	Environment, lifestyle
Hepatitis B	Environment, lifestyle
Helicobacter pylori, intestinal parasites, and malaria	Environment, lifestyle
Trichuris trichuria, Enterobius vermicularis, Entamoeba histolytica, Dientamoeba fragilis, Ascaris lumbricoides, and Schistosoma mansoni	Environment, lifestyle
Lactase deficiency	Genetic

- Other common infectious diseases are hepatitis B, *Helicobacter pylori*, intestinal parasites, and malaria. Many arrivals come from refugee camps in countries other than Somalia. HIV and hepatitis C are relatively uncommon. Almost all Somali immigrants have antibodies to hepatitis A and do not need vaccination against this disease.
- *H. pylori* infection is common, although of uncertain relationship to common upper gastrointestinal symptoms. **Antibodies to *H. pylori* in the blood do not indicate current infection. A better indicator is the presence of *H. pylori* DNA in the stool. Screen Somali refugees for hepatitis B, *H. pylori*, intestinal parasites, and malaria.**
- Intestinal parasites may persist for years after immigration. Usually these do not cause symptoms but should be treated for public health reasons. Common infections involve *Trichuris trichuria, Enterobius vermicularis, Entamoeba histolytica, Dientamoeba fragilis, Ascaris lumbricoides,* and *Schistosoma mansoni*. Stool tests will also reveal many nonpathogenic organisms such as *Endolimax nana, Entamoeba coli, Entamoeba hartmanii, Iodamoeba buetchlii,* and *Chilomastix mesnili*. These do not need treatment. There is some controversy about whether *Blastocysitis hominis* is pathogenic; generally, treatment is recommended only if the patient is symptomatic or immunosuppressed.
- Little is written about drug metabolism specific to Somali. Lactase deficiency occurs in adulthood. **Assess for lactose intolerance on intake interviews.**

28.1.5 High-Risk Health Behaviors

- Rates of obesity and diabetes have increased dramatically, especially among women adjusting to availability of food in Western countries and decreased physical activity climates. **Encourage indoor physical activities such as swimming, working out at gyms, and dancing when appropriate.**
- The Muslim religion prohibits alcohol consumption, and few Somali immigrants use alcohol. **If you ask about alcohol, explain "we ask these questions of everybody."**

- More of a problem is *khat*, a vegetable stimulant chewed like tobacco that produces euphoria. *Khat* can be obtained from leaves, shoots, or twigs of an evergreen shrub found in Ethiopia, Kenya, and Yemen but not in Somalia. It can cause psychosis and is classified as a controlled substance by the United Nations. Possession in the US is a felony. It loses potency within days after harvest, so imported *khat* must be used promptly. People who use *khat* regularly tend to gather in houses where they sit together for many hours talking with each other, abandoning family and work obligations. *Khat* can cause erectile dysfunction. In pregnancy, it is associated with low birth weight.
- Promiscuity, intravenous drug use, and HIV are relatively uncommon. **If you ask about promiscuity, *khat*, intravenous drug use, or HIV, explain "we ask these questions of everybody."**
- Teenagers are more oriented toward school and family life and less likely to engage in premarital sex, gangs, weapons, and drugs than their American counterparts.

28.1.6 Nutrition

- As in many cultures, the main meal of the day is an important opportunity for a family to gather and talk. In urban areas, the main meal is usually in early afternoon. In rural areas, it's in the evening after work.
- In Somalia, common foods are rice, pasta, bread, potatoes, corn, beans, and sugar. Meat products include goat, lamb, beef, camel, and fish. Fruits include banana, mango, papaya, grapefruit, oranges, lemons, grapes, and dates. Beverages include milk from goats, cows, and camels; tea and coffee; and soda.
- A typical breakfast includes *angela*, a flour product resembling pizza, to which oil or sauce can be added, and tea. At mid-day, the meal is rice or pasta with sauce and vegetables including onions and potatoes, often with goat meat and bananas. For dinner, a mixture of beans, corn, oil, and sugar, with milk is consumed.
- Somali food is less spicy than Ethiopian, Middle Eastern, or Indian cuisine.
- Adapting to American culture and the ready availability of food has produced an epidemic of obesity and diabetes, especially in Somali women who were very physically active in their home country and are now living indoors in cold northern climates.
- **Initiate nutritional assessments and food preferences on intake assessments.**
- Vitamin D deficiency may occur in winter months related to lack of sunlight from being indoors and wearing a headscarf.
- Everyone fasts from sunup to sundown during the lunar month of **Ramadan**. Some religious individuals may select additional days of fasting at any time. **Adjust medication schedules to coincide with nonfasting hours. See Chap.** 8 **for Ramadan schedules.**

- Meat slaughtered by a butcher of faith is called *halal*. The butcher is ideally Muslim, but Christianity and Judaism are respected religions, and butchers from these faiths are acceptable. Pork of any kind is forbidden. **Provide *halal* meats when possible; otherwise, ask the patient's family to bring food from home. Some online services have frozen *halal* foods available.**

28.1.7 Pregnancy and Childbearing Practices

- Somali families tend to be large by American and European standards. Children and pregnancy are invariably considered a precious gift. Infertility is a psychological burden that causes marked suffering and elicits sympathy. Abortion is prohibited.
- Contraception for the purpose of spacing children is allowed by Islam. Caesarian sections are avoided because of the fear of preventing subsequent normal pregnancy and delivery.
- During pregnancy, women are expected to eat well and avoid heavy exertion or lifting. The custom of working until the due date seems strange to many Somali women. However, Somalis tend to be practical and adapt to new situations and expectations.
- **Incorporate healthy traditional practices into care.**
- After childbirth, *afartanbah* is observed, a period of 40 days during which the woman stays home, eats well, is assisted by her family and neighbors, and abstains from sexual intercourse.
- Breast-feeding for as long as possible up to 2 years is a religious obligation. Supplementation with milk or formula and other food as the child grows is allowed.
- A naming ceremony is held between 2 weeks and 40 days.

28.1.8 Death Rituals

- When a patient appears to be dying, Somali families come together to provide support for each other and the dying person. This may include many members of the extended family. A religious person such as the ***imam*** from the local mosque may visit the person in the hospital, offer prayers, and help mobilize community support. **Assist family to obtain an *imam* and make arrangements for family to be at the bedside of the dying patient.**
- As faithful Muslims, Somalis believe that Allah will determine how long a person will live. For this reason, discussing end-of-life care or advance directives is taboo and may be misunderstood. In general, impending death is not discussed directly with the patient. "Allah will decide, and I will follow" is a common expression. **When the need arises to discuss end-of-life care, hospice, or palliative care, approach the family indirectly.**

- After death, the funeral home cooperates with the mosque in preparing the body, washing it, wrapping it in white cloth, and delivering it to the mosque for a prayer service. Bodies are never cremated, always buried. In Somalia, bodies are buried without coffins; the idea of being "in a box" may be an uncomfortable one.

28.1.9   Spirituality

- Essentially all Somalis are devout Sunni Muslims who believe that Allah cares for them in their daily lives, and follow instructions in the Qur'an regarding prayer and fasting, caring for the poor, and kindness to strangers.
- In Islam, no priests mediate between the believer and God, but there are religious leaders including teachers, preachers, and mosque officials.
- In the US, Somali attend formal services at mosques which serve as centers of community life as well as religion. Prayers *(salat)* are performed five times daily, in any location, facing Mecca. These prayers take about 3 min apiece and involve kneeling and touching the forehead to the ground. On Fridays, men are required to attend the local mosque to listen to the *khadiib* who gives a sermon called *khudba*. Women may go to mosque but attendance is not required. **Plan care around prayer times. Assist patients and families with prayer as necessary. Contact a chaplain for assistance.**
- Strict fasting occurs during the lunar month of Ramadan. Ramadan advances each year within the Western calendar. During this time, all people older than 14 years refrain from taking food or drink from sunrise to sundown. Sometimes this is interpreted to include medication, including insulin. Exceptions can be made in the event of illness, travel, menstruation, pregnancy, etc. At the end of this month comes a feast called *Eid al-Fidri*, a celebration in which everyone joins together to give thanks, make a special prayer, wear their best clothes, and dine out. **Adjust meal and medication schedules to accommodate for Ramadan.**
- Another feast day, *Eid al-Adha*, coincides with the annual pilgrimage to Mecca *(haj)*. Every Muslim is expected to make *haj* once in his or her life.
- There is a strong sense of sharing and being part of a faithful community. Individualism is regarded as self-centered. Somali culture does not encourage celebrating birthdays. Many refugees will give January 1 of the year they were born when confronted by immigration authorities or Western medical records systems.

28.1.10   Health-Care Practices

- Somali sometimes seek healing from traditional and religious healers first. But they do not hesitate to go to a health-care provider when symptoms persist or are alarming. This includes seeking care for mental health problems. Older individuals who feel well may not understand the concept of preventive medicine, but usually accept advice when the reasons are explained to them. **Integrate non-harmful traditional health-care practices with allopathic practices and treatments.**

- Somali patients tend to listen, ask questions, and value medical advice. Seeking care and keeping appointments are usually not barriers to care; lack of insurance may be a barrier.
- Affording prescription medication is a problem for many patients; prescribing less expensive generic drugs is a discreet way to help. Newly arrived persons will appreciate assistance in navigating the complex American health-care system. **Recommend generic medications when feasible and assist patients and families in navigating the health-care system. Refer to social services and community agencies as necessary.**
- Traditional Somali healing practices include ritual reading or reciting of verses from the Qur'an; creating superficial skin burns on the abdomen, chest, or head; superficial cutting of the skin (especially used for jaundice in children suffering from hepatitis A); *mingis* (or *saar*), which consists of ceremonial drumming, singing, and dancing around the sick person; and herbs that may be consumed or used as a poultice. Such practices are guided by a traditional healer, usually an older man called an ***alaqad***, and may include participation by a large group of people. The patient is expected to provide food and clothing for the gathering, and to pay the *alaqad* a sum of money. Some traditional healers specialize in setting fractures or trephine procedures for subdural bleeding.
- In Somalia, prepubertal girls are circumcised by removing the clitoris, labia minora, and labia majora. The vaginal orifice is then narrowed by sutures, leaving room for menstruation. At the time of marriage, these sutures are removed. In another variety of this practice, only the clitoris is removed. While these practices seem extreme to Western medical providers, they are accepted as normal by many Somali women who value chastity and tradition. In the US, this practice is illegal. In response, some families arrange for their children to travel to Africa for circumcision. **Carefully explain that female circumcision is illegal in the US as well as in most other countries.**
- There are no restrictions on accepting blood products when needed.
- Pain is a common complaint, including the unusual pattern of pain on one half of the body. Pain may be a culturally acceptable way to express sadness or social and psychological discomfort. **Ascertain if the pain is physical or emotional in origin.**
- Many Somali immigrants have significant depression or post-traumatic stress disorder related to war, family disruption, and long years in refugee camps. Symptoms include depressed mood, insomnia, rumination, social isolation, lack of trust, and a host of somatic symptoms for which investigations do not reveal a specific cause. **Antidepressants are generally more helpful than insight psychotherapy.**
- Many patients are uncomfortable with the word "depression" and respond better to simpler explanations like too much stress, not enough sleep, or a reaction to the past. There is a sense that people should bear up under their symptoms, that

Allah would not ask a person to bear more than he or she is capable of tolerating.
- **Avoid the use of the word "depression" on first encounter because it may elicit feelings of weakness or guilt. Approach mental health issues cautiously and indirectly until trust is established, unless the situation is an emergency.**

28.1.11 Health-Care Providers

- In urban Mogadishu, Western medicine was practiced by European standards before the civil war, supplemented by traditional practices. Somali Western medicine and traditional practices as complementary. **Integrate traditional health-care providers and traditional treatments with Western health-care providers and treatments.**
- Respect for medical education and expertise is the rule; doctors and nurses are valued as professionals. A health-care career is a common choice for young people. Somali integrate easily into Western medical systems, displaying medical values such as patience, tactfulness, and kindness. **Encourage Somali children and older adults seeking a second career to think about the health professions.**

28.2 Reflective Exercises

1. What is the primary language spoken by Somalis? Where might you find an interpreter for this group?
2. How would you greet a Somali male patient? A Somali female patient?
3. Explain decision making practices among the Somali. What implications does this have for your nursing care?
4. What are the primary non-communicable health conditions among Somalis? What might you do to help decrease some of them?
5. Name at least three infections conditions common among Somali immigrants. What are the implications for your initial assessment and follow up care?
6. Explain the name format for Somali men. For Somali Women.
7. Identify infant feeding beliefs among Somali families. What educational advice would you offer?
8. What is the dominant religious practice among Somalis? How does this affect your care?
9. How is depression seen among Somalis? What is your approach to addressing depression?
10. Identify several traditional health-care practices. How might you incorporate traditional healers with allopathic treatments?

Bibliography

Adair R, Greminger A, Post B (2006) Differences between insured and uninsured patients in South Minneapolis. Minn Med 89:46–47

Adair R, Nwaneri O (1999) Communicable disease in African immigrants in Minneapolis. Arch Intern Med 159:83–85

Adair R, Jama Y (2013) People of Somali heritage. In: Purnell L (ed) Transcultural health care: a culturally competent approach, 4th edn. F.A. Davis, Philadelphia

CIA (2016) World factbook: Somalia. Accessed from https://www.cia.gov/library/publications/the-world-factbook/geos/so.html

Epstein R, Fiscella K, Gipson T, Volpe E, Jean-Pierre P (2007) Caring for Somali women. Implications for clinician-patient communication. Patient Educ Couns 66(3):337–345

Global Affairs Canada (2014) Cultural information - Somalia. Accessed from https://www.international.gc.ca/cil-cai/country_insights-apercus_pays/ci-ic_so.aspx?lang=eng

Gerristen AAM, Bramsen I, Deville W, van Willigen LHM, Joven JE, Ploeg HM (2006) Physical and mental health of Afghan, Iranian, and Somali asylum seekers and refugees living in the Netherlands. Soc Psychiatry Med 41(1):18–26

Lewis T (2016) Somali cultural profile. Ethnomed. Accessed from https://ethnomed.org/culture/somali/somali-cultural-profile

Pavlish CL, Noor S, Brandt J (2010) Somali immigrant women and the American health care system: discordant beliefs, divergent expectations and silent worries. Soc Sci Med 71(2):353–361

Somalis in Minnesota (2016) Culture care connection. Accessed from http://www.culturecareconnection.org/matters/diversity/somali.html

Steinman L, Doescher M, Keppel G, Pak-Gorstein S, Graham E, Haq A, Johnson DB, Spicer P (2009) Understanding infant feeding beliefs and practices and preferred nutrition education and health provider approaches: an exploratory study with Somali mothers in the US. Matern Child Nutr 6(1):67–88

Chapter 29
People of Turkish Heritage

29.1 Overview and Heritage

Türkiye, as it is written in Turkish, means "land of Turks." Referred to as a geographic, religious, and cultural crossroads, the Republic of Turkey is situated at the geographic intersection of Europe, Asia, the Middle East, and Africa. One-30th of Turkey lies in Europe and is referred to as Thrace. The remainder, located in Asia, is commonly called Asia Minor or Anatolia. Because of its geopolitical location and its cultural and religious ties, Turkey remains strategically important to the West and is a strong ally of the US.

Most people of Turkish descent in the US live in New York, California, New Jersey, and Florida. Many come from the elite and upper-middle classes. Education is highly valued in Turkey, and significant numbers of Turks in the US hold advanced degrees. Great diversity exists among Turks in the US, and their beliefs and practices vary depending on acculturation, individuality, and the variant characteristics of culture (see Chap. 1).

Although Westernization had begun before independence, Turkey's former president, Mustafa Kemal Atatürk, became synonymous with Westernization and secularism. During his presidency from 1923 to 1938, he initiated many reforms, including banning the fez, outlawing polygamy, instituting marriage as a civil contract, abolishing communal law for ethnic minorities, removing Islam as the state religion, promoting nationalism and pride, instituting educational and cultural reforms, making surnames obligatory, changing the weekly day of rest from Friday to Sunday, and electing 17 female deputies to the National Assembly. Atatürk died on November 10, 1938, but he is still revered as the father of Turkey.

© Springer Nature Switzerland AG 2019
L. D. Purnell, E. A. Fenkl, *Handbook for Culturally Competent Care*,
https://doi.org/10.1007/978-3-030-21946-8_29

29.1.1 Communications

- Turkish, spoken by 90% of the population, does not distinguish gender pronouns, such as "he" from "she" or "her" from "his." However, Turkish does distinguish a formal from an informal "you," signifying the importance of status in Turkish society.
- Pronouncing Turkish names can be troublesome for Americans. Until 1928, Turkish was written in Arabic script, but under Atatürk's direction, a Turkish alphabet was developed based on Latin script. The Turkish alphabet is much like the English alphabet. It does not have a "w" or an "x," and additional sounds are symbolized by an "i" without a dot; a "⇨," an "ö," and a "ü" with accents; and an "☺" and a "ç" with a cedilla, symbolizing "sh" and "ch," respectively.
- Speaking in loud voices is common, signifying excitement or involvement in a discussion. More than one person may speak at the same time or to interrupt another person; this is not necessarily considered rude. However, someone of lower status should not interrupt someone of higher status. **Do not interpret speaking in a loud voice as anger.**
- Group affiliation is valued over individualism in Turkish society. Identity may be determined by family membership or group, school, and work associations. Turks generally do not desire much privacy and tend to rely on cooperation between family and friends, although competition between groups can be fierce.
- For women, expressions of anger are usually acceptable only within same-sex friendships and kinship networks or toward those of lower social status. Generally, women are not free to vent their anger toward their husbands or other powerful men.
- Touching, holding hands, and patting one another on the back are acceptable behaviors between same-sex friends and opposite-sex partners. Likewise, personal space is closer between same-sex friends than opposite-sex partners; physical proximity is valued as a sign of emotional closeness.
- **Touch, when necessary, is allowed and expected from health-care providers. However, strict Muslims may not shake hands or touch members of the opposite sex, especially if they are not related.**
- Eye contact may be used as a way of demonstrating respect. When interacting with someone of higher status, one is expected to maintain occasional eye contact to show attention; however, prolonged eye contact may be considered rude or may be interpreted as flirting. **Maintain eye contact with patients, demonstrating respect and truthfulness.**
- Turkish people tend to dress formally; men wear suits rather than sports jackets and slacks on social occasions. Women tend to dress modestly, wearing skirts and dresses rather than slacks. More traditional Muslim women may wear very modest clothing and cover their heads with either a black or a colorful print scarf. However, styles continue to change, and denim jeans and casual dress are becoming common among young people for less formal occasions. **Expect a wide variety of dress among Turks, depending on religiosity, acculturation, and age.**

- Turks tend to display emotions openly, such as happiness, disgust, approval, disapproval, and sadness, through facial expressions and gestures. "No" is indicated by raising the eyebrows or lifting the chin slightly while making a snapping or "tsk" sound. Appreciation may be expressed by holding the tips of the fingers and thumb together and kissing them and is commonly used to express appreciation for food.
- Turkish take pride in keeping their homes immaculately clean; one is expected to remove one's shoes inside the home. Most Turkish hosts in Turkey and many in the US offer slippers to their guests. Whether wearing shoes or not, showing the sole of one's foot is considered to be offensive. Women are expected to sit modestly with knees together and not crossed. **Home health-care providers should look for cues and ask if they should remove their shoes when entering the home of a Turkish family.**
- In business relationships and for health-care appointments, punctuality among Turkish Americans is gaining in importance. **Explain the necessity of being punctual for health-care appointments.**
- Titles are used to show respect and acknowledge status. Strangers are always greeted with their title, such as *Bey* (Mr.), *Hanim* (Mrs., Miss, or Ms.), *Doktor*, or *Profesör*. Members of the family are also addressed using specific titles that recognize relationships, such as *agbi* (older brother or older close male friend), *amca* (uncle or elderly male relative or stranger), *abla* (older sister or older close female friend), *teyze* (maternal aunt or older female relative or older female stranger), and *yenge* (wife of a brother or paternal uncle). **Greet patients formally with a title or Mr., Mrs., or Ms.**

29.1.2 Family Roles and Organization

- In a very traditional Turkish home, the father is considered the absolute ruler. Less traditional families show more equality between spouses, especially in nuclear families in which the wife is well educated and works outside the home. Modern women tend to be more Westernized than some of their Middle Eastern or Muslim counterparts.
- **Ascertain the family spokesperson before eliciting health-care decisions. Accept family decision-making patterns without judgment.**
- Once children enter school, they are expected to study hard, show respect, and obey their elders, including older siblings. Traditionally, children are not allowed to act out or talk back to their superiors. Light corporal punishment is generally acceptable. **Explain legalities of child abuse in the US.**
- Circumcision is a major rite of passage for a male child. This is a time of celebration within the extended family, and newly circumcised boys are honored with gifts. Traditionally, boys can be circumcised up to the age of about 12 years, although the modern trend is to perform the circumcision in the hospital shortly after birth.

- Sexual interaction is strongly discouraged among youth and the unmarried, especially for women, for whom virginity is a strong cultural value.
- Although financial independence is valued in Turkish culture, independence from the family is not encouraged. Adult children, especially men, remain an integral part of their parents' lives, and parents expect their children to care for them in their old age.
- Parents or other family members are often consulted before major decisions are made.
- **Arranging for a family conference with the family spokesperson is a useful tactic in order to obtain compliance with health prescription and health teaching.**
- Young people generally live in their parents' home until they are married, unless school or work necessitates other arrangements. Although some marriages in Turkey are still arranged, this practice is not as common among Turks in the Western world.
- Individuals are socialized to take care of parents when self-care becomes a concern. Grandparents play a significant role in raising their grandchildren, especially if they live in the same home.
- In many Turkish families, aunts, uncles, cousins, and in-laws form the extended family. A cooperative relationship includes sharing child care, labor, and food when necessary.
  Use the extended family arrangement for health teaching and home care.
- Divorce is becoming more common but remains socially undesirable, especially for women. Widows are generally taken care of by their late husband's family; depending on their age and socioeconomic background, they may have the option to remarry.
- Homosexuality is beginning to be received "at a distance." Most Turks would be hesitant to associate themselves with the gay community. **Do not disclose same-sex relationships to family members or others.**

29.1.3 Workforce Issues

- Türkiye is a group-oriented culture with a workplace that may be more team oriented.
- Turkish relationship orientation may lead to dependence on personal contacts and networks to accomplish tasks. Developing relationships and networks may appear as nepotism or as too much socializing from the American perspective.
- **Orient American nurses to the Turkish perspective on relationships and the Turkish perspective of relationship to American nurses.**
- Turkish employees expect an authoritative relationship between superior and subordinates. Indirect criticism is expected and appreciated to "save face." A Turk may be highly offended if openly criticized, especially if done in front of other people.
- **Do not criticize Turks in front of other employees.**

- They may be reticent about asking questions for fear of exposing a lack of knowledge. **Ask employees to demonstrate a procedure to ensure understanding.**
- Turks handle differences of opinion in a brisk and clear-cut manner.
- Because most Turkish immigrants speak English, language barriers in the workplace may be only subtle.
- Turks perceive that aggressive face-to-face confrontation may cause relationships to deteriorate; therefore, the dominant means of conflict resolution is collaboration reinforced by compromise and forcing. Compromise and avoidance behaviors are more likely among peers, while accommodation behaviors are used with superiors.
- Türkiye is known for its high-power distance (the psychological and emotional distance between superiors and subordinates), respect for authority, centralized administration, and authoritarian leadership style.

29.1.4   Biocultural Ecology

- The Turkish population is a mosaic in terms of appearance, complexion, and coloration. Appearances range from light-skinned with blue or green eyes to olive or darker skin tones with brown eyes. Mongolian spots, usually found at or near the sacrum, are common among Turkish babies and should not be confused with bruising.
- Table 29.1 lists commonly occurring health conditions for Turks in three categories: lifestyle, environment, and genetics. See Chap. 2.
- Endemic goiter associated with iodine deficiency is a major health problem in Turkey despite iodine prophylaxis. **Screen newer immigrants for helminthiasis and goiter.**

Table 29.1 Health conditions common to Turks

Health conditions	Causes
Helminthiasis	Environment, lifestyle
Goiter	Environment, lifestyle
Behçet's disease	Genetic
Thalassemia	Genetic
Cardiovascular disease	Genetic, environment, lifestyle
Diabetes mellitus	Genetic, environment, lifestyle
Hepatitis	Environment, lifestyle
Parasitic diseases	Environment, lifestyle
Malaria	Environment, lifestyle
Tuberculosis	Environment, lifestyle
Rickets	Environment, lifestyle
Lactase deficiency	Genetic, environment, lifestyle
Sickle cell disease	Genetic, environment

- **Assess for lactose intolerance as part of the intake interviews, and help patients identify alternative sources of calcium. Assess Turkish patients for β-thalassemia before prescribing medications.**
- Turks in Turkey have high rates of occupational injuries. **Provide education regarding safety issues in occupational health.**
- Turks are at high-risk for multiple cancers because of the Chernobyl accident in Russia in 1986; this includes children born after this period as well.

29.1.5 High-Risk Health Behaviors

- Cigarette smoking is widespread in Turkey and tends to start at an early age. Turkey, a major producer of tobacco in the world, has instituted very limited anti-tobacco activities. **Encourage patients to cease smoking, and help them access smoking cessation programs. Start teaching the hazards of smoking in elementary school.**
- Turks tend to consume less alcohol than Americans or Europeans, perhaps as a result of the Muslim culture, which discourages more than moderate alcohol use.

29.1.6 Nutrition

- Food choices are varied and tend to provide a healthy, balanced diet. Turkish cooking is not terribly spicy, and is prepared artfully and fastidiously.
- While a typical family dinner may be simple, guests are generally served a bountiful array of dishes. More food is always better, and dinner guests may have difficulty finishing everything on their plates because Turkish hostesses may relentlessly offer to replace what has been eaten. Polite guests refuse the first offer.
- Breakfast is typically a simple meal of *beyaz peynir* (white feta cheese), olives, tomatoes, eggs, cucumbers, toast, jam, honey, and Turkish tea. Turks typically eat their evening meal at about 8 p.m. **Take mealtimes into consideration when teaching Turkish American patients about medication therapies.**
- Soups range from light to substantial. Hors d'oeuvres include a great variety of small dishes, either hot or cold, such as stuffed grape leaves in olive oil, olives, Circassian chicken with walnut sauce, dried mackerel, roasted chickpeas, or a savory cheese pastry fried until crispy. Hors d'oeuvres may be accompanied by *rakí*, a traditional anisette liquor distilled from grapes served with water over ice and drunk slowly.
- Salads include lettuce, tomatoes, cucumbers, onions, and other raw vegetables, with a dressing of olive oil and lemon juice or vinegar. Turks generally prepare meat in small pieces in combination with other vegetables, potatoes, or rice. Famous Turkish cuisine includes *köfte*, small spicy meatballs, and *kebab*, skewered beef or lamb and vegetables.

- The Turkish people invented yogurt, which is an essential part of the Turkish diet and is generally served with hot meals rather than as cold breakfast food.
- Rice and *börek* are important parts of Turkish culinary tradition. *Börek* is made by wrapping *yufka* (thin sheets of flour-based dough) around meat, cheese, or spinach and then frying or baking until the dough is flaky.
- Turkish desserts fall into four categories: rich and sweet pastry, such as *baklava*; puddings; cooked fruits; and fresh fruits. Most meals are concluded with fresh fruit and coffee or tea.
- Turkish *kahve*, is famous for its dark, thick, sweet taste. *Ayran*, a mixture of yogurt and milk, is the national cold drink and is drunk by children and adults alike. The Muslim religion requires abstinence from eating pork and drinking alcohol, but not all Muslims abstain, depending on their degree of religious practice.
- **Given the diversity among food options for in the US, provide dietary counseling according to the individual's unique food choices and practices.**
- The Islamic tradition of **Ramazan (Ramadan** in Arabic countries) is a month of fasting observed by practicing Muslims throughout the world. During Ramazan, one is not allowed to eat or drink anything from sunrise to sunset. Observance of this tradition varies, from some not observing to others who observe the ritual strictly.
- Sunni Muslims start practicing Ramazan at age 10 or 11 years, and some believe that women have the duty to fast even during pregnancy and the postnatal period. Alevi Muslims do not require fasting for men or women. Generally, pregnant and postpartum women, travelers, and those who are ill are excused from fasting but may be required to make up lost time at a later date. **Health teaching strategies for Turks in America should include the recognition and prevention of dehydration, bloating, constipation, and fatigue during periods of Ramazan fasting (see Chap. 8).**
- Fasting also can cause a variety of digestive problems and may endanger the health of a pregnant or postnatal woman and her baby. **Provide factual information regarding fasting and the potential effect on some health conditions.**
- Molasses and *baklava*, *lokom* (Turkish delight), tahini, and honey and nuts and raisins are believed to increase strength and sexual vigor.
- Fruits, especially bananas, oranges, tangerines, and apples, are brought to convalescing people, helping them to regain their strength and aiding in the healing process.
- Chicken soup is a common remedy for cold and flu symptoms. An *ebe* (a traditional midwife or healer in Turkey) relies on various herbs and home remedies to heal patients. *Ebegömeci* (a spinach-like leaf or herb) may be prepared for topical or oral use to treat inflammation, infection, and sometimes infertility.
- Tea, cinnamon, hot sugar water, ginger, mint, and various roots are used separately or in various combinations to treat rheumatism, low blood pressure, intestinal gas, or colds and flu. Nettles may be used topically for rheumatism, arthritis, and varicose veins.
- A folk remedy for diabetes is boiling olive leaves and, after refrigeration, drinking the juice.

- *Lapa*, a watery rice mixture with a gruel-like texture, or a boiled potato may be used to treat diarrhea and is followed by yogurt to replace the natural flora of the intestines.
- **Ask Turkish patients if they are using folk dietary practices, and incorporate them into prescription therapies. Because malnutrition may be a significant problem among economically disadvantaged Turks immigrating to the US, consider extensive nutritional assessments for more recent immigrants.**

29.1.7 Pregnancy and Childbearing Practices

- Before the 1960s, the sale of contraceptives and birth control education were prohibited in Turkey. Abortion and sterilization, except for medical reasons, were illegal until 1983. Currently, common methods of contraception used by Turkish women include withdrawal, intrauterine devices, and birth control pills. **Given the history of family planning in Turkey, be open in culturally congruent ways to discussions of family planning with Turks in the US.**
- Motherhood is accorded great respect, and pregnant women are usually made comfortable in any way possible, including satisfying their cravings. Pregnant women may continue their daily activities or work as long as they are comfortable.
- Many pregnant women take prenatal vitamins, drink a lot of milk, and apply salves such as Vaseline to avoid stretch marks. Light exercise, such as walking, is encouraged, but weather conditions often hamper such efforts because Turks generally tend to avoid wet or cold weather, fearing its ill effects on one's health.
- Most Turks prefer hospitals for physician-assisted child delivery. The husband and the birth mother's father may be present during the birthing process. Expressions of discomfort and pain are quite acceptable.
- The postpartum period can last up to 40 days. Light exercise is encouraged during this period, and bathing, an important part of the Muslim tradition, is strongly encouraged.
- A special food called *lošgusalik* is served to the postpartum woman. *Lošgusalik* is a sweet sherbet-like foodstuff that is prepared by dissolving *lošgusalik* beads in hot water. This high-carbohydrate mixture increases the woman's strength. Postpartum women drink hot soups and other fluids such as milk, especially when breast-feeding. Most Turks value breast-feeding, which is commonly but modestly practiced.
    Encourage incorporation of nonharmful post-partum practices with Western interventions.
- A small blue bead called a *nazar boncuk*, believed to protect the child from the "evil eye," is usually placed on the child's left shoulder to protect the child from the evil angel whispering in the left ear. Other traditional practices include placing iron under the baby's mattress to protect against anemia, tying a yellow ribbon to the crib to ward against jaundice, and placing a red bow on the crib to

distract any envy or negativity. **Do not remove protective beads or ribbons from the infant's bedside.**
- Swaddling, a common practice, has been linked to congenital hip dislocation, pneumonia, and upper respiratory infections among young infants. **Explain the harmful effect of swaddling.**

29.1.8 Death Rituals

- Turkish Muslims do not generally practice cremation because of their belief that the body must remain whole. After death, the body is washed in a ritual manner and wrapped in a white sheet.
- Muslim Turks are placed in the grave with the right side of the body facing Mecca.
- The traditional mourning period is 40 days, during which time traditional women may wear black clothes or a black scarf.
- Although Muslim Turks believe in the afterlife, death is always an occasion of great sorrow and mourning. An expression of sympathy to one who has just lost someone to death is *Basiniz saŠg olsun* (May your head be healthy), hoping that one is not overwhelmed with grief.

29.1.9 Spirituality

- Most Turks are Sunni Muslim, with a minority from the Alevi Muslim group. Most Turks who emigrate to the West tend to be very moderate Muslims.
- Traditional prayer is practiced five times each day and can take place anywhere, as long as one is facing the holy city of Mecca. A special small rug called *seccade* is used for praying in places other than the *cami*, mosque. When entering the *cami*, shoes are always removed, and women must cover their heads. Men and women go to separate parts of the *cami* for prayer. One prepares for prayer by ritual cleansing, which, at minimum, includes washing the face, ears, nostrils, neck, hands to the elbow, and feet and legs to the knee, three times each.
- **Caregivers may need to make special arrangements so that Muslims can practice their religious obligations when they are in a health-care facility.**
- Spiritual leaders or healers are sought for assistance with relationships or emotional problems. A *muska*, a paper inscribed by a *hoca* (spiritual teacher) with a prayer in Arabic, is wrapped in fabric and hidden in the home or worn by the person seeking help. *Turbe* and *yatir* are the practice of going to the saints' graves to pray about wishes, mental or emotional problems, or fertility problems. *Tesbih* (small beads traditionally used for praying) now take a more secular meaning and are often referred to as "worry beads." **Religious or folk items should not be removed because they provide comfort for the patient, and removal may increase anxiety.**

29.1.10   Health-Care Practices

- Most Turks rely on Western medicine and highly trained professionals for health and curative care. However, remnants of traditional beliefs continue. A common explanation for the cause of illness is an imbalance of hot and cold. For example, diarrhea is thought to come from too much cold or heat; pneumonia results from extreme cold.
- Many Turks have a traditional fatalistic worldview, which is more common among those living in extended households who believe "You cannot change what God has written." **Incorporate factual information for the entire family regarding disease causation and treatment in patient education.**
- Terminally ill patients are generally not told the severity of their condition. Informing a patient of a terminal illness may take away hope, motivation, and energy that should be directed toward healing, or it may cause the patient additional anxiety related to the fear of dying and concern for those being left behind. Furthermore, it may be believed that no one can second-guess Allah. **Take cues from the patient, and discuss the patient's condition with the family before disclosing the severity to the patient.**
- In general, women are responsible care-taking of the ill in the home. The person who is respected as the most educated has primary input into decisions about health care. **Determine the family spokesperson for decision-making on health issues, and incorporate group interventions whenever possible.**
- Turkey has high rates of consumption of over-the-counter antibiotics and painkillers; aspirin is commonly used as a panacea for a variety of ailments, including gastric upset. Turks, especially those who have difficulty affording the services of a physician, commonly consult a pharmacist before visiting a physician. **Assess Turkish American patients for their use of over-the-counter medications to prevent conflicting or potentiating effects with prescription medications.**
- The concept of the "evil eye" is prevalent. It is a cultural inclination not to speak too well of one's health for fear that one may incur misfortune through others' *nazar* (envy). So pervasive is this concept that taxi drivers and medical doctors alike respect the *nazar boncuk*, a blue bead that is used as protection from the evil eye. When describing an illness, one avoids using oneself or another person as an example for fear that it may invite the illness or condition upon that person.
- *Kolonya* (cologne) is part of a traditional practice that crosses religious and secular lines. Originally derived from the religious value of cleanliness, cologne is sprinkled on the hands of guests before and after eating to provide cleanliness and a fresh lemon scent. Inhaling from a cloth or handkerchief doused with cologne may be used for relief from motion sickness. In the hospital, patients may offer cologne to a physician or nurse prior to examination. **Respect this custom as appropriate.**

- Other home remedies include using rubbing alcohol or a wet cloth to bring down a fever and warming the back to treat coughing.
- Turkish culture allows freedom to express pain through either emotional outbursts or verbal complaints. **Accept a wide range of pain expression among Turkish patients.**
- Although stigma is attached to mental illness, many families seek treatment or care for the patient at home. The most common reasons given for mental illness include discord among family members, marital and love problems, gossip, and other familial problems. Social causes include financial inadequacies, societal disorders, and a "struggle with life." Men more frequently blame social causes for mental illness, and women often give psychological causes. **Address somatization among Turkish American patients with mental health concerns, and encourage expression of their feelings. Incorporate group interventions whenever possible.**
- During hospitalization, *refakatçí* refers to the person who stays overnight with the patient, providing emotional and physical support and comfort. A show of concern and compassion for the patient eases his or her fears and reduces loneliness. Family members may also attend to physical needs such as bathing. A balanced healthy diet is considered essential to regaining one's health; Turks frequently bring food from home.

  Encourage family-assisted care as well as having families bring food from home.
- While blood transfusions are gaining acceptance, Turkish people usually prefer to receive blood from family members. Muslims traditionally prefer that the body remain intact after death, a belief that may conflict with organ donation.

29.1.11 Health-Care Providers

- Although Turkish people are inclined toward Western health-seeking behaviors, medical care tends to be holistic. Great value is placed on emotional well-being, especially as it affects physical well-being. Physicians may be "adopted" as members of their patients' families, and it is common to give gifts (usually food) to the physician as an expression of gratitude.
- Generally, treating someone of the opposite sex is not an issue among Turks. **Ask each patient for preference on gender issues in health care.**
- Physicians, and to a lesser extent nurses and midwives, have historically been held in very high esteem. Patients rarely question the authority of physicians, but the notion of obtaining a second opinion is gaining popularity.

29.2 Reflective Exercises

1. Explain the Turkish language and now it might be difficult for Americans to pronounce Turkish names. What might you do to ensure you pronounce a name correctly?
2. Give some examples of the collectivistic Turkish family versus the individualistic culture of European Americans. What implications does this have for patient education?
3. Identify communication practices related to touch among Turks? How does this vary among the women and men and their religion and tradition?
4. Identify at least genetic conditions common among Turks. Which of them have a lifestyle component?
5. Describe the traditions of *Ramazan*? How would you change your practice to accommodate a person who practices *Ramazan*?
6. What function does somatization have in the Turkish culture? How would you address a patient who is somaticizing?
7. What is *nazar boncuk* and what is its function?
8. Identify two folk remedies and for what conditions are they used?
9. Posit why newer Turkish immigrants may easily assimilate into the dominant European American culture?
10. As a home health nurse, you are visiting a Turkish patient in their home. What do you need to consider before entering the home?

Bibliography

CIA (2016) World factbook: Turkey. https://www.cia.gov/library/publications/the-world-factbook/geos/tu.html

Fassaert T, Hesselink AE, Verhoeff AP (2009) Acculturation and use of health care services by Turkish and Moroccan migrants: a cross-sectional population-based study. BMC Public Health 9. http://bmcpublichealth.biomedcentral.com/articles/10.1186/1471-2458-9-332

Herich B (2011) Introduction—beyond state Islam: religiosity and spirituality in contemporary Turkey. Eur J Turk Stud. https://ejts.revues.org/4527

Kara B (2017) Self-related health and associated factors in older Turkish adults with type 2 diabetes: a pilot study. J Transcult Nurs 28(1):40–48

Karakas F, Sarigollu E, Kavas MJ (2015) Discourses of collective spirituality and Turkish Islamic ethics: an inquiry into transcendence, connectedness, and virtuousness in Anatolian tigers. J Bus Ethics 129(4):811–822

Ozbassaran F, Ergul S, Temel AB, Aslan GG, Coban A (2011) Turkish nurses' perceptions of spirituality and spiritual care. J Clin Nurs 20:21–22

Serdar U (2011) Traditional fold medicine in the Turkish folk culture. Turk Stud 6(4):317–327

Thobaben M, Kuguoglu S (2013) People of Turkish heritage. In: Purnell L (ed) Transcultural health care: a culturally competent approach (chapter on DavisPlus Web site), 4th edn. F.A. Davis Company, Philadelphia

Yaman A, Mesman J, van Ijzendoom MH, Bakermans-Kranenbutg MJ, Linting M (2010) Parenting in an individualistic culture with a collectivistic cultural background: the case of Turkish immigrant families with toddlers in the Netherlands. J Child Fam Stud 19(5):617–628

Chapter 30
People of Vietnamese Heritage

30.1 Overview and Heritage

Most Vietnamese living in the US have arrived since 1975. Their departures from Vietnam were often precipitous and tragic. Escape attempts were long, harrowing, and, for many, fatal. Survivors were often placed in squalid refugee camps for years. Many departed Vietnam in small, often unseaworthy and overcrowded vessels in hopes of reaching a non-Communist port. Half died during their journey. Many were forcibly repatriated to Vietnam or eventually returned voluntarily; others continued to languish in camps. The 1979 Orderly Departure Program provided safe and legal exit for the Vietnamese to reunite with family members already in the US (referring specifically to the US and not North, South, or Central America). In 1987, The Amerasian Homecoming Act provided for the entry of former South Vietnamese military officers, other political detainees, children of American servicemen, and Vietnamese women and their close relatives. The Vietnamese in America differ substantially, depending on acculturation, individuality, and the variant characteristics of culture as described in Chap. 1. Vietnamese place a high value on education and accord scholars an honored place in society. The teacher is highly respected as a symbol of learning and culture.

30.1.1 Communications

- Vietnamese is a single distinctive language with northern, central, and southern dialects, all of which can be understood by individuals speaking any one of these dialects. Vietnamese is the only language of the Asian mainland that is regularly written in the Roman alphabet.
- All words in Vietnamese consist of a single syllable, although two words are commonly joined with a hyphen to form a new word. Each vowel can be spoken

© Springer Nature Switzerland AG 2019
L. D. Purnell, E. A. Fenkl, *Handbook for Culturally Competent Care*,
https://doi.org/10.1007/978-3-030-21946-8_30

in five or six tones that may completely change the meaning of the word. One perennial stumbling point is that the word for "blue" and "green" is the same. **Do not give directions such as take one "blue" pill at a specified time. Instead, provide the name and dosage of the medication.**

- Many individuals cannot easily give a blunt "no" as an answer because they feel that such an answer may create disharmony. A "yes" response, rather than expressing a positive answer or agreement, may simply reflect an avoidance of confrontation or a desire to please the other person. **Ask open-ended questions, and have the patients demonstrate rather than verbalize their understanding of treatments.**

- The terms "hot" (*duong*) and "cold" (*am*), rather than expressing physical feelings associated with fever and chills, may actually relate to other conditions associated with perceived bodily imbalances. **See the sections on Nutrition and Health-Care Practices in this chapter for a fuller understanding of the hot and cold theory. Provide an interpreter as needed.**

- English skills may not be sufficient to communicate in psychiatric interviews, which are usually carried out at a highly abstract level. **Observe patients for behavioral cues, use simple sentences, paraphrase words with multiple meanings, and avoid metaphors and idiomatic expressions. Explain all points carefully. Approach patients in a quiet, unhurried manner.**

- Self-control, another traditional value, encourages keeping to oneself. Expressions of disagreement that may irritate or offend another person are avoided. Individuals may be in pain, distraught, or unhappy yet rarely complain to health-care providers.

- Expressions of emotion are considered a weakness that may interfere with self-control. Negative emotions and expressions may be conveyed by silence or a reluctant smile. A smile may express joy or convey stoicism in the face of difficulty or may indicate an apology for a minor social offense. A smile may also be a response to a scolding to show sincere acknowledgment for the wrongdoing. **Clarify the meaning of a smile.**

- The head is a sacred part of the body that should not be touched. To place one's feet on a desk is considered offensive. **If it is medically necessary to touch the head, provide an explanation, and ask permission. Do not place feet on a table or desk.**

- Looking another person directly in the eyes may be deemed disrespectful.

- To signal for someone to come by using an upturned finger is considered a provocation, usually used to call a dog; waving the hand is considered more proper. **Do not point when beckoning patients; call them by name.**

- Hugging and kissing are not seen outside the privacy of the home. Men greet each other with a handshake, but they do not shake hands with a woman unless she offers her hand first. **Greet men with a handshake. Men should wait for the woman to extend her hand for a greeting.**

- Two men or two women can walk hand in hand without implying sexual connotations.

- Women may be reluctant to discuss sex, childbearing, or contraception when men are present.
- Traditional Vietnamese prefer more distance during personal and social relationships than the European American culture.
- Not keeping appointments may relate to not understanding oral or written instructions or not knowing how to use the telephone. The more acculturated understand the significance of punctuality.
- Little attention is given to one's precise age. Birth dates may pass unnoticed, with everyone celebrating birthdays together during the Lunar New Year (*Tet*) in January or February. A person's age is calculated from the time of conception; children are considered to be a year old at birth and gain a year each *Tet*. A child born just before *Tet* could be regarded as 2 years old when only a few days old. Because of the difficulty in determining age, many immigrants use January 1 as a date of birth for official records.
- Most names consist of a family name, a middle name, and a given name of one or two words, always written in that order. There are relatively few family names, with Nguyen (pronounced "nwin") and Tran accounting for more than half of all Vietnamese names. Other common family names are Cao, Dinh, Hoang, Le, Ly, Ngo, Phan, and Pho. There are relatively few middle names, with Van being used regularly for men and Thi (pronounced "tee") for women. A typical woman's name is Tran Thi Thu. That is how she would write or give her name if requested. She would expect to be called simply Thu or sometimes *Chi* (sister) Thu by friends and family. In other situations, she would expect to be addressed as *Cô* (Miss) or *Ba* (Mrs.) Thu. If married to a man named Nguyen Van Kha, the proper way to address her would be as Mrs. Kha, but she would retain her full three-part maiden name for formal purposes. The man would always be known as *Kha* or *Ong* (Mr.) Kha. Some Vietnamese American women have adopted their husband's family name. Children always take the father's family name.
  Ask women their preferred name as well as their legal name.

30.1.2 Family Roles and Organization

- The family is the main reference point for the individual throughout life, superseding obligations to country, religion, and self. The traditional family is patriarchal, with an extended family structure. Men deal with matters outside the home. A wife is expected to be dutiful and respectful toward her husband and his parents. Women often make family health-care decisions.
- Children are expected to be obedient and devoted to their parents and to worship their memory after death. The eldest son is usually responsible for rituals honoring the memory and invoking the blessings of departed ancestors. A son's obligations and duties to his parents may assume a higher value than those to his wife or children.

- For the first 2 years of life, mothers primarily care for their children; thereafter, grandmothers and others take on much of this responsibility.
- Exposure of the younger generation to American culture can become a source of conflict with considerable family strain when adolescents are influenced by the perceived American values of individuality, independence, self-assertion, and egalitarian relationships. **Counsel with parents and children together and discuss cultural values to help them reach satisfactory outcomes.**
- Unattached male refugees may join *pseudofamilies*, households made up of close and distant relatives and friends who share accommodations, finances, and companionship.
    Pseudofamilies **form an important source of social support in the refugee communities.**
- Differing sexual orientation is difficult for Vietnamese to face because being gay or lesbian brings shame upon the family, causing many gays and lesbians to remain closeted. **When questioning a gay or lesbian person about his or her sexual activity, an interpreter unknown to the family is an absolute requirement. Do not disclose gay and lesbian relationships to family members.**
- Traditional Vietnamese are class-conscious and rarely associate with individuals at different levels of society. Respect is accorded to people in authoritative positions who are well educated or otherwise successful or who have professional titles.

30.1.3   Workforce Issues

- Priority is given to the concerns of the family, rather than to those of the employer. **Explain the necessity of punctuality in the American workforce.**
- The Vietnamese are highly adaptable and adjust their work habits to meet requirements.
- Most Vietnamese respect authority figures with impressive titles, achievements, education, and a harmonious work environment.
- Confucian cultural background results in conformity and reluctance to undertake independent action.
- Many fear losing their job if they speak out about inequities; they are likely to be taken advantage of by some more unscrupulous employers. **Allow extra time, provide visually oriented instructions, and provide programs that enhance communications to promote increased harmony in the workplace.**
- Many individuals cannot easily give a blunt "no" as an answer because they feel that such an answer may create disharmony. A "yes" response, rather than expressing a positive answer or agreement, may simply reflect an avoidance of confrontation or a desire to please the other person. **Ask open-ended questions and have employees demonstrate rather than verbalize their understanding of procedures.**

- Looking another person directly in the eyes may be deemed disrespectful. **Do not assume that lack of eye contact means the person is not listening, does not care, or is being less than truthful.**
- **Do not take offense if the Vietnamese employee stands at a greater distance from how you are accustomed.**

30.1.4 Biocultural Ecology

- Skin color ranges from pale ivory to dark brown. Mongolian spots, bluish discolorations on the lower back of a newborn child, are normal hyperpigmented areas in many Asians. **Assess for oxygenation and cyanosis in dark-skinned Vietnamese by examining the sclera, conjunctiva, buccal mucosa, tongue, lips, nail beds, palms of the hands, and soles of the feet. Assess for patches of melanin in the buccal mucosa and the conjunctiva for petechiae and rashes. Assess for jaundice by observing for yellow discoloration of the conjunctiva.**
- Most Vietnamese are small in physical stature. Adult women average 5 ft tall and weigh 80–100 lb. Men average a few inches taller and weigh 110–130 lb.
- Typical physical features include inner eye folds that make the eyes look almond shaped, sparse body hair, and coarse head hair. They also have dry earwax, which is gray and brittle. People with dry earwax have few apocrine glands, especially in the underarm area, and thus produce less sweat and associated body odor.
- Children are small by U.S. standards. American growth charts do not provide adequate assessments for evaluating the physical development of Vietnamese children. Standing, walking, and language skills begin at a slightly later age in Vietnamese children, but they rapidly catch up with European American norms by the age of 1½ to 2 years.
- Table 30.1 lists commonly occurring health conditions for Vietnamese in three categories: lifestyle, environment, and genetics. See Chap. 2.
- Betel-nut pigmentation may be found in some adults, resulting from the practice of chewing *chau* (betel leaves), which is common among older women for its narcotic effect on diseased gums. Some elderly women lacquer their teeth, believing that it strengthens the teeth and symbolizes beauty and wealth.
- Cancer and other problems common to this group may be associated with the widespread application of chemical agents during the Vietnam War. Explain the importance of cancer screening. **Assess patients for tuberculosis, parasitism, anemia, malaria, and hepatitis B.**
- Melioidosis and paragonimiasis may mimic tuberculosis. Microcytosis may be misdiagnosed as iron deficiency, and inappropriate treatment with iron may be initiated. Erythrocytic microcytosis is a reflection of the presence of thalassemia or of hemoglobin E trait. **Screen for these disorders in people with findings consistent with tuberculosis but with a negative purified protein derivative response.**

Table 30.1 Vietnamese
health conditions

Health conditions	Causes
Cervical cancer	Environment, lifestyle
Depression and post-traumatic stress syndrome	Environment, lifestyle
Tuberculosis	Environment, lifestyle
Hepatitis	Environment, lifestyle
Malaria	Environment, lifestyle
Parasitosis	Environment, lifestyle
Thalassemia	Genetic
Liver cancer	Environment, lifestyle
Trichinosis	Environment, lifestyle
Nasopharyngeal cancer	Environment, lifestyle
Lactase deficiency	Genetic
Leprosy	Genetic

- Most Vietnamese people are slow metabolizers of alcohol, making them more sensitive to the adverse effects of alcohol as expressed by facial flushing, palpitation, and tachycardia. They are twice as sensitive to the effects of propranolol on blood pressure and heart rate, experience a greater increase in heart rate from atropine, and require lower doses of benzodiazepines, diazepam, and alprazolam. Because of their increased sensitivity to the sedative effects of these drugs, many require lower doses of imipramine, desipramine, amitriptyline, and clomipramine. They are less sensitive to cardiovascular and respiratory side effects of analgesics (for example, morphine) but are more sensitive to their gastrointestinal side effects.
- **Most Asians require lower doses of neuroleptics. Because Vietnamese people are considerably smaller than most European Americans, medication dosages may need to be reduced.**

30.1.5   High-Risk Health Behaviors

- Some newer immigrants have never heard of cancer; among those who have heard of it, some believe it is contagious. Many women have never heard of or had a Pap test. Many have never performed breast self-examination or had a mammogram. **Explain the necessity of Pap tests, mammograms, and breast self-exams.**
- High rates of gastrointestinal cancer may be due to asbestos that, in some parts of the world, is used in the process of "polishing" rice. **Remind patients that polished, imported rice should always be washed.**
- Many do not know that cigarette smoking can cause cancer. **Explain the adverse effects of cigarette smoking and encourage cessation, starting in elementary school.**

- Trichinosis risk is 25 times greater in Southeast Asian refugees than in the general population. This increased risk is related to undercooking pork and purchasing pigs directly from farms. **Screen newer immigrants for the possibility of trichinosis if symptomatology warrants.**

30.1.6   Nutrition

- Meal preparation is precise and may occupy much of the day. Celebrations and holidays involve elaborately prepared meals.
- The normal daily caloric intake of the Vietnamese is approximately two-thirds that of average Americans. White or polished rice is the main staple in the diet, providing up to 80% of daily calories. Other common foods are fish (including shellfish), pork, chicken, soybean curd (tofu), noodles, various soups, and green vegetables. Preferred fruits are bananas, mangoes, papayas, oranges, coconuts, pineapples, and grapefruit. Soy sauce, garlic, onions, ginger root, lemon, and chili peppers are used as seasoning. Rice and other foods are commonly served with *nuoc mam* (a salty, marinated fish sauce). A meal typically consists of rice, *nuoc mam*, and a variety of other seasonings, green vegetables, and sometimes meat cut into slivers. Chicken and duck eggs may be used.
- Most Vietnamese prefer white bread, particularly French loaves and rolls, and pastry. A regular dish is *pho*, a soup containing rice noodles, thinly sliced beef or chicken.
- Food preparation includes *com chien* (fried rice) and *thit bo xau ca chua* (beef fried with tomatoes). *Cha gio* (pronounced "cha-yuh") is a combination of finely chopped vegetables, mushrooms, meat, or bean curd that is rolled into delicate rice paper and deep-fried. Stir-frying, steaming, roasting, and boiling are the preferred methods of cooking. Hot tea is the usual beverage.
- A predominant aspect of the traditional Asian system of health maintenance is the principle of balance between two opposing natural forces, known as *am* (cold) and *duong* (hot) in Vietnamese. The terms have nothing to do with temperature and are only partly associated with seasoning. Rice, flour, potatoes, most fruits and vegetables, fish, duck, and other things that grow in water are considered cold. Most other meats, fish sauce, eggs, spices, peppers, onions, candies, and sweets are hot. Tea is cold, coffee is hot, water is cold, and ice is hot.
- Illness or trauma may require therapeutic adjustment of hot-cold balance to restore equilibrium. Hot foods and beverages, used to replace and strengthen the blood, are preferred after surgery or childbirth. During illness, certain foods, such as *chao* (light rice gruel) mixed with sugar or sweetened condensed milk and a few pieces of salty pork cooked with fish sauce, are consumed in greater quantity. Fresh fruits and vegetables are usually avoided, being considered too cold. Water, juices, and other cold drinks are restricted. **Nutritional counseling**

should take into consideration these factors and other aspects of the usual Vietnamese diet, because advice to simply eat certain kinds of American foods may be ignored.
- Most adults and many children suffer from lactose intolerance, which may cause problems in schools, other institutional settings, and adoptive families. **Use of substitute milk products that use soybeans for those who are lactose intolerant.**

30.1.7   Pregnancy and Childbearing Practices

- Women have high fertility rates (average of six pregnancies, with four live births) and commonly have children until their early to middle 40s. Many know little about contraception. Abortions are commonly performed in their homeland because pregnancy outside of marriage is considered a disgrace to the family. More acculturated individuals practice some form of birth control. **Avoid family planning issues on the first encounter; but such information is usually well received on subsequent visits.**
- Prescriptive food practices for a healthy pregnancy include noodles, sweets, sour foods, and fruit and exclude fish, salty foods, and rice. After birth, to restore equilibrium and provide adequate warmth to the breast milk, women consume soups with chili peppers, salty fish and meat dishes, and wine steeped with herbs.
- Foods are also classified as *tonic* and *wind*. Tonic foods include animal protein, fat, sugar, and carbohydrates; they are usually also hot and sweet. Sour and sometimes raw and cold foods are classified as *antitonic*. Wind foods, often classified as cold, include leafy vegetables, fruit, beef, mutton, fowl, fish, and glutinous rice. It is considered critical to increase or decrease foods in various categories to restore bodily balances upset by unusual or stressful conditions such as pregnancy. While the balance of foods may be followed, the terminology is not used consistently.
- During the first trimester, the expectant mother is considered to be in a weak, cold, and antitonic state. Therefore, she should correct the imbalance by eating and drinking hot foods, such as ripe mangoes, grapes, ginger, peppers, alcohol, and coffee. To provide energy and food for the fetus, she is prescribed tonic foods, including a basic diet of steamed rice and pork.
- Cold foods, including mung beans, green coconut, spinach, and melon, and antitonic foods, such as vinegar, pineapple, and lemon, are avoided during the first trimester. In the second trimester, the pregnant woman is considered to be in a neutral state. Cold foods are introduced, and the tonic diet is continued.
- During the third trimester, when the woman may feel hot and suffer from indigestion and constipation, cold foods are prescribed, and hot foods are avoided or strictly limited. Tonic foods, which are believed to increase birth weight, are restricted to reduce the chances of a large baby. Wind foods are generally avoided throughout pregnancy, as they are associated with convulsions, allergic

reactions, asthma, and other problems. This regimen may appear more complex and restrictive than it actually is in practice. Most women use it only as a general guide, commonly restricting, rather than totally abstaining from the restricted foods. A great variety of food, including rice, many kinds of vegetables and fruits, various seasonings, and certain meats and fish are generally permissible throughout pregnancy. **Ask women at each stage of their pregnancy about preferred foods before initiating dietary recommendations. Dispel any myths, and recommend foods that are congruent with beliefs.**

- Many women do not seek medical attention until the third trimester because of cost, fear, or lack of perceived need. Women who are generally better educated seek early prenatal care. For obstetric and gynecologic matters, they tend to feel more comfortable with a female physician or midwife.
- Women maintain physical activity to keep the fetus moving and to prevent edema, miscarriage, or premature delivery. Prolonged labor may result from idleness. An undesirably large baby may result from afternoon napping.
- Restrictive beliefs include avoiding heavy lifting and strenuous work and raising the arms above the head, which pulls on the placenta, causing it to break. Sexual relations late in pregnancy may cause respiratory stress in the infant.
- Many consider it taboo for pregnant women to attend weddings or funerals.
- Some pregnant women look at pictures of happy families and healthy children, believing that it helps give birth to healthy babies.
- Anesthesia during labor and delivery is usually accepted. Once in labor, the woman tries to maintain self-control and may even smile continuously. Her period of labor is usually short, and there may be no warning of impending delivery.
- **Because the head is considered sacred, neither that of the mother nor of the infant should be touched or stroked. Removal of vernix from the infant's head can cause distress. If it is necessary to touch the head for medical treatment, an explanation is essential. Stress the importance and necessity of invasive procedures.**
- Because body heat is lost during delivery, women avoid cold foods and beverages and increase consumption of hot foods to replace and strengthen their blood. Ice water and other cold drinks are usually not welcome and most raw vegetables, fruits, and sour items are consumed in lesser amounts. Prescriptive foods include steamed rice, fish sauce, pork, chicken, eggs, soups with chili or black peppers, other highly seasoned and salty items, wine, and sweets.
- Because water is *am* (cold), women traditionally do not fully bathe, shower, or wash their hair for a month after delivery. Some women have complained that they were adversely affected by showering shortly after delivery in American hospitals; others, however, welcomed the opportunity to shower and seemed willing to give up some traditional practices.
- Postpartum women avoid drafts and strenuous activity; wear warm clothing; stay in bed, indoors, or both for about a month; and some avoid sexual intercourse for months. Many use hot water bottles or electric blankets to combat the cold forces of childbirth.

- Other women in the family assume responsibility for the baby's care. In Vietnam, husbands would never be present at their child's delivery; this may also occur in the US and elsewhere. **Do not interpret the mother's inactivity and dependence on others as apathy or depression. Do not praise the baby lest jealous spirits steal the infant. Do not cut the child's hair or nails for fear that this might cause illness.**
- The infant is generally maintained on a diet of milk for the first year, with the introduction of rice gruel at around 6 months.
- Some women discard colostrum and feed the baby rice paste or boiled sugar water for several days. This does not indicate a decision against breast-feeding. **Explain the immune properties of colostrum.**
- After the milk comes in, the mother and baby benefit from the hot foods consumed by the mother for the first month. Then, however, a conflict arises: the mother believes that hot foods benefit her health but that cold foods ensure healthy breast milk. Having the mother change from breast-feeding to formula can easily solve this dilemma. However, if the mother cannot afford formula, she may use fresh milk or rice boiled with water, which may result in anemia and growth retardation. **Reinforce the benefits of colostrum.**

30.1.8   Death Rituals

- Most Vietnamese accept death as a normal part of life. Ancestors are commonly honored and worshipped and are believed to bestow protection on the living.
- Most individuals have an aversion to hospitals and prefer to die at home. Some believe that a person who dies outside the home becomes a wandering soul.
- Many do not want to artificially prolong life and suffering, but it may still be difficult for relatives to consent to terminating active intervention, which might be considered contributing to the death of an ancestor who shapes the fate of the living. **Having a family conference and providing factual information is appreciated.**
- Clergy visitation is usually associated with Sacrament of the Sick; rites influenced by Catholicism can actually be upsetting to hospitalized patients. Sending flowers may be startling, as flowers are usually reserved for the rites of the dead. **Do not place flowers in the patient's room until getting permission from the patient or family. Contact religious leaders only with the permission or request of the patient or family.**
- Families gather around the body of a deceased relative and express great emotion. Traditional mourning practices include the wearing of white clothes for 14 days, the subsequent wearing of black armbands by men and white headbands by women, and the yearly celebration of the anniversary of a person's death.
- Few families consent to autopsy unless they know and agree with the reasons for it. Cremation is an acceptable practice to some families. **Explain legalities of autopsy.**

30.1.9 Spirituality

- Most practice Buddhism, either solely or combined with other religions. Other major religions practiced are Confucianism and Taoism; a few are Christians, most of whom are Catholic. Animism is found mainly among people from the highland tribes. Many believe that deities and spirits control the universe and that the spirits of dead relatives continue to dwell in the home.
- Spirituality is also gained by participating in ethnic community activities. **Partner with ethnic activities to gain trust among Vietnamese.**

30.1.10 Health-Care Practices

- Good health is achieved by having harmony and balance with the two basic opposing forces, *am* (cold, dark, female) and *duong* (hot, light, male). An excess of either force may lead to discomfort or illness. *Am* (cold) represents factors that are considered negative, feminine, dark, and empty, whereas *duong* (hot) represents those that are positive, masculine, light, and full. Proper balance of these two life forces ensures the correct circulation of blood and good health. If the balance is not proper, life is short.
- Diseases and debilitating conditions result from either cold or hot influences. For example, diarrhea and some febrile diseases are due to an excess of cold, whereas pimples and other skin problems result from an excess of hot. Countermeasures involve using foods, medications, and treatments that have properties opposite those of the problem and avoiding foods that would intensify the problem. Asian herbs are cold, and Western medicines are hot. **Incorporate traditional practices into allopathic care.**
- Common Vietnamese medical practices are discussed in Table 30.2.
- Many Vietnamese try home remedies, allowing the condition to become serious before seeking professional assistance. Once a physician or nurse has been consulted, the Vietnamese are usually quite cooperative and respect the wisdom and experience of health-care providers.
- The belief that life is predetermined is a deterrent to seeking health care. Diagnostic tests may be baffling, and invasive procedures are frightening. The prospect of surgery can be terrifying. **Explain procedures in simple, realistic terms, and reinforce information as necessary.**
- Hospitalization is considered a last resort and is acceptable only in case of emergency when everything else has failed.
- Naturalistic explanations for poor health include eating spoiled food and exposure to inclement weather. Natural elements are associated with bad weather and cold drafts, and cause problems such as the common cold, mild fever, and headache. Countermeasures involve dietary, herbal, hygienic, and simple medical practices. Collectively, these measures are categorized as *thuoc nam*, the

Table 30.2 Traditional Vietnamese medicine

- ***Cao gio*** (literally, "rubbing out the wind") is used for treating colds, sore throats, flu, sinusitis, and similar ailments. An ointment or hot balm oil is spread across the back, chest, or shoulders and rubbed with the edge of a coin (preferably silver) in short, firm strokes. This technique brings blood under the skin, resulting in dark ecchymotic stripes, so the offending wind can escape. Dermabrasion may provide a portal for infection. Do not interpret these ecchymotic areas as evidence of abuse
- ***Be bao*** or ***bar gio*** (skin pinching) is a treatment for headache or sore throat. The skin of the affected area is repeatedly squeezed between the thumb and forefinger of both hands, as the hands converge toward the center of the face. The objective is to produce ecchymoses or petechiae. Do not interpret these ecchymotic areas as evidence of abuse
- ***Giac*** (cup suctioning), another dermabrasive procedure, is used to relieve stress, headaches, and joint and muscle pain. A small cup is heated and placed on the skin with the open side down. As the cup cools, it contracts the skin and draws unwanted hot energy into the cup. This treatment leaves marks that may appear as large bruises. Do not interpret these ecchymotic areas as evidence of abuse
- ***Xong*** (an herbal preparation) relieves motion sickness or cold-related problems. Herbs or an agent such as Vicks VapoRub is put into boiling water, and the vapor is inhaled. Small containers of aromatic oils or liniments are sometimes carried and inhaled directly
- **Moxibustion** is used to counter conditions associated with excess cold, including labor and delivery. Pulverized wormwood or incense is heated and placed directly on the skin at certain meridians
- Acupuncture, acupressure, and acumassage relieve symptomatic stress and pain (see Chap. 10 for a description of these healing practices)
- Balms and oils, such as Red Tiger Balm, available in Asian shops, are applied to affected areas for relief of bone and muscle ailments
- Herbal teas, soups, and other concoctions are taken for various problems, generally in the sense of using cold measures to overcome hot illnesses. Eating organ meats such as liver, kidneys, testes, brains, and bones of an animal is said to increase the strength of the corresponding human part. Two additional practices are consuming gelatinized tiger bones to gain strength and taking powdered rhinoceros horn to reduce fever

traditional southern medicine of Vietnam, and *thuoc bac*, the more formal northern or Chinese medicine.

- Supernaturalistic causes, such as gods, spirits, or demons, may cause illness. Illness may be considered a punishment for offending such an entity or violating some religious or moral code.
- A "weak heart" may refer to palpitations or dizziness, a "weak kidney" to sexual dysfunction, a "weak nervous system" to headaches, and a "weak stomach or liver" to indigestion.
- Loss of blood from any route is feared, and some may refuse to have blood drawn for laboratory tests. The patient may complain, although not to the health-care provider, of feeling weak for months. A patient may fear that any body tissue or fluid removed cannot be replaced and that the body suffers the loss in this life as well as into the next life. **Explain that blood is naturally replenished with good nutrition.**
- Many believe that Western medicine is very powerful and cures quickly, but few understand the risks of over- or underdosages. The concept of long-term medica-

tion for chronic illnesses and acceptance of unpleasant side effects and increased autonomic symptoms are not congruent with traditional notions of safe and effective treatment of illnesses. Some people politely accept a prescription but may not fill it. Even if they have filled it, they may not take the medicine, or they may adjust the dosage without telling the health-care provider. **Extensive education, repetition of instructions, and home visitations are necessary to ensure adherence.**

- **Persistent reminding, as part of an overall effort to improve communication and information dissemination, has been suggested as the best way to encourage Vietnamese women to undergo regular cancer screening and follow-up treatment.**
- **Include family members in all major treatment decisions regarding physical and mental health.**
- Common problems that pose barriers to health care are discussed in Table 30.3.
- Fatalistic attitudes and the belief that problems are punishment may reduce the amount of complaining and expression of pain among those who view endurance as an indicator of strong character. Many Vietnamese accept pain as part of life and attempt to maintain self-control as a means of relief. **Explain that accepting pain medication will hasten the healing process.**
- In times of distress or loss, they often complain of physical discomforts, such as headaches, backaches, or insomnia.
- Mental illness is believed to result from offending a deity and brings disgrace to the family. A shaman may be enlisted to help, and additional therapy is sought only with the greatest discretion and often after a dangerous delay. **Partner with traditional healers to refer patients early in the diagnostic stage.**
- Emotional disturbance is usually attributed to possession by malicious spirits, bad luck of familial inheritance, or, for Buddhists, bad karma accumulated by misdeeds in past lives.

Table 30.3 Barriers to health care for Vietnamese people

• Subjective beliefs and the cost of health care
• Lack of a primary provider
• Differences between Western and Asian health-care practices
• Caregivers' judgment of Vietnamese as deviant and unmotivated because of noncompliance with medication schedules, diagnostic tests, follow-up care, and failure to keep appointments
• Inability to communicate effectively in the English language with recent immigrants, who lack confidence in their ability to communicate their needs; failure of providers to communicate
• Avoidance of Western practitioners out of fear that traditional methods will be criticized
• Fear of conflicts and ridicule, resulting in loss of face
• Lack of knowledge of the available resources
• Fear of stigmatization, difficulty in locating agencies that can provide assistance without distorted professional and cultural communication, and unwillingness to express inner feelings

- The term *psychiatrist* has no direct translation in Vietnamese and may be interpreted to mean nerve physician or specialist who treats crazy people. The nervous system is sometimes considered the source of mental problems, with neurosis being thought of as "weakness of the nerves" and psychosis as "turmoil of the nerves." A person with a mental disability may be stigmatized by the family and society, which can jeopardize the ability of relatives to find marriage partners. People with mental disabilities are usually harbored within their families unless they become destructive; then they may be admitted to a hospital. **Do not disclose mental health conditions outside the family. Use the depression scale developed by Kinzie and Associates and Buchwald when working with Vietnamese patients.**
- People with a physical disability are commonly accepted and are treated well and cared for by their families.
- Many women who have abnormal Pap smears fail to return for follow-up care, thereby contributing to the shockingly high incidence of cervical cancer among Vietnamese women. Many women fail to comprehend the severity of the situation and the potential for recovery if regular treatment begins early enough. **To increase follow-up visits and care, carefully explain the problems that may result if they do not follow up after an abnormal Pap smear. Explain that lack of symptoms or pain may be only temporary. Persistent reminding, as part of an overall effort to improve communication and information dissemination, has been suggested as the best way to encourage Vietnamese women to undergo regular cancer screening and follow-up treatment.**
- Because many believe that the body must be kept intact even after death, they are averse to blood transfusions and organ donation. Those who prefer cremation will donate body parts under certain circumstances.
- Women may not want to discuss sexual problems, reproductive matters, and birth control techniques until after an initial visit and confidence has been established in the health-care provider. **Pelvic examinations on unmarried women should not be made on the first visit or without careful advance explanation and preparation. When such an examination is necessary, the woman may want her husband present. If possible, the health-care provider and an interpreter should both be female.**

30.1.11   Health-Care Providers

- Four kinds of traditional and folk healers exist in Vietnam, as shown in Table 30.4. Many consult one or more of these healers in an attempt to find a cure. **Partner with traditional healers as a means for social support and treatment adherence.**
- Acknowledgment and support of traditional belief systems are important in building a trusting relationship. Traditional healers often provide the Vietnamese with necessary social support.

Table 30.4 Folk practitioners

• Asian physicians who are learned individuals who use herbal medication and acupuncture
• Informal folk healers who use special herbs and diets as cures based on natural or pragmatic approaches. The secrets of folk medicine are passed down through the generations
• Spiritual healers, some of whom have a specific religious outlook, and others who have powers to drive away malevolent spirits
• Magicians or sorcerers who have magical, curative powers but no communication with the spirits

- Adults, particularly young and unmarried women, are more comfortable with health-care providers of the same gender, especially for obstetrical and gynecological conditions. **Provide female health-care providers whenever possible for gynecological conditions.**
- Some Vietnamese are not accustomed to female authority figures and may have difficulty relating to women as professional health-care providers.
- While many individuals have great respect for professional, well-educated people, they may be distrustful of outside authority figures. Most have come to America to escape oppressive authority. A common suspicion is that divulging personal information for a medical history may jeopardize their legal rights. Respect and mistrust are not mutually exclusive concepts.

30.2   Reflective Exercises

1. Explain the naming format for Vietnamese men and women.
2. When does a traditional Vietnamese have a birthday? What might this mean for child development in accordance with the U.S. Standards and expectations?
3. What are traditional Vietnamese family roles and decision making?
4. Identify several common health conditions resulting from refugee camps.
5. Identify several common health conditions not related to being in a refugee camp. What can the health-care provider do to help decrease these health problems?
6. Explain the *am* and *duong* concepts of nutrition. What might you do if you are unfamiliar with the foods in each category?
7. How do *am* and *duong* practices vary with pregnancy and postpartum practices?
8. How are the principles of *am* and *duong* displayed in illnesses?
9. How are mental illnesses seen among the Vietnamese? As a health-care provider, what can you do to improve Vietnamese's' assess to mental health professionals?
10. Explain the five different traditional Vietnamese medical practices.

Bibliography

CIA (2016) World factbook. https://www.cia.gov/library/publications/the-world-factbook/geos/vm.html

Global Health: Vietnam (2016) Centers for Disease Control and Prevention. https://www.cdc.gov/globalhealth/countries/vietnam/

Gordon S, Bernadett M, Evans D, Shapiro NB, Dang L (2006) Vietnamese culture: influences and implications for health care. Molina Institute for Cultural Competency. http://www.molina-healthcare.com/providers/nm/medicaid/resource/PDF/health_nm_vietnameseculture-influenc-esandimplicationsforhealthcare_materialandtest.pdf

LaBorde P, Duong B (2010) Vietnamese cultural profile. Ethnomed. https://ethnomed.org/culture/vietnamese/vietnamese-cultural-profile

Mattson S (2013) People of Vietnamese heritage. In: Purnell L (ed) Transcultural health care: a culturally competent approach (chapter on DavisPlus Web site), 4th edn. F.A. Davis Company, Philadelphia

Thai C (2003) Traditional Vietnamese Medicine: historical perspective and current usage. Ethnomed. https://ethnomed.org/clinical/traditional-medicine/traditional-vietnamese-medicine-historical-perspective-and-current-usage

Thai H (n.d.) Traditional post-partum practices among the Vietnamese and Chinese. https://ethnomed.org/clinical/pediatrics/post-partum-viet-chin-brochure.pdf

Vietnamese guide: a look at Vietnamese language, culture, customs, and etiquette (2016) Commisceo Global. http://www.commisceo-global.com/country-guides/vietnam-guide

Vietnamese pregnancy and birth practices (2016) http://www.vietnam-culture.com/zones-28-1/Pregnancy-and-Birth-Customs.aspx

Glossary

aagwachse Amish folk illness, referred to in English as livergrown, with symptoms of abdominal distress believed to be caused by too much jostling, especially occurring in infants during buggy rides.

Abnemme Amish folk illness characterized by "wasting away"; usually affects infants or young children who seem to be too lean and not active.

Abwaarde Amish term for ministering to someone by being present and serving when someone is sick in bed.

Acadia Part of the Canadian Maritime Provinces.

adab Egyptian term for politeness.

afatanbah Somali term for after childbirth.

alaqad Somali traditional healer.

Allah Greatest and most inclusive of the names of God. Arabic word used to describe the God worshipped by Muslims, Christians, and Jews.

am Pervasive force in Vietnamese traditional medicine, associated with cold conditions and things that are dark, negative, feminine, and empty.

Anabaptist Adherent of the radical wing of the Protestant Reformation who espouses baptism of adult believers.

antyesti Hindu equivalent of last rites.

Arabic Semitic language of the Arabs.

arwah Egyptian term for spirits.

Ashkenazi Descended from Eastern Europe and Russia.

!Ay bendito! Frequently used Puerto Rican phrase expressing astonishment, surprise, lament, or pain.

Ayurveda Traditional Asian Indian medicine.

baten Iranian term for inner self.

be bao or bat gio Vietnamese folk practice in which the skin is pinched in order to produce ecchymosis and petechiae; practiced to relieve sore throats and headaches.

boat people Haitian or Cuban immigrants who arrive in small boats; usually of undocumented status.

© Springer Nature Switzerland AG 2019
L. D. Purnell, E. A. Fenkl, *Handbook for Culturally Competent Care*,
https://doi.org/10.1007/978-3-030-21946-8

Boricua Puerto Rican term used with great pride; name given to Puerto Rico by the Taino Indians.

botanica Traditional Cuban or other Spanish store selling a variety of herbs, ointments, oils, powders, incenses, and religious figurines used in Santería.

Brauche Folk healing art common among Pennsylvania Germans.

Braucher Amish practitioner of brauche, a folk healer.

bris or brit milah Ritual circumcision of a male Jewish child.

caida de la mollera Condition of fallen fontanelle, believed to occur because the infant was withdrawn too harshly from the nipple; common among some Spanish-speaking populations.

Cami Turkish word for mosque.

cao gio Vietnamese practice of placing ointments or hot balm oil across the chest, back, or shoulders and rubbing with a coin; used to treat colds, sore throats, flu, and sinusitis.

catimbozeiros Portuguese word for sorcerer; can be a folk practitioner.

Chasidic (or Hasidic) Ultra-Orthodox Jewish sect.

Chiac or ciac French dialect used in New Brunswick.

cho Haitian word for cold.

Chondo-Kyo Korean naturalistic religion that combines Confucianism, Buddhism, and Daoism.

choteo Cuban term for a lighthearted attitude, involving teasing, bantering, and exaggeration.

comadre Portuguese term for godmother.

compadre Portuguese term for godfather.

confianza Hispanic term for trust developed between individuals; essential for effective communication and interpersonal interactions in health-care settings.

Conservative Jewish term for the religious group between Reform and Orthodox in terms of religious practice.

contadini Italian term for peasants.

cornicelli Italian charm with little red horns worn for good luck.

Cultural awareness Appreciation of the external signs of diversity such as the arts, music, dress, and physical characteristics.

Cultural competence Having the knowledge, abilities, and skills to deliver care congruent with the patient's cultural beliefs and practices. (See Chap. 1 for a more extensive definition.)

Cultural sensitivity Having to do with personal attitudes and not saying things that may be offensive to someone from a cultural or ethnic background different from the health-care provider's background.

Culture Totality of socially transmitted behavior patterns, arts, beliefs, values, customs, lifeways, and all other products of human work and thought characteristics of a population of people that guides their worldview and decision making. Patterns may be explicit or implicit, are primarily learned and transmitted within the family, and are shared by the majority of the culture.

curandeiro Portuguese folk practitioner whose healing powers are divinely given.

curandero Traditional folk practitioner common in Spanish-speaking communities; treats traditional illness not caused by witchcraft.

Deitsch/Duetsch Pennsylvania German (sometimes incorrectly anglicized as Pennsylvania Dutch); American dialect derived from several uplands and Alemanic German dialects, with an admixture of American English vocabulary.

demut German term for humility, a priority value for the Amish, the effects of which may be seen in details such as the height of the crown of an Amish man's hat, as well as in very general features such as the modest and unassuming bearing and demeanor usually shown by Amish in public. This behavior is reinforced by frequent verbal warnings against its opposite, hochmut, pride or arrogance, which is to be avoided.

descensos Spanish term for fainting spells.

doule Haitian word for pain.

duong Vietnamese force used in traditional health practice, associated with things positive, masculine, light, and full.

Eid Arabic, Iranian, and Somali term for celebration of a feast—for example, Eid Gorgan (day/feast ending pilgrimage to Mecca); Eid Fetr (last day of the month of Ramadan).

el ataque/ataque de nervios Hyperkinetic spasmodic activity common in Spanish-speaking groups. The purpose is to release strong feelings or emotions. The person requires no treatment, and the condition subsides spontaneously. It is an expression of deep anger or depression.

endropi Greek term for shame.

espiritista (espiritualista) Spanish or Portuguese folk practitioners who receive their talent from "God"; treat conditions believed to be caused by witchcraft.

Falasha Black Jews originating from Africa.

Farsi The national language of Iran.

Fatalism Acceptance that occurrences in life are predetermined by fate and cannot be changed by human beings.

fret Haitian word for cold.

Gaelic The language spoken in Ireland.

garm Iranian term for hot.

Generalization Reducing numerous characteristics of an individual or group of people to a general form that renders them indistinguishable. Generalizations have to be validated by the individual.

geophagia The practice of eating clay and nonfood substances

giac Vietnamese dermabrasive procedure performed with cup suctioning.

giagia Greek term for grandma.

Gullah Creole language spoken by African Americans who reside on or near the islands off the coasts of Georgia and the Carolinas.

halal The lawful—that which is permitted by Allah; also, the term used to describe ritual slaughter of meat.

haram The unlawful—that which is prohibited by Allah; anyone who engages in what is prohibited is liable to incur punishment in the hereafter (as well as legal punishment in countries that incorporate Islamic law into legal codes).

Hasidic Jewish ultra-Orthodox sect.

Hasidic (or Chasidic) Ultra-Orthodox Jewish sect.

hijab Modest covering of a Muslim woman; conceals the head and the body, except for the hands and face, with loosely fitting, nontransparent clothing.

hilot Filipino folk healer and massage therapist.

Hochmut Amish term for pride and arrogance.

honor Spanish term for goodness or virtue; can be diminished or lost by an immoral or unworthy act.

hot-and-cold theory Hispanic concept that illness is caused when the body is exposed to an imbalance of hot and cold; foods are also classified as hot or cold.

hwa-byung Korean traditional illness that occurs from repressing anger or other strong emotions.

Ihteram Egyptian Arabic word for respect.

il mal occhio Italian term for the evil eye.

imam Muslim leader of the prayer; usually the most learned member of the local Islamic community.

Indian Health Service Federal agency that has the responsibility for providing health services to Native Americans.

Indochinese Individuals originating from Vietnam, Cambodia, or Laos.

Islam Monotheistic religion in which the supreme deity is Allah; according to Muslim belief, God imparted his final revelations—the Holy Qur'an—through his last prophet, Mohammed, thereby completing Judaism and Christianity.

issei First-generation Japanese immigrant.

itami Japanese term for pain.

jenn Egyptian term for the devil.

Jinn Arabic term for demons.

kaddish Jewish prayer said for the dead.

kampo Japanese term for East Asian or Chinese medical practices and botanical therapies.

karma Hindu term for actions performed in the present life and the accumulated effects from past lives.

kashrut or kashrus Jewish laws that dictate which foods are permissible under religious law.

Koran See Qur'an.

kosher Kashrut laws in the Jewish religion.

Ladino Guatemalan term for Mestizo (mixed American Indian and Spanish)

Latino(a) Person from Latin America.

laub Hmong dish made with raw pork and vegetables and spices.

lien Vietnamese concept that represents control over and responsibility for moral character.

maghi Italian word for witch.

magissa Greek folk healer.

mal ojo or mal de ojo Spanish term for the evil eye, a hex condition with unspecific signs and symptoms believed to be caused by an older person admiring a younger person; condition can be reversed if the person doing the admiring touches the person being admired.

matiasma Greek term for the evil eye.

Métis People of mixed Native American and European, especially French Canadian, heritage.

mien Vietnamese concept based on wealth and power.

mohel Ritual circumciser in the Jewish faith.

morita therapy Indigenous Japanese school of psychotherapy.

Moslem See Muslim.

moxibustion Vietnamese health-care practice in which pulverized wormwood is heated and placed directly on the skin at specified meridians to counter conditions associated with excess cold.

Muslim Person who follows the Islamic faith, the world's second-largest religion.

naharati Iranian term for generalized distress.

Naikan therapy Japanese indigenous therapy focusing on bodily illnesses that are emotionally induced.

nervioso(a) Hispanic term used to describe signs and symptoms of nervousness, anxiety, sadness, and grief.

nevra Greek folk illness.

Nihon/Nippon Japanese name for Japan.

nisei Japanese term for the second generation of an immigrant family.

Nuyoricans Puerto Ricans born in New York.

Old Order Amish Most conservative and traditionalist group among the followers of Jacob Ammann; today simply called Amish, but technically known as Old Order Amish Mennonite to distinguish them from other related Amish and Mennonite groups.

Oppression Haitian ailment related to asthma; describes a state of anxiety and hyperventilation.

oreo derogatory term for "black on the outside and white on the inside".

Paj ntaub (pan dow) Form of embroidery that Hmong women do to decorate their clothing and make historical story cloths.

pappou Greek word for grandfather.

parve Jewish term used for foods that are neutral and can be eaten with meat or milk products.

Pasah Dai Dialect in southern Thailand.

Pasah Isaan Dialect in northeastern Thailand.

Pasah Nua Dialect in northern Thailand.

philptimo Greek term for respect.

Polonia Communities heavily occupied by Polish immigrants and descendants of Polish nationals. Also the medieval name for Poland.

Ponos Greek word for pain.

practika Greek herbal remedies.

Pseudofamilies Vietnamese households made up of close and distant relatives and friends that share accommodations, finances, and fellowship.

pu tong hua Recognized language of China.

qi One of five substances or elements of traditional Chinese medicine; encompasses the foundation of the energy of the body, environment, and universe; includes all sources and expenditures of energy.

Qur'an or Koran Muslim holy book; believed by Muslims to contain God's final revelations to humankind.

rabbi Jewish religious leader.

Ramadan or Ramazan The ninth month of the Islamic year during which Muslims are required to fast during daylight hours for 30 days.

Reconstructionism Mosaic of the three main branches of Judaism; is an evolving religion of the Jewish people; seeks to adapt Jewish beliefs and practices to the needs of the contemporary world.

Reform Liberal or Progressive Judaism.

respeto Hispanic term denoting respect; refers to the qualities developed toward others such as parents, the elderly, and educated people who are expected to be honored, admired, and respected.

Sabra Jew who was born in Israel.

sansei Japanese term for the third generation of an immigrant family.

Santería 300-year-old Afro-Cuban religion that syncretizes Roman Catholic elements with ancient Yoruba tribal beliefs and practices.

santero Practitioner of Santería.

sard Iranian term for cold.

sensei Japanese term for master; used to address teachers, physicians, or those in seniority in a corporate setting.

Sephardic Jewish term for being descended from Spain, Portugal, the Mediterranean, Africa, or Central or South America.

sheikhs The most learned individuals in an Islamic community.

Shinto Indigenous religion of Japan.

simpatia Spanish term for smooth interpersonal relationships; characterized by courtesy, respect, and the absence of harsh criticism or confrontation.

sobador Spanish folk practitioner, similar to a chiropractor, who treats illnesses and conditions affecting the joints and musculoskeletal system.

Solidarnosc "Solidarity" Union of interests, purposes, and sympathies promoting fellowship with Polish nationals.

stereotype Oversimplified conception, opinion, or belief about some aspect of an individual or group of people.

Synagogue, temple, or shul Jewish house of worship.

ta'arof Iranian ritual expressing courtesy.

tae-kyo Korean term, literally fetus education, with the objective being health and well-being of the fetus and the mother through art, beautiful objects, and a serene environment.

Tagalog Filipino national language.

Tesbih Turkish small beads traditionally used for praying, now take a more secular meaning and are often referred to as worry beads.

Tet Asian Lunar New Year; celebrated in January or February.

Torah Five books of Moses; referred to in the Jewish faith.

treyf Jewish term for forbidden or unclean.

tribe Native American social organization comprising several local villages, bands, districts, lineages, or other groups who share a common ancestry, language, and culture.

Unani Traditional medicine practiced in Arab countries and Indochinese such as Vietnam.

Variant cultural characteristics Determine a person's adherence to beliefs and values of his or her dominant culture. Includes nationality, race, color, gender, age, religious affiliation, educational status, socioeconomic status, occupation, military experience, political beliefs, urban or rural residence, enclave identity, marital status, parental status, physical characteristics, sexual orientation, gender issues, and reason for migration (sojourner, immigrant, or undocumented status).

velorio Spanish term for a wake; a festive occasion following the burial of a person.

vendouses Greek practice of cupping.

voudou or voodoo Vibrant religion born from slavery and revolt; the term means sacred in the African language of Fon.

yang In Chinese belief system, one of two opposing principles of the balance of life; can be either a single phenomenon or a state of being of a phenomenon. See yin.

yarmulke Jewish head covering worn by men.

yerbero See jerbero.

Yiddish Language often spoken by elderly Jews.

yin In Chinese belief system, one of two opposing principles of the balance of life; can be either a single phenomenon or a state of being of a phenomenon. See yang.

zaher Iranian term for public persona.

zar Egyptian transmeditative ceremony.

Zhong guo The Chinese name for China and means "middle kingdom."